Climate-Smart Rice Breeding

Akansha Singh • Shravan Kumar Singh
Jiban Shrestha
Editors

Climate-Smart Rice Breeding

Springer

Editors
Akansha Singh
Genetics and Plant Breeding
Institute of Agricultural Sciences, Banaras Hindu University
Varanasi, Uttar Pradesh, India

Shravan Kumar Singh
Genetics and Plant Breeding
Institute of Agricultural Sciences, Banaras Hindu University
Varanasi, Uttar Pradesh, India

Jiban Shrestha
Nepal Agricultural Research Council (NARC), National Plant Breeding and Genetics Research Centre (NPBGRC)
Lalitpur, Nepal

ISBN 978-981-97-7097-7 ISBN 978-981-97-7098-4 (eBook)
https://doi.org/10.1007/978-981-97-7098-4

This Springer imprint is published by the registered company Springer Nature Singapore Pte Ltd.
The registered company address is: 152 Beach Road, #21-01/04 Gateway East, Singapore 189721, Singapore

Preface

The challenges posed by climate change on a global scale have necessitated a significant shift in agricultural practices, especially in the cultivation of rice, which is a staple food for more than half of the world's population. As climate patterns become increasingly unpredictable and extreme, the sustainability and productivity of rice farming are under significant threat. In response to this urgent need, the book *Climate-Smart Rice Breeding* presents innovative breeding techniques and climate-resilient strategies to address these challenges. This book offers a comprehensive examination of the advancements and methodologies in breeding rice varieties that can withstand the adverse effects of climate change. The 14 well-presented chapters in the book integrate knowledge from the use of several advanced crop breeding technologies, including Marker-Assisted Backcross Breeding (MABB), genomic-assisted breeding, genomic selection, haplotype breeding, speed breeding, and genome editing technologies, in rice improvement. Additionally, it covers the application of Artificial Intelligence and Machine Learning in crop phenotype prediction, which is essential for climate-smart breeding aimed at addressing food security. This book presents a holistic perspective on climate-smart rice breeding by integrating traditional methods with cutting-edge technologies like genomic selection, CRISPR-Cas9 gene editing, and marker-assisted selection. Each chapter is carefully crafted to combine theoretical knowledge with practical applications, offering insights into the latest research, breeding techniques, and field practices. By demonstrating the synergy between these approaches, the book highlights how they can be utilized to develop new rice varieties that are not only high-yielding but also resilient to climatic stresses.

The development of climate-smart rice varieties is not just a scientific endeavor; it is a crucial step towards ensuring food security, reducing poverty, and promoting sustainable agriculture. This book aims to empower its readers with the knowledge and tools necessary to contribute to this vital mission. The book *Climate-Smart Rice Breeding* will be an essential resource for students, researchers, breeders, and policy-makers that will deepen their understanding and inspire innovative solutions to the challenges posed by climate change in rice improvement.

In conclusion, I extend my gratitude to all the contributors, my teachers, students, my friends and family whose work has been instrumental in the creation of this book. It is my hope that *Climate-Smart Rice Breeding* will not only inform but also motivate collaborative efforts towards building a resilient and sustainable future for rice cultivation in the face of climate change.

Varanasi, Uttar Pradesh, India — Akansha Singh
Varanasi, Uttar Pradesh, India — Shravan Kumar Singh
Lalitpur, Nepal — Jiban Shrestha

Contents

Editors and Contributors

About the Editors

Akansha Singh is presently working as DBT BioCARe Scientist in the Department of Genetics and Plant Breeding, Institute of Agricultural Sciences, Banaras Hindu University, India. She has also worked as Postdoctoral Fellow, Mumbai University and as Associate Professor, Department of Genetics and Plant Breeding, College of Agriculture, Parul University, India. She has authored numerous national and international publications in journals of repute. She is one of the breeders of a popular rice variety "Malaviya Manila Sinchit Dhan-1," released and notified by the Government of India in March, 2024.

Shravan Kumar Singh is working as senior professor and Head of the Department of Genetics and Plant Breeding, Institute of Agricultural Sciences, B.H.U., Varanasi, India and has more than 34 years of teaching and research experiences. He had been Vice President of the Indian Society of Genetics and Plant Breeding during 2020–2022 at national level. He has been honored with Fellow of the Indian Society of Genetics and Plant Breeding as well as NARES-IRRI Fellowship Award. He has developed many crop varieties and five of them have been released/notified by the Govt. of India and many rice varieties are in the pipeline of release. He has authored more than 130 research publications in journal of repute. He has produced 12 Ph.D. and 40 M.Sc. students as supervisor/co-supervisor.

Jiban Shrestha is presently working as Scientist (S2) (Plant Breeding and Genetics) at Nepal Agricultural Research Council (NARC), National Plant Breeding and Genetics Research Centre (NPBGRC), Khumaltar, Lalitpur, Nepal. He is plant breeder and is currently working in rice and maize breeding works. He has published more than 300 journal articles and 5 books. He is senior editor working in editorial works over 12 years. He has been editing/reviewing articles from reputed publishers. Till date he had reviewed more than 550 articles and edited more than 150 articles for renowned journals. He has provided scientific article writing and data analysis training to students and researchers at national and international universities.

Contributors

Karthikeyan Adhimoolam Subtropical Horticulture Research Institute, Jeju National University, Jeju City, Jeju, South Korea

Abhishek Anand Division of Agricultural Bioinformatics, ICAR-Indian Agricultural Statistics Research Institute, New Delhi, India

T. Anandhan Department of Plant Breeding and Genetics, Pandit Jawaharlal Nehru College of Agriculture and Research Institute, Karaikal, Puducherry (U.T), India

J. Bonipas Antony Centre for Plant Breeding and Genetics, Tamil Nadu Agricultural University, Coimbatore, Tamil Nadu, India

Rachana Bagudam Department of Genetics and Plant Breeding, Research Fellow, ICRISAT, Hyderabad, Telangana, India

P. S. Basavaraj ICAR-National Institute of Abiotic Stress Management, Baramati, Maharashtra, India

Narender Kumar Bharat Department of Seed Science and Technology, Dr Yashwant Singh Parmar University of Horticulture and Forestry, Nauni, Solan, Himachal Pradesh, India

H. A. Bhargavi ICAR-Indian Institute of Horticultural Research, Bengaluru, Karnataka, India

Prastuti Bhattacharyya Genetics and Plant Breeding, Institute of Agricultural Sciences, Banaras Hindu University, Varanasi, Uttar Pradesh, India

Sadras Bhavana Department of Crop Physiology & Plant Pathology, College of Agriculture, Kaveri University, Gowraram, Telangana, India

Ritik Digamber Bisane Genetics and Plant Breeding, Institute of Agricultural Sciences, Banaras Hindu University, Varanasi, Uttar Pradesh, India

Neeraj Budhlakoti Division of Agricultural Bioinformatics, ICAR-Indian Agricultural Statistics Research Institute, New Delhi, India

Sheshanka Dugyala Department of Crop Physiology & Plant Pathology, College of Agriculture, Kaveri University, Gowraram, Telangana, India

Geetha Govind Department of Agricultural Biotechnology, College of Agriculture, Hassan, University of Agricultural Sciences, Bangalore, Karnataka, India

B. S. Harshitha Central Sericultural Research and Training Institute, Berhampore, West Bengal, India

Girish Kumar Jha Division of Agricultural Bioinformatics, ICAR-Indian Agricultural Statistics Research Institute, New Delhi, India

B. T. Krishnaprasad Department of Agricultural Biotechnology, College of Agriculture, Hassan, University of Agricultural Sciences, Bangalore, Karnataka, India

Vemula Anjula Krishna Genetics and Plant Breeding, Institute of Agricultural Sciences, Banaras Hindu University, Varanasi, Uttar Pradesh, India

Vikas Kulkarni AICRP on Sunflower, MARS, Raichur, Karnataka, India

Dharmalingam Kumaresan Department of Rice, Centre for Plant Breeding and Genetics, Tamil Nadu Agricultural University, Coimbatore, Tamil Nadu, India

M. Karthik Kumar Central Sericultural Research and Training Institute, Mysuru, Karnataka, India

Sakthi Anand Muthazhagu Kuppuraj Department of Plant Breeding and Genetics, Pandit Jawaharlal Nehru College of Agriculture and Research Institute (PAJANCOA & RI), Karaikal, Puducherry (U.T), India

C. Subha Lakshmi ICAR-Indian Institute of Millets Research, Rajendranagar, Hyderabad, Telangana, India

Jai Prakash Lal Genetics and Plant Breeding, Institute of Agricultural Sciences, Banaras Hindu University, Varanasi, Uttar Pradesh, India

Puja Mandal Department of Genetics and Plant Breeding, Centre for Plant Breeding and Genetics, Tamil Nadu Agricultural University, Coimbatore, Tamil Nadu, India

Swaminathan Manonmani Department of Rice, Centre for Plant Breeding and Genetics, Tamil Nadu Agricultural University, Coimbatore, Tamil Nadu, India

Dwijesh Chandra Mishra Division of Agricultural Bioinformatics, ICAR-Indian Agricultural Statistics Research Institute, New Delhi, India

K. Krishna Mohan Department of Botany, Government Degree College, Khairatabad, Telangana, India

Tushar Arun Mohanty Department of Genetics and Plant Breeding, Centre for Plant Breeding and Genetics, Tamil Nadu Agricultural University, Coimbatore, Tamil Nadu, India

Tamilzharasi Murugesan Department of Plant Breeding and Genetics, Pandit Jawaharlal Nehru College of Agriculture and Research Institute (PAJANCOA & RI), Karaikal, Puducherry (U.T), India

Aavula Naveen Division of Genetics, ICAR-Indian Agricultural Research Institute, New Delhi, India

S. K. Pooja Department of Agricultural Biotechnology, College of Agriculture, Hassan, University of Agricultural Sciences, Bangalore, Karnataka, India

Amrit Prasad Poudel Genetics and Plant Breeding, Institute of Agricultural Sciences, Banaras Hindu University, Varanasi, Uttar Pradesh, India

Ameena Premnath Department of Plant Biotechnology, Centre for Plant Molecular Biology and Biotechnology (CPMB&B), Tamil Nadu Agricultural University (TNAU), Coimbatore, Tamil Nadu, India

R. Pushpam Department of Forage, CPBG, TNAU, Coimbatore, Tamil Nadu, India

B. Rajanna AICRP on Groundnut, MARS, Raichur, Karnataka, India

Thota Joseph Raju Department of Seed Science and Technology, College of Agriculture, Hassan, University of Agricultural Sciences, Bangalore, Karnataka, India

Bharathi Raja Ramadoss Centre for Plant Breeding and Genetics (CPBG), Tamil Nadu Agricultural University (TNAU), Coimbatore, Tamil Nadu, India

Plenty Unlimited Inc., Laramie, WY, USA

Shobica Priya Ramasamy Department of Genetics and Plant Breeding, Centre for Plant Breeding and Genetics (CPBG), Tamil Nadu Agricultural University (TNAU), Coimbatore, Tamil Nadu, India

M. Raveendran Director of Research, Tamil Nadu Agricultural University, Coimbatore, Tamil Nadu, India

U. S. Sree Vathsa Sagar Centre for Plant Breeding and Genetics, Tamil Nadu Agricultural University, Coimbatore, Tamil Nadu, India

Ambothi Rathnasamy Sakthi Department of Plant Biotechnology, Centre for Plant Molecular Biology and Biotechnology (CPMB&B), Tamil Nadu Agricultural University (TNAU), Coimbatore, Tamil Nadu, India

Shubham Rajaram Salunkhe Department of Plant Biotechnology, Centre for Plant Molecular Biology and Biotechnology (CPMB&B), Tamil Nadu Agricultural University (TNAU), Coimbatore, Tamil Nadu, India

Subburaj Samuthirapandi Centre for Plant Breeding and Genetics, Tamil Nadu Agricultural University, Coimbatore, Tamil Nadu, India

Venna Santhosh Department of Genetics and Plant Breeding, Institute of Agricultural Sciences, Banaras Hindu University, Varanasi, India

Aravindan Shanmugam ICAR-Central Institute for Cotton Research, Regional Station, Coimbatore, Tamil Nadu, India

Nitesh Kumar Sharma Division of Agricultural Bioinformatics, ICAR-Indian Agricultural Statistics Research Institute, New Delhi, India

Parshant Kumar Sharma Genetics and Plant Breeding, Institute of Agricultural Sciences, Banaras Hindu University, Varanasi, Uttar Pradesh, India

Akansha Singh Genetics and Plant Breeding, Institute of Agricultural Sciences, Banaras Hindu University, Varanasi, Uttar Pradesh, India

Akashdeep Singh Department of Agronomy, Chaudhary Sarwan Kumar Himachal Pradesh Krishi Vishavydyalaya, Palampur, Himachal Pradesh, India

Faculty of Agriculture, School of Agricultural Sciences, Nirwan University, Jaipur, Rajasthan, India

Lopamudra Singha Genetics and Plant Breeding, Institute of Agricultural Sciences, Banaras Hindu University, Varanasi, Uttar Pradesh, India

Anil Kumar Singh Department of Genetics and Plant Breeding, Institute of Agricultural Sciences, Banaras Hindu University, Varanasi, India

Shravan Kumar Singh Genetics and Plant Breeding, Institute of Agricultural Sciences, Banaras Hindu University, Varanasi, Uttar Pradesh, India

Ramalingam Suresh Department of Rice, Centre for Plant Breeding and Genetics, Tamil Nadu Agricultural University, Coimbatore, Tamil Nadu, India

Manonmani Swaminathan Department of Rice, Centre for Plant Breeding and Genetics, Tamil Nadu Agricultural University, Coimbatore, Tamil Nadu, India

Anandhan Tamilselvan Department of Plant Breeding and Genetics, Pandit Jawaharlal Nehru College of Agriculture and Research Institute (PAJANCOA & RI), Karaikal, Puducherry (U.T), India

Department of Genetics and Plant Breeding, Centre for Plant Breeding and Genetics (CPBG), Tamil Nadu Agricultural University (TNAU), Coimbatore, Tamil Nadu, India

M. Tamilzharasi Department of Plant Breeding and Genetics, Pandit Jawaharlal Nehru College of Agriculture and Research Institute, Karaikal, Puducherry (U.T), India

Anubhav Thakur Department of Seed Science and Technology, Dr Yashwant Singh Parmar University of Horticulture and Forestry, Nauni, Solan, Himachal Pradesh, India

Ashish Kumar Tiwari Department of Genetics and Plant Breeding, College of Agriculture, IGKV, Raipur, Chhattisgarh, India

Jitendra Kumar Tiwari Department of Genetics and Plant Breeding, Rani Lakshmi Bai Central Agricultural University, Jhansi, Uttar Pradesh, India

J. Umabalan Department of Plant Breeding and Genetics, Pandit Jawaharlal Nehru College of Agriculture and Research Institute, Karaikal, Puducherry (U.T), India

D. Umamaheswari Department of Plant Breeding and Genetics, Pandit Jawaharlal Nehru College of Agriculture and Research Institute, Karaikal, Puducherry (U.T), India

S. N. Vasudevan Department of Seed Science and Technology, College of Agriculture, Hassan, University of Agricultural Sciences, Bangalore, Karnataka, India

Vengadessan Vedachalam Department of Plant Breeding and Genetics, Pandit Jawaharlal Nehru College of Agriculture and Research Institute (PAJANCOA & RI), Karaikal, Puducherry (U.T), India

V. Vengadessan Department of Plant Breeding and Genetics, Pandit Jawaharlal Nehru College of Agriculture and Research Institute, Karaikal, Puducherry (U.T), India

Sakthivel Viswabharathy Centre for Plant Molecular Biology and Biotechnology, Tamil Nadu Agricultural University, Coimbatore, Tamil Nadu, India

Climate Change Impact on Rice Production and Breeding for Climate Resilient Rice

1

Ramalingam Suresh, Aravindan Shanmugam, Sakthivel Viswabharathy, J. Bonipas Antony, Subburaj Samuthirapandi, and Swaminathan Manonmani

Abstract

Human society is contending with a substantial environmental issue in the shape of global warming. IPCC's sixth assessment report indicates that high-emission climate change scenarios are expected to render around 10% of the currently suitable cropland unusable by the middle of this century. Rice, among the food crops, holds paramount significance in the context of global food security. Until now, annual rice production has matched the requirements due to advancements in understanding the crop's physiological and biochemical processes, the genetics of key traits, and the rapid transition and refinement of breeding tools. Rice production must be sustained in the coming years due to the crop's increased susceptibility to severe biotic and abiotic challenges, particularly unexpected drought and submergence, caused by changing climatic circumstances. Increased events of drought and submergence, seawater intrusions and inland salinity pose serious threats to rice production. Hence, breeding for climate-resilient rice varieties becomes paramount to ensure future rice demands. In the recent past, mechanisms

R. Suresh (✉) · S. Manonmani
Department of Rice, Centre for Plant Breeding and Genetics, Tamil Nadu Agricultural University, Coimbatore, Tamil Nadu, India
e-mail: suresh.r@tnau.ac.in

A. Shanmugam
ICAR-Central Institute for Cotton Research, Regional Station, Coimbatore, Tamil Nadu, India

S. Viswabharathy
Centre for Plant Molecular Biology and Biotechnology, Tamil Nadu Agricultural University, Coimbatore, Tamil Nadu, India

J. B. Antony · S. Samuthirapandi
Centre for Plant Breeding and Genetics, Tamil Nadu Agricultural University, Coimbatore, Tamil Nadu, India

A. Singh et al. (eds.), *Climate-Smart Rice Breeding*,
https://doi.org/10.1007/978-981-97-7098-4_1

underlying climate resilience in rice have been unraveled, which accelerate the discovery of several resistant genes/QTL regions against various stresses. Ultimately, marker-assisted breeding programs were intensified by the researchers across the globe. Further, establishing well-informed regulatory frameworks concerning new breeding technologies and enhanced collaboration in sharing genetic resources, genomic data, and bioinformatics proficiency between developed and developing countries will be pivotal in effectively addressing the multifaceted challenges arising from rapidly evolving climatic conditions and escalating food demands. The upcoming obstacles to long-term sustainable rice production in the face of global climate change are covered in this chapter, along with contemporary breeding techniques for climate-resilient rice.

Keywords

Temperature stress · Drought · Submergence · Salinity · Modern breeding

1.1 Introduction

Global warming, the phrase used to describe the alarming rise in land surface and ocean temperatures worldwide, results from the continuous accumulation of greenhouse gases, especially CO_2. This process, which raises the frequency and intensity of heat waves, cold snaps, dry spells, and flooding, is what leads to climate change (Von der Gathen et al. 2021; Van Houtan et al. 2021; Rivero et al. 2022; Singh et al. 2023). Most cropping areas would be exposed to increased air temperature by 2080 (Battisti and Naylor 2009). Plant breeders now have a significant problem in meeting human food demands due to unpredictable weather patterns, the world's ever-increasing population and other pressing issues, including the diminishing of arable land and water quality. Many factors, such as rising temperatures, melting glaciers, water scarcity, and devastating floods brought on by climate change, have a significant influence on rice production.

Rice is grown over 160 Mha, yielding about 493 MT of milled rice annually. It nourishes billions globally due to its remarkable adaptability, thriving in various environments from high altitudes to below-sea-level regions. Asia dominates rice production, led by China (212.84 million tons) and followed by India, Bangladesh, Indonesia, and Vietnam. By 2050, there will be a 2–3 °C increase in global temperature (IPCC 2014). This pattern of small increments in the average temperature causes adverse effects on rice production. The IPCC predicts that by the end of this century, the mean temperature will have increased by 1.1–6.4 °C, resulting in a 3–17% yield loss for rice crops for every 1–2 °C increase in temperature (Zhao Chuang et al. 2017).

In South and Southeast Asia, rice cultivation faces severe challenges due to waterlogging and water scarcity. It is expected that flood-prone areas receive more frequent extreme events of heavy precipitation even in the off-season, too, while drought-prone areas result in extreme drought due to climate change. Flash droughts are the swift start of drought for a short period with highly intensified irrecoverable impacts on crop production (Pendergrass et al. 2020).

India is thought to be among those countries most susceptible to drought stress, with at least one drought spell occurring every 3 years (Gupta et al. 2020). From 1870 to 2018, 18 severe meteorological droughts and 16 intense hydrological droughts occurred for over 6 months (Mishra et al. 2021). Dryness is expected to peak in the twenty-first century due to a significant rise in evapotranspiration, and during 2071–2100, drought incidences are expected to be the highest in India (Gupta et al. 2020). Due to waterlogging, flooding, and ineffective irrigation, salinity is predicted to become significant in land degradation in this climate change era (Dagar and Yadav 2017).

Crops of the next generation should be efficient in water usage and produce sustainable yields without consuming much water. They should also be adaptable to varying environments (Pareek et al. 2020). Therefore, it is imperative that efforts be directed towards creating rice varieties that are resistant to climate change. Modern molecular breeding techniques combined with conventional plant breeding methods have greatly expanded the pool of enhanced rice cultivars that can withstand climatic change. Moreover, the cost of genotyping has decreased in the past few decades due to the rapid development of marker and sequencing technology. The declining cost of genotyping has led to the discovery of major and minor quantitative trait loci (QTLs) for diverse abiotic stresses through many genetic mapping studies. This chapter focused on the advancements made towards contemporary breeding techniques concerning changing climate to obtain insight into the possibility of generating climate-smart rice through modern breeding approaches and maintaining global food security.

1.2 Impact of Climate Change on Rice Production

Climate change exposes crops to several abiotic pressures over their life cycle, resulting in a loss of agricultural productivity (Fig. 1.1). Around 64% of worldwide crop-growing areas are affected by drought spells, while 13% by submergence, 6% by salinity, and 15% and 57% by acidic conditions and cold, respectively (Cramer et al. 2011). Rice is commonly exposed to abiotic conditions such as severe drought, high salt, submergence, low and high temperature as a typical glycophyte.

1.2.1 Heat Stress

Rice is a tropical crop that can thrive in hot conditions and is naturally tolerant to high temperatures. Nonetheless, increasing temperature made rice vulnerable to heat stress, as the temperature goes beyond the critical point. Temperatures exceeding extremes are particularly damaging during critical stages of crop development, particularly during reproduction (Prasad et al. 2006). Based on NOAA data, the rate

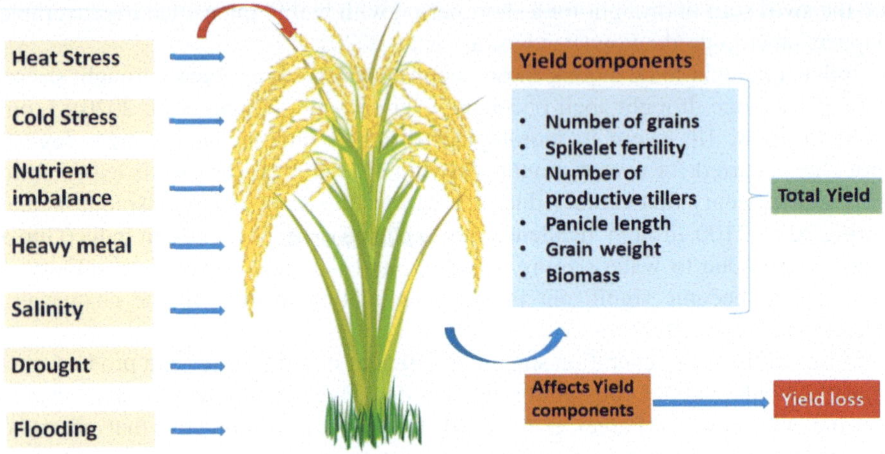

Fig. 1.1 Impact of abiotic stress on rice crop

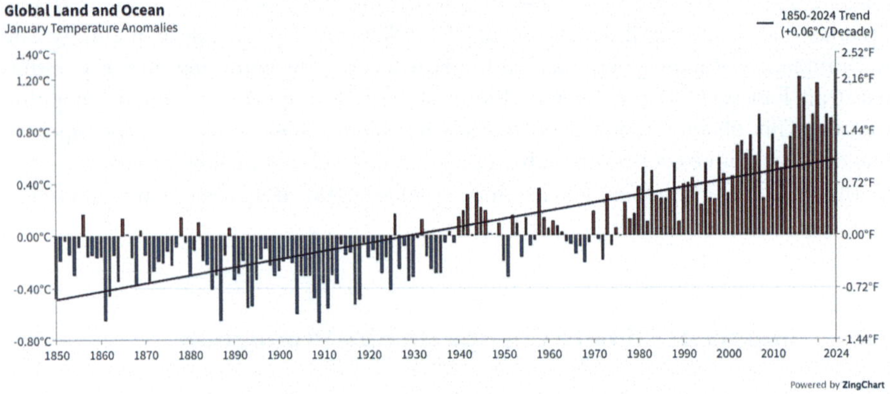

Fig. 1.2 Global temperature regime from 1850–2024 (NOAA, 2024)

of increase in Earth's temperature since 1981 has been 0.32 °C each decade, which is more than twice as fast as in earlier years (Fig. 1.2).

An incessant temperature increase makes heat stress a major hindrance to rice production. High temperatures throughout many developmental phases of rice, from meristem growth to the grain filling stage, are predicted to reduce rice yield by 40% (Jagadish et al. 2013). The high temperature throughout the reproductive stage reduces pollen fertility by preventing the spikelet from absorbing water. Less germination and extension of the pollen tubes impacted the vitality of the pollen, which in turn led to spikelet sterility and a decrease in grain yield (Yang et al. 2022; Zafar et al. 2022). Temperature increases at the booting stage led to severe effects, such as poor milling characteristics, grain volume reduction and amylose content (Zhen et al. 2019; Mohammed and Tarpley 2011). The economic value of rice is largely dependent on its grain quality while grain's chalkiness might have a detrimental effect on its value by increasing grain breakage during milling (Sharma and Khanna 2019). Heat stress causes increased chalkiness in heat-sensitive lines and reduces amylose content, which leads to poor milling quality (Jagadish et al. 2015).

El Nino Southern Oscillation (ENSO), Indian Ocean Dipole (IOD), and North Atlantic Oscillation (NAO), three climate prediction models, are conceivable processes that might concurrently alter agricultural yield in many breadbaskets (Anderson et al. 2023; Anderson et al. 2020; Najafi et al. 2020; Deepti Singh et al. 2019). This model can predict climate change at different scales, from field to global, and highlights the practical application of climate predictions in agriculture (Hammer et al. 2001). A decline in rice production of up to 10% in South Asia by 2030 and 10–25% in developing nations by 2080 has been projected by a number of climate prediction models (Lobell et al. 2008). Climate change may have an impact on 20 Mha of rice-growing land in Asian countries. This comprises South Asia's 14% rice productivity, East Asia and the Pacific's 10%, and sub-Saharan Africa's 15% (Senguttuvel et al. 2022). Changes in temperature of one degree Celsius have an impact on the yield of rice crops worldwide at a rate of 3.2% (Zhao Chuang et al. 2017; Senguttuvel et al. 2022). However, depending on the season and variety, a 1 °C rise in night-time temperatures might result in a 7–10% loss in rice productivity (Senguttuvel et al. 2022). Similarly, yield may drop by as much as 7–8% when the daytime temperature increases from 28 °C to 34 °C (Korres et al. 2017).

Around 16% of the area used to cultivate rice is expected to be exposed to temperatures over the threshold range for at least 5 days during the reproductive period by 2030 (Senguttuvel et al. 2022; Gourdji et al. 2013). A 1 °C increase in temperature might cause 5–15% sterility, impacting 62% of rice cultivable areas. Major rice cultivable regions in South Asia showed more than 30% sterility with a 3 °C temperature increase (Jagadish et al. 2015; Senguttuvel et al. 2022). Extreme heat periods (>35 °C) during critical stages can significantly impact rice yields as a result of decreased head rice recovery and increased spikelet sterility (Djanaguiraman et al. 2020). Reduced pollen germination, limited pollen tube development, and failure of anther dehiscence are the causes of the reduced yields (Jagadish et al. 2010).

1.2.2 Cold Stress

Originating in tropical and subtropical locations, rice is a warm-weather crop that needs temperatures between 20–40 °C to thrive (Sridevi and Chellamuthu 2015). Temperature patterns alter due to global climate change, making temperate and subtropical regions colder. The low-temperature sensitivity of modern rice cultivars also limits cultivation at higher latitudes and altitudes in South and Southeast Asia (Kaw and Khush 1986). Low-temperature stress, also known as cold stress, which is separated into freezing (0 °C) and chilling (0 °C) stress, affects rice crop development, productivity, and geographic distribution (Zhu et al. 2007; Zhang et al. 2014). At every stage of the rice crop's growth, from germination to maturity, low temperatures significantly decrease production (Andaya and Mackill 2003). The low temperature directly affects crop germination, seedling vigor and crop establishment at early stages. Low seedling numbers, decreased tillering, and increased mortality of plants during the seedling stage are other consequences (Aghaee et al. 2011). When compared to cold-tolerant cultivars, the effects of cold stress during the flowering stage result in an 89.7% rise in spikelet sterility and a 97% decrease in pollen fertilization efficiency (Susanti et al. 2019).

Rice is a temperature-sensitive crop requiring an ideal temperature level of 25–30 °C for its growth and development (Lv et al. 2019). Temperatures dropping below 15 °C can adversely affect rice germination and seedling growth, while temperatures between 17 °C to 20 °C during critical stages can also impact development (Lv et al. 2019; Liu et al. 2018). Cold stress can particularly affect seed germination in rice, highlighting the importance of traits like low-temperature germination (LTG) in breeding for rice cultivation in temperate regions using direct seeding methods (Lv et al. 2019; Sales et al. 2017). Cold stress during the early development stage can cause reduced growth rate, tillering, chlorosis, and lower seedling vigor (Moraes de Freitas et al. 2019). On the other hand, panicle branching, spikelet production, and pollen fertility may all decrease during the booting stage (Shimono et al. 2016). Often cold stress, combined with overcast skies and rain. In rice, these circumstances can impair photosynthetic capacity, flowering and fertilizing rate, and seed setting rate during the reproductive development stage (Suh et al. 2010; Zhou et al. 2010).

Plant cells are able to perceive alterations in the cytoskeleton's organization and membrane fluidity brought on by cold stress. This triggers the activation of second messenger signals, viz., calcium ions, abscisic acid and reactive oxygen species (ROS). Subsequently, the expression of genes sensitive to cold and mechanisms for cold tolerance is regulated by downstream signal transduction pathways that involve kinases, transcription factors, and other components (Guo et al. 2018; Zhu 2016). Elevated levels of ROS and MDA accumulation in rice led to cellular oxidative damage, impairing metabolism during cold stress (Xie et al. 2009; Nakashima et al. 2007). Nonetheless, rice has evolved defense mechanisms against and adaptations to cold stress. For example, proline aids in stabilizing protein synthesis, is accumulated by rice plants during cold stress. This procedure aids in preserving rice cells' ideal function (Zhang et al. 2014).

1.2.3 Drought Stress

Rice is damaged throughout its life cycle by drought stress (Barik et al. 2019), which includes delayed blooming, decreased partitioning, dry matter accumulation, photosynthetic capability, metabolic constraints, and oxidative damage to chloroplasts (Farooq et al. 2009). In Asia, 8 Mha and over 34 Mha of upland and lowland rice, respectively, are impacted by drought stress every year (Singh et al. 2016). In India, 60% of rice-growing areas fall under an irrigated ecosystem, whereas 40% is under a rainfed ecosystem (Anonymous 2016). Rice under such rainfed conditions is vulnerable to abiotic conditions, viz., drought, submergence, high temperature and salinity. In recent days, water has become scarce and less available for rice cultivation even in the traditional irrigated areas due to poor discharge of water, rapid depletion of ground water and frequent failure of monsoon rains due to changing climate. However, area under rainfed rice cultivation is steadily increasing from 80 thousand hectares in 1990's to 2.5 lakh hectares in 2021 in delta districts and is predicted to rise even more in the coming years. Research on developing drought tolerant rice cultivars has lagged behind, even while the area utilized for dry and semi-dry paddy cultivation has increased. Therefore, farmers are forced to cultivate varieties which are breed for irrigated ecosystem, which leads to significant yield loss during dry spells.

On average, the global rice production decreases by nearly 18 MT annually (O'Toole 2004). Drought affects more than one-third of the world's agricultural land. In Asia alone, about 42 Mha (8 Mha and 34 Mha in upland and lowland rice) are thought to be drought-prone (Singh et al. 2016). In India, the annual rice yield suffers to a tune of about 15.0% due to drought. Of the total area of 43 Mha of rice cultivation, at least 17 Mha is affected mainly by drought, leading to 20–40% decreased productivity. Each year is different regarding the extent and severity of the drought, and it is influenced by a number of variables, including temperature, frequency of rainfall, evaporation, and soil moisture (Hao et al. 2018; Oladosu et al. 2019). Hence, it is difficult to predict the onset of drought, its end, and its severity, as well as to know the time for management practices. Many farmers who produce rice under irrigated conditions have profited from the "green revolution," but its influence on rainfed rice output has been very small (Evenson and Gollin 2003). The yield gap between rainfed and irrigated rice in Asia increased from 1.7 (1960s) to 3.6 t ha^{-1} (1990s) (Maclean 2002), which is mainly due to poor research on increasing yield under drought stress. Drought is now considered a major constraint of immediate concern due to changes in rainfall patterns, rising global temperatures, and climate change. Because the water available for agriculture worldwide is decreasing, it is crucial to boost rice's resistance to drought. Climate change, global temperature increase, and rainfall patterns made drought a major constraint of immediate concern. Hence, it is important to develop drought-tolerant varieties due to the world's declining water supply for agriculture.

One of the most drought-sensitive plants is rice, which is regarded as a semi-aquatic and water-loving plant because of its shallow root system, thin cuticular wax, rapid stomata closure, and leaf senescence under mild water stress (Sahebi et al. 2018) and limited capacity to extract water from deeper layers (Kondo et al. 2000). The extent of damage caused by drought mainly depends on the duration of water stress and the stage of the crop in which drought occurs. Drought conditions at the vegetative and reproductive stages reduce yield and productivity; however, the latter affects the most (Singh et al. 2022; Majhi et al. 2022). While late-season dryness impacts crop maturity, early-season drought impacts seedling transplanting and direct-seeded rice growth (Bunnag and Pongthai 2013).

Drought stress results in a variety of negative effects, such as delayed blooming, low dry matter accumulation, reduced photosynthetic capacity, and increased spikelet sterility. Furthermore, given current climate change forecasts and a drop in water supplies, the impact of drought on rice production in Asia is expected to be increasingly severe. Traditional varieties suited to the different rice-growing habitats were grown in these regions before the Green Revolution. However, in the years after the Green Revolution, a few high-yielding, fertilizer-responsive cultivars that were tailored to the irrigated environment took the place of these types. These cultivars have significant production losses even in moderate stress because they were never selected for drought resistance (Kumar et al. 2008). However, compared to other cereals, rice is unusual in that it has a broad and varied area under cultivation, is grown using diverse techniques (direct-seeded upland, transplanted lowland), and has an inherent diversity that makes it tolerant to drought (Kumar 2014).

Assessment of drought stress is commonly based on morphological, physiological, and biochemical reactions. These responses are typified by a decline in the plant's water content, a reduction in cell elongation and growth, the closure of stomata, a reduction in gaseous exchange, and disruption of enzyme-catalyzed activities. Severe droughts also cause significant disruptions to photosynthesis and metabolism, eventually leading to the plant's death (Ozga et al. 2017). According to Anjum et al. (2017), drought stress inhibits cell division rather than cell expansion. This also affects a wide range of physiological and biochemical functions, including respiration, photosynthesis, ion absorption, growth stimulants, carbohydrates, source-sink connections, and nutrient metabolism (Fahad et al. 2017). To develop high-yielding varieties of plants under drought circumstances, it may be necessary to understand how plants react physiologically to changes in drought resistance.

1.2.4 Submergence/Water Logging Stress

After heat and drought stress that is impacting agricultural production, flooding is unquestionably the third most significant barrier to achieving high productivity. Annually, 5.2 Mha of rice growing area was periodically damaged by flooding (Oladosu et al. 2020). During the northwest monsoon season alone, 0.3 million acres of rice land were subjected to flooding in India. Over 100 million people's livelihoods are at risk due to the annual unscheduled submergence of 22 Mha of rice-cultivable areas (Barik et al. 2020). Heavy rainfall-induced floods in South and Southeast Asia can occur at any growth stage and impact an estimated 10–15 Mha annually, causing an approximate loss of US$1 billion in economic production (Dey and Upadhyaya 1996). Though rice is a semi-aquatic plant, extreme flooding and complete or partial submergence for a prolonged period lead to complete yield loss. Rice plants encounter many obstacles in their development and survival while submerged. It inhibits oxygen intake, decreases the rate of gas diffusion, and limits anaerobic metabolism. Additionally, the presence of turbid floodwater reduces light accessibility, which in turn reduces photosynthesis (Pradhan et al. 2019).

From a scientific standpoint, crops' growth and developmental processes are profoundly impacted by stagnant, waterlogged, or submerged conditions. For almost half the world's population, rice is a staple diet, particularly in Asian nations where flooding is frequent. Roughly 30% of the world's rice is grown on lowlands that get rain (Yuan et al. 2021). The present climate emergency is reshaping global precipitation patterns, leading to unpredictable floods and droughts, which pose significant risks to agricultural productivity and food security.

When submerged, rice plants face a range of physiological, biochemical, and molecular changes, such as Oxygen deprivation, anaerobic metabolism, nutrient imbalance, hormonal regulation and genetic variation that can severely impact their growth, development, and yield. Asia dominates rice production, led by China (212.84 million tons) and followed by India, Bangladesh, Indonesia, and Vietnam (FAO 2021). Assuming no changes in fertility and mortality rates, the world's population has the potential to surpass 17 billion by 2100 (Nations U 2019). To have enough rice by 2050, we need to grow 70% more despite limited land, water, and energy (Siddiq and Vemireddy 2021).

The scientific consensus, supported by studies such as Trenberth (2011), GISTEMP (Gistemp 2018) and Cai et al. (2014), suggests that climate change will disrupt precipitation patterns, increasing the likelihood of droughts or floods while global temperatures rise and sea levels elevate. Developing robust and adaptive rice varieties is essential to effectively addressing the effects of climate change in response to these issues (Zhang et al. 2023). Ranked as the third most significant constraint among various stresses, submergence is one of the main factors in limiting rice production and productivity (Setter et al. 1997).

1.2.5 Salinity Stress

Soils with high salt content are frequent in nature, particularly in coastal, semi-arid, and desert regions. These naturally occurring soils are caused by a variety of processes, such as the weathering of salt-bearing rocks or mineral deposits, drainage restrictions brought on by hardpan or topographic factors, subsurface salinity water near the surface, marine intrusion, volcanic activity, arid climate, and aeolian transport of salt from saline inland and coastal areas. In recent days, agricultural land has become more salinized and sodified due to unsustainable farming methods, water scarcity and climate changes. It is estimated that more than a billion hectares of land exhibit saline or sodic characteristics. Roughly 6% of the arable land worldwide exhibits salt stress (Organization FaA 2008). Between a quarter and a third of irrigated areas, totaling around 70 million hectares, are affected by salt stress (Zaman et al. 2018). Earlier studies reported that around half of the global population was affected by soil salinity. An estimated 5.02 billion people on Earth are expected to be affected by seawater intrusion and inundation by the year 2050 (Liu et al. 2020). Salt stress hinders seed germination, limits root and shoot growth, decreases the number of tillers per plant, reduces grain yield per panicle, and lowers pollen viability. It also causes delays in seed setting and the occurrence of sterile spikelets, lowers the amount of dry matter accumulation overall, and results in poor leaf area (Singh et al. 2008; Thomson et al. 2010). By the end of the twenty-first century, salt-affected soils may spread significantly, forcing more farmers to modify their agricultural practices to keep their land viable in the face of water scarcity and soil salinity.

Rice, being highly sensitive to salt, experiences significant reductions in growth and yield when subjected to 30 mmol L^{-1} NaCl (3 dS m^{-1}) salinity (Lutts et al.

1995). In rice cultivation, the seedling and reproductive stages are more prone to salt sensitivity than the germination and vegetative stages. Previous research indicated that modern high-yielding rice varieties can experience up to a 12% yield reduction at salinity levels exceeding 3 dS m^{-1}, with a 50% reduction observed at 6 dS m^{-1} (Linh et al. 2012).

Three major effects on rice plants due to the accumulation of salts have been reported, including decreased water potential, impaired nutrient uptake, and direct toxicity of sodium (Na^+) and chlorine (Cl^-) (Flowers and Flowers 2005). A yield reduction is attributed to several factors, such as altered transpiration, water relations, stomatal conductance, nutritional imbalances, and oxidative damage resulting from salinity stress. Salinity lowers rice production via modifying morpho-physiological and biochemical processes. Furthermore, it accelerates the senescence of reproductive organs and decreases source-sink activity, biomass accumulation, and photosynthetic activity, all of which have adverse effects on yield and its component traits (Khataar et al. 2018). Comparably, throughout the reproductive phase, variations in water potential led to a decrease in the thickness of the flag leaf and vascular tissue, as well as in the size of mesophyll cells, cell elongation, and epidermal cells. These changes, in turn, impact the flag leaf turgidity, leaf area, yield potential and assimilate synthesis. Salt tolerance, a complex quantitative trait, is primarily influenced by interactions among multiple genes (Chinnusamy et al. 2005).

1.3 Breeding Approaches for Climate Smart Rice

1.3.1 Direct Seeded Rice

The phenomenon of climate change, alongside the issue of water scarcity, presents notable barriers to agricultural productivity. Each year, 3853 km^3 of groundwater is extracted, primarily for agricultural irrigation (69%), followed by industry (19%) and municipalities (12%) (Quilloy et al. 2021). Asia produces and consumes almost 90% of the world's rice, meeting three-fourths of the caloric needs of 520 million Asians (Rathna Priya et al. 2019). Global rice production is projected to surge, with a 26% rise by 2035 and a 50% increase by 2050, driven by population growth, where India and China contribute 22% and 28% of the total, respectively (Shekhawat et al. 2020).

Direct seeding involves sowing seeds directly in the field instead of transplanting seedlings raised in a nursery. Three primary methods of direct seeding include Wet direct-seeded rice (wet-DSR), Dry direct-seeded rice (dry-DSR), and Water seeding (Quilloy et al. 2021). Dry-DSR scatters dry seeds on untilled soil, while Wet-DSR plants pre-germinated seeds on puddled soil are categorized as aerobic or anaerobic. Water seeding involves dispersing pre-germinated seeds over flooded or non-flooded fields (Kumar and Ladha 2011; Sandhu et al. 2021; Kumar et al. 2021, 2023).

The advancement of direct-seeded rice (DSR) has emphasized improving key physiological traits like early seedling vigor, anaerobic germination (AG) and seed longevity (SL). In direct-seeded rice (DSR) systems, these characteristics are vital for effective crop establishment, promoting vigorous growth and outcompeting weeds. Anaerobic germination (AG), the capacity of seeds to sprout and thrive in low to no oxygen conditions, is crucial in direct-seeded rice (DSR) systems. It mitigates risks during early crop stages and aids in weed control. While many modern rice varieties lack underwater germination ability, AG is vital for breeding DSR-suited varieties. Various landraces, such as Ma Zhan (red), Nanhi, Khao Hlan On, Khaiyan, and Kalarata, exhibit tolerance to flooded germination. These tolerant landraces span major rice groups, with indica and aus types being prominent in genetic research and breeding programs (Quilloy et al. 2021).

The seed life cycle encompasses germination, dormancy, and fertilization, with dormancy described as the incomplete germination of a viable seed under optimal conditions (Bewley 1997). Further, Seed longevity refers to a seed's capacity to remain viable post-harvest, often assessed by germination percentage. It varies among rice varieties due to genetic and environmental factors. Factors such as ambient conditions during harvest maturity, seed development, presence of pathogens, mechanical damage, post-harvest storage conditions, and moisture content also play a role (Quilloy et al. 2021; Chhabra and Singh 2019; Rehman Arif et al. 2012). High temperature and moisture can reduce seed longevity and viability (Quilloy et al. 2021; Mb 1999). Anaerobic conditions during direct seeding reduce seed longevity, hamper crop establishment, and increase weed competition, underscoring the importance of resilient varieties with strong seedling vigor for rapid emergence and early growth (Quilloy et al. 2021; Diwan et al. 2013).

1.3.2 Green Super Rice

Over the last twenty years, researchers in breeding and biotechnology have been dedicated to enhancing crop productivity by developing resistance to pests, diseases and environmental stresses (Ali et al. 2021). This is possible by altering plant structures and integrating specific traits into desired plant varieties using both traditional and marker-assisted breeding techniques (Janaki Ramayya et al. 2021; Dixit et al. 2020; Sandhu et al. 2019).

Multiple genes regulate the ability to withstand abiotic stresses like drought and salinity, potentially affecting grain yield components and interacting with mechanisms similar to those of similar traits. Selecting appropriate donors for different target traits poses a challenge. Even though rice cultivars tolerant to environmental stresses are usually bred by crossing with tolerant landraces, concerns arise due to challenges in breaking undesirable linkages via conventional methods (Ali et al. 2021).

The Green Super Rice (GSR) breeding strategy was initiated in 2008 at IRRI to efficiently develop rice lines that possess tolerance to multiple stresses and enhanced nutrient and water use efficiency (Ali et al. 2021). The aim is to achieve high genetic gains for various targeted ecosystems. The GSR approach focuses on developing superior rice varieties that can thrive under reduced inputs of irrigation water, fertilizer and pesticide. These varieties are developed for growth in less favorable conditions, leading to lower greenhouse gas emissions and enhanced grain nutrient content (Ali et al. 2021; Wing et al. 2018).

High throughput phenotyping and genotyping in Green Super Rice (GSR) breeding aids in identifying promising rice varieties and understanding genetic and physiological trait mechanisms (Ali et al. 2021; Wing et al. 2018; Yu et al. 2020). By employing this breeding strategy, researchers have been able to quickly introgress the desired target traits with high precision and minimal genetic drag.

1.3.2.1 Key Feature of GSR

The GSR project, as outlined by Zhang et al. (2018), focuses on five key areas:

1. Theoretical and technical frameworks for GSR breeding.
   - Population, individual, trait, and genome levels.
   - Integration of diverse germplasm resources.
2. Platforms for selection across the entire genome.
   - Derived from global rice functional genomics research.
   - Creation of comprehensive breeding databases and gene chips.
3. Creation of new genetic resources.
   - Combining genes for environmentally beneficial traits.
   - Improved resistance to pests, diseases, and environmental stresses.
   - Enhanced water and nutrient-use efficiencies.
4. Developing new GSR cultivars.
   - Inbred and hybrid varieties.
   - Incorporation of diverse green traits.
   - Improved grain yields and quality.
5. High-yield cultivation and field management techniques for GSR.
   - Assessment of green traits.
   - "Resource-saving and environment-friendly" cultivation management systems
   - Wide planting of GSR varieties.

1.3.3 Marker-Assisted Breeding for Climate Resilience

QTL region identification enhances precise gene transfer into elite cultivars with flanking markers. Marker-assisted selection is a valuable approach for improving the rice crop against various pests, diseases, and environmental stresses. In recent times, marker-assisted backcrossing (MABC) has gained significant traction as a promising approach for developing stress-tolerant rice cultivars. MABC aims to transfer genes/QTLs from donor parents to popular elite mega varieties, enabling the creation of stress-tolerant varieties with the genetic characteristics of the mega variety. Marker-assisted selection (MAS) is a highly accurate and efficient technique compared to conventional breeding methods. Further, developing trait-specific introgression lines by incorporating QTLs from diverse donors with elite backgrounds can provide elite pre-breeding material for varietal development. To improve submergence tolerance in rice cultivars, marker-assisted backcross breeding has been employed to transfer specific segments of chromosome 9, ranging from 0.8 to 7.8 Mbp, which include the *Sub1* locus. Farmers did not widely accept the identified submergence-tolerant lines because of their low crop yield and inferior grain quality. As a result, the *Sub1* QTL was incorporated into the widely popular variety, Swarna.

Scientists at the IRRI (International Rice Research Institute) utilized Marker-Assisted Backcrossing (MAB) to incorporate the *Sub1* gene into the widely grown and high-yielding Swarna variety (Neeraja et al. 2007). This achievement served as a pivotal step in the development of flood-resistant versions of several major high-yielding varieties. Assessments on *Sub1* introgressed varieties, compared to their original counterparts, showed that integrating Sub1 did not negatively impact agronomic traits such as growth, maturity, grain yield, and quality under normal conditions (Neeraja et al. 2007; Sarkar et al. 2006, 2009; Singh et al. 2009). A thin gas layer on hydrophobic plant leaves facilitates underwater gaseous exchange, controlled by the wax synthesis gene *LGF1* (Leaf Gas film). Further, combining *Sub1* mega varieties with the *LGF1* allele will boost flood tolerance. Future work should explore genetic traits for enhanced survival and post-flood recovery.

Introgression of major-effect drought-resistant QTLs, such as *qDTY1.1*, *qDTY2.1*, etc., into well-known rice varieties typically results in a 10–30% genetic gain and increased yield of 150–500 kg ha^{-1} over their original parents. To ensure a substantial economic impact for rice farmers, the yield increase of 1.0 t/ha is crucial, making the sole introgression of a single QTL for drought tolerance less likely to yield optimal results in breeding drought-tolerant cultivars (Swamy and Kumar 2013).

The most environmentally and economically viable method to effectively combat salt stress is to develop superior salt-tolerant rice cultivars. In 1939, Sri Lanka first introduced the salt-tolerant rice variety known as Pokkali. Using marker-assisted backcross breeding, Salt-tolerant QTL was introduced into mega varieties in several nations such as Pusa Basmati 1121 (Babu et al. 2017) and PB6 (Singh et al. 2011) in India, as well as BR11 and BRRI dhan28 (Rahman et al. 2008) in Bangladesh. Rassi (West Africa) (Bimpong et al. 2016), AS996 and BT7 (Vietnam) were also developed for salt tolerance. It is challenging to produce rice cultivars resistant to salt stress for commercial cultivation by introducing one or many genes. To enhance the salt tolerance of important rice varieties, multiple genes resistant to salinity stress should be incorporated through marker-assisted gene pyramiding. The efficient and speedy transfer of these genes into mega varieties is another challenge that needs to be solved. More work has to be done on cloning genes with respect to saline stress, comprehending the regulatory processes underlying genes responsive to salt stress, and investigating the possibility of many genes being introgressed simultaneously with stable inheritance to combat the changing climate.

Near-isogenic lines containing *qHTSF4.1* showed consistent increases in heat tolerance (spikelet fertility) throughout all the backcross experiments (Ye et al. 2021). Ye et al. (2021) reported that in BC_5F_2 generations, NILs with QTL *qHTSF4.1* (44.6 ± 13.1%) had higher spikelet fertility than the female parent IR64 (19.4 ± 8.4%) and the lines without the QTL (27.1 ± 9.6%). To enhance tolerance to cold stress, QTLs conferring resistance to cold stress at different growth stages are important to be identified and pyramided using marker-assisted breeding methods. The researchers revealed that cold tolerance is a highly complex trait dependent on epistatic interaction among several unlinked genes. Therefore, complementing functional alleles and exploiting hidden diversity could help to increase tolerance for complex traits like cold tolerance. Developing trait-specific introgression lines by incorporating QTLs from diverse donors with elite backgrounds can provide elite pre-breeding material for varietal development. These achievements were pivotal in developing several abiotic stress-tolerant cultivars, gaining popularity among farmers (Table 1.1).

Table 1.1 Rice varieties in India developed for climate resilience through marker-assisted breeding

S. No	Recurrent parent	Donor parent	Improved recurrent parent	Combination of QTLs/genes	Improved traits	Reference
1.	Swarna	Swarna-*Sub1* for submergence IR91659-54-35 (N22/Swarna) for *qDTY1.1* IR 81896-B-B-195 (Apo/Swarna) *qDTY2.1* and *qDTY3.1*	IR96322-34-223-B-1-1-1 1 (released as CR Dhan 801 in India)	*qDTY1.1 + qDTY2.1 + qDTY3.1 + Sub1*	Drought and submergence	Sandhu et al. (2019)
			IR 96321-1447-651-B 1-1-2 (released as a Bahuguni dhan-1 in Nepal)	*qDTY1.1 + qDTY3.1 + sub 1*	Drought and submergence	
			IR 94391-131-358-19-B-1-1-1 released as a Bahuguni dhan-2 in Nepal	*qDTY3.1 + Sub1*	Drought and submergence	
2.	Swarna + drought	IRBB60 (*Xa4, xa5, xa13, and Xa21*, IRBL9 (*Pi9*), Rathu Heenati (*Bph3* and *Bph17*) Abhaya (*Gm4*) Aganni (*Gm8*)	IL1	*Pi9 + Xa4+ xa5 + Xa21+ Bph17+ Gm8 + qDTY1.1 + qDTY3.1*	Drought, blast, BLB, brown plant hopper and gall midge	Dixit et al. (2020)
			IL7	*Pi9 + Xa4 + xa5+ Bph3 + Gm4+ qDTY1.1 + qDTY3.1*	Drought, blast, BLB, and gall midge	
			IL6	*Pi9+ Xa4 + xa5+ Xa21+ Bph3+ Bph17+ Gm4+ Gm8+ qDTY1.1+ qDTY3.1*	Drought, blast, BLB, brown plant hopper and gall midge	

(continued)

3.	IR64	N22 (*qDTY1.2, qDTY3.2, qDTY4.1, qDTY6.1* and *qDTY12.1*), and two heat-tolerance QTLs (*qSSPF10* and *qHT6*) N22 has two WUE-related QTLs (i.e., *Mrl9a* and *SLA9.1*)	PL425	*%N11.1+ SLA9.1 + ci/Ca2.1+ qWue7a + qDTY3.2 + qDTY6.1 + qSSPF10+ qHT6 + LP1 + qFLW4.1+ qFLL9*	Water use efficiency, drought, heat tolerance and grain yield	Withanawasam et al. (2022)
			PL130	*Mr19a + qWue7a + qDTY3.2+ qDTY4.1+ qDTY1.2 + qDTY12.1+* $qSSP_F10$*+ qHT6 + LP1 + qFLW4.1*	Water use efficiency, drought, heat tolerance and grain yield	
			PL76	*Mr19a + ci/Ca2.1+ qDTY3.2+ qDTY6.1 + qDTY1.2+* $qSSP_F10$*+ LP1 + GW5 + qFLW4.1+ qFLL9*	Water use efficiency, drought, heat tolerance and grain yield	
4.	Naveen	IRBL9 (*Pi9*), IRBB60 (*Xa4, xa5, xa13* and *Xa21*), however, they targeted only *Xa21*gene from IRBB60 Aganni (*Gm8*), IR96321-1447-561-B-1(*qDTY1.1*), IR87707–445-B (*qDTY2.2* and *qDTY4.1*)	MABC9	*Gm8 + Pi9 + Xa21 + qDTY1.1 + qDTY2.2 + qDTY4.1*	Drought, blast, BLB and gall midge	Janaki Ramayya et al. (2021)
			MABC1, MABC2 and MABC3	*Gm8 + Xa21 + qDTY1.1 + qDTY2.2 + qDTY4.1*	Drought, BLB and gall midge	
			MABC10	*Gm8 + Xa21 + qDTY1.1 + qDTY2.2 + qDTY4.1*	Drought, BLB and gall midge	
			MABC12	*Gm8 + Pi9 + Xa21 + qDTY1.1 + qDTY2.2*	Drought, blast, BLB and gall midge	

(continued)

Table 1.1 (continued)

S. No	Recurrent parent	Donor parent	Improved recurrent parent	Combination of QTLs/genes	Improved traits	Reference
5.	IR 91648-B-1-B-3-1	IR96322-34-223-B (*DTY1.1* + *qDTY2.1* + *qDTY3.1* and *Sub1*) IR74371-46-1-1 (*qDTY12.1*) WHD-1S-75-1-127 (*Pi9*) Tadukan (*Pita2*) IRBB60 (*Xa4, xa5, xa13* and *Xa21*) IRBB23 (*Xa23*) Abhaya (*Gm4*) Rathu Heenathi (*Bph3* and *Bph17*)	IR18L112	*qDTY1.1* + *qDTY2.1* + *qDTY3.1* + *qDTY12.1* + *Sub1* + *GM4* + *Pi9* + *xa5* + *Xa21* + *Bph17*	Drought, submergence, gall midge, blast, brown plant hoppers and bacterial leaf blight	Yadav et al. (2021)
			IR18L1137	*qDTY1.1* + *qDTY2.1* + *qDTY3.1* + *qDTY12.1* + *Sub1* + *GM4* + *Pi9* + *Xa21* + *Xa23* + *Bph17*	Drought, submergence, gall midge, blast, brown plant hoppers and bacterial leaf blight	
			IR18L1156	*qDTY1.1* + *qDTY12.1* + *Sub1* + *Pi9* + *Gm4* + *Xa4* + *xa5* + *Xa21* + *Xa23* + *Bph3*	Drought, submergence, gall midge, blast, brown plant hoppers and bacterial leaf blight	
6.	Reeta (CR Dhan 401)	Swarna-*Sub1* variety bearing *Sub1*, *Pup1*, and *GW5* QTLs, with agronomic traits submergence tolerance, low P (phosphorus) deficiency tolerance, and grain size, respectively	RSS45-83-27-54, RSS 45-83-27-235, RSS 45–83–27-415, RSS 45–83- 27-526, and RSS45–83–27-648	*Sub1* + *Pup1* + *GW5*	Submergence tolerance, low P (phosphorus) deficiency-tolerance, and grain size	Barik et al. (2023)
7.	Improved white Ponni (IWP)	Apo (*qDTY1.1*, *qDTY2.1*), Pokkali FL478 (*Saltol*), and FR13A (*Sub1*)	BIL 3-11-9-2, BIL 3-11-11-1, and BIL 3–11–11–2	*qDTY1.1* + *qDTY2.1* + *Saltol* + *Sub1*	Drought, salinity and submergence	Muthu et al. (2020)

(continued)

8.	Dongan	Indica-type Myanmar landrace for anaerobic germination (*AG1* and *AG2*)	SR35884-25- 1 – 1-B, SR35883-1-2 – 1-B, SR35884–60– 3 – 1-B and SR35884–66– 3 – 3-B	*AG1* + *AG2*	Anaerobic germination	Kim et al. (2019)
9.	Maudamani	Swarna-*Sub1* carrying *Sub1* QTL for submergence tolerance and yield component QTLs *SCM2* and *GW5*	MSS128-102-97-117 and MSS128-102-97-436 MSS128–102- 97-68, MSS128–102–97-335, MSS128–102–97-601, MSS128–102–97-617, and MSS128–102–97-613	*Sub1* + *SCM2* + *GW5*	Submergence and yield components	Pandit et al. (2021)
10.	Lalat (possess *Xa4*)	IR96321-1447-561-B 1(*qDTY1.1, qDTY3.1*), IR74371-46-1-1 (*qDTY12.1*), IRBL9 (*Pi9*) and IRBB60 (*Xa4, xa5, xa13, Xa21*)	IR144974-161-213-27 47-B	*qDTY1.1* + *qDTY3.1* + *qDTY12.1* + *Pi9* + *xa5* + *xa13* + *Xa21*	Drought, blast and BLB	Singh et al. (2022)
			IR144974-161-176-35 51-B	*qDTY3.1* + *qDTY12.1* + *Pi9*+ *Xa4* + *xa5* + *xa13* + *Xa21*	Drought, blast and BLB	
			IR143207-131-917-21 4-5-63-B	*qDTY1.1* + *qDTY3.1* + *qDTY12.1* + *Pi9* + *Xa4* + *xa5* + *Xa21*	Drought, blast and BLB	
11.	Improved Tapaswini (harbours *Xa4*, *xa5*, *xa13* and *Xa21*)	WHD-1S-75-1-127 (*O. Minuta* derivative-*Pi9*) and C1O1A51(*Pi2*), Kavya (*Gm1*), and Abhaya (*Gm4*) FR13A (*Sub1*), FL478 (*saltol*)	ITGP1, ITGP2, ITGP4, ITGP5	*Xa4* + *xa5* + *xa13* + *Xa21* + *Gm1* + *Gm4* + *Pi2* + *Pi9* + *Sub1* + *Saltol*	Submergence, salinity, blast, BLB and gall midge	Das et al. (2018)
			ITGP7	*Xa4* + *xa5* + *xa13* + *Xa21* + *Gm1* + *Gm4* + *Sub1* + *Saltol*	Submergence, salinity, blast, BLB and gall midge	

1.4 Conclusion

Creating commercially viable rice cultivars resistant to multiple stresses warrants efficient gene transfer, advanced breeding techniques, and a comprehensive understanding of regulatory processes. Leveraging the success of submergence tolerance and drought tolerance in farmers' fields, a parallel approach is undertaken for various abiotic stresses in high-yielding mega varieties, encountering challenges due to the complex genetic basis of these traits. Harmonized, interdisciplinary initiatives are crucial to cultivating advanced rice varieties, seamlessly blending tolerance to diverse abiotic stresses in vulnerable regions, and ensuring sustained production amidst unpredictable future climates. Further, harnessing molecular tools and next-generation sequencing paves the way for the strategic development of climate-resilient rice cultivars. Launching a global rice phenotyping network through GRiSP aligns with the genomic revolution, fostering the discovery of diverse gene pools for inducing multiple environmental stresses in various elite rice cultivars across the rice ecologies.

Author Contributions RS, conceptualized the framework of the content and sub-titles, RS, SM, and AS designed the work. RS, AS, SV and JBA gathered the literature related to the chapter. AS, SV, JBA, SS, SM and RS participated and wrote the chapter. RS, AS, SV, and SM reviewed and polished the manuscript. All authors contributed to the article and approved the submitted version.

References

Aghaee A, Moradi F, Zare-Maivan H, Zarinkamar F, Irandoost HP, Sharifi P (2011) Physiological responses of two rice (Oryza sativa L.) genotypes to chilling stress at seedling stage. Afr J Biotechnol 10:7617–7621

Ali J, Anumalla M, Murugaiyan V, Li Z (2021) Green super Rice (GSR) traits: breeding and genetics for multiple biotic and abiotic stress tolerance in rice. In: Rice improvement: physiological, molecular breeding and genetic perspectives. Springer International Publishing, Cham, pp 59–97

Andaya V, Mackill D (2003) QTLs conferring cold tolerance at the booting stage of rice using recombinant inbred lines from a japonicax indica cross. Theor Appl Genet 106:1084–1090

Anderson W, Han E, Baethgen W, Goddard L, Muñoz Á, Robertson A (2020) The madden-Julian oscillation affects maize yields throughout the tropics and subtropics. Geophys Res Lett 47:e2020GL087004

Anderson W, Baethgen W, Capitanio F, Ciais P, Cook BI, da Cunha CG, Goddard L, Schauberger B, Sonder K, Podestá G (2023) Climate variability and simultaneous breadbasket yield shocks as observed in long-term yield records. Agric For Meteorol 331:109321

Anjum SA, Ashraf U, Zohaib A, Tanveer M, Naeem M, Iftikhar Ali IA, Tahira Tabassum TT, Usman Nazir UN (2017) Growth and developmental responses of crop plants under drought stress: a review

Anonymous (2016) Statistical handbook. Directorate of Economics and Statistics

Babu NN, Krishnan SG, Vinod K, Krishnamurthy S, Singh VK, Singh MP, Singh R, Ellur RK, Rai V, Bollinedi H (2017) Marker aided incorporation of Saltol, a major QTL associated with seedling stage salt tolerance, into Oryza sativa 'Pusa basmati 1121'. Front Plant Sci 8:41

Barik SR, Pandit E, Pradhan SK, Mohanty SP, Mohapatra T (2019) Genetic mapping of morpho-physiological traits involved during reproductive stage drought tolerance in rice. PLoS One 14:e0214979

Barik J, Kumar V, Lenka SK, Panda D (2020) Assessment of variation in morpho-physiological traits and genetic diversity in relation to submergence tolerance of five indigenous lowland rice landraces. Rice Sci 27:32–43
Barik SR, Moharana A, Pandit E, Behera A, Mishra A, Mohanty SP, Mohapatra S, Sanghamitra P, Meher J, Pani DR (2023) Transfer of stress resilient QTLs and panicle traits into the Rice variety, Reeta through classical and marker-assisted breeding approaches. Int J Mol Sci 24:10708
Battisti DS, Naylor RL (2009) Historical warnings of future food insecurity with unprecedented seasonal heat. Science 323:240–244
Bewley JD (1997) Seed germination and dormancy. Plant Cell 9:1055
Bimpong IK, Manneh B, Sock M, Diaw F, Amoah NKA, Ismail AM, Gregorio G, Singh RK, Wopereis M (2016) Improving salt tolerance of lowland rice cultivar assi' through marker-aided backcross breeding in West Africa. Plant Sci 242:288–299
Bunnag S, Pongthai P (2013) Selection of rice (Oryza sativa L.) cultivars tolerant to drought stress at the vegetative stage under field conditions. Am J Plant Sci 4:1701
Cai W, Borlace S, Lengaigne M, Van Rensch P, Collins M, Vecchi G, Timmermann A, Santoso A, McPhaden MJ, Wu L (2014) Increasing frequency of extreme El Niño events due to greenhouse warming. Nat Clim Chang 4:111–116
Chhabra R, Singh T (2019) Seed aging, storage and deterioration: an irresistible physiological phenomenon. Agric Rev 40:234–238
Chinnusamy V, Jagendorf A, Zhu JK (2005) Understanding and improving salt tolerance in plants. Crop Sci 45:437–448
Cramer GR, Urano K, Delrot S, Pezzotti M, Shinozaki K (2011) Effects of abiotic stress on plants: a systems biology perspective. BMC Plant Biol 11:1–14
Dagar J, Yadav R (2017) Climate resilient approaches for enhancing productivity of saline agriculture. J Soil Salinity Water Qual 9:9–29
Das G, Rao GJ, Varier M, Prakash A, Prasad D (2018) Improved Tapaswini having four BB resistance genes pyramided with six genes/QTLs, resistance/tolerance to biotic and abiotic stresses in rice. Sci Rep 8:2413
Deepti Singh DS, Seager R, Cook B, Cane M, Ting MingFang TM, Cook E, Davis M (2019) Climate and the global famine of 1876–78. J Clim 31(23):9445–9467
Dey M, Upadhyaya H (1996) Yield loss due to drought, cold and submergence tolerance. Rice research in Asia: progress and priorities. International Rice research Institute in Collaboration with CAB international, UK
Diwan J, Channbyregowda M, Shenoy V, Salimath P, Bhat R (2013) Molecular mapping of early vigour related QTLs in rice. Res J Biol 1:24–30
Dixit S, Singh UM, Singh AK, Alam S, Venkateshwarlu C, Nachimuthu VV, Yadav S, Abbai R, Selvaraj R, Devi MN (2020) Marker assisted forward breeding to combine multiple biotic-abiotic stress resistance/tolerance in rice. Rice 13:1–15
Djanaguiraman M, Narayanan S, Erdayani E, Prasad PV (2020) Effects of high temperature stress during anthesis and grain filling periods on photosynthesis, lipids and grain yield in wheat. BMC Plant Biol 20:1–12
Evenson RE, Gollin D (2003) Assessing the impact of the green revolution, 1960 to 2000. Science 300:758–762
Fahad S, Bajwa AA, Nazir U, Anjum SA, Farooq A, Zohaib A, Sadia S, Nasim W, Adkins S, Saud S (2017) Crop production under drought and heat stress: plant responses and management options. Front Plant Sci 8:1147
FAO (2021) FAOSTAT. Food and Agriculture Organization of the United Nations, Rome, Italy
Farooq M, Wahid A, Kobayashi N, Fujita D, Basra SM (2009) Plant drought stress: effects, mechanisms and management. Sustain Agric:153–188
Flowers T, Flowers S (2005) Why does salinity pose such a difficult problem for plant breeders? Agric Water Manag 78:15–24
Gistemp T (2018) GISS surface temperature analysis (GISTEMP). NASA Goddard Institute for Space Studies. Accessed on the internet at https://data.giss.nasa.gov/gistemp on 23
Gourdji SM, Sibley AM, Lobell DB (2013) Global crop exposure to critical high temperatures in the reproductive period: historical trends and future projections. Environ Res Lett 8:024041

Guo X, Liu D, Chong K (2018) Cold signaling in plants: insights into mechanisms and regulation. J Integr Plant Biol 60:745–756
Gupta V, Kumar Jain M, Singh VP (2020) Multivariate modeling of projected drought frequency and hazard over India. J Hydrol Eng 25:04020003
Hammer G, Hansen J, Phillips J, Mjelde J, Hill H, Love A, Potgieter A (2001) Advances in application of climate prediction in agriculture. Agric Syst 70:515–553
Hao Z, Hao F, Singh VP, Zhang X (2018) Changes in the severity of compound drought and hot extremes over global land areas. Environ Res Lett 13:124022
IPCC (2014) Mitigation of climate change. In: Contribution of working group III to the fifth assessment report of the intergovernmental panel on climate change, vol 1454, p 147
Jagadish S, Muthurajan R, Oane R, Wheeler TR, Heuer S, Bennett J, Craufurd PQ (2010) Physiological and proteomic approaches to address heat tolerance during anthesis in rice (Oryza sativa L.). J Exp Bot 61:143–156
Jagadish KS, Craufurd P, Shi W, Oane R (2013) A phenotypic marker for quantifying heat stress impact during microsporogenesis in rice (Oryza sativa L.). Funct Plant Biol 41:48–55
Jagadish S, Murty M, Quick W (2015) Rice responses to rising temperatures–challenges, perspectives and future directions. Plant Cell Environ 38:1686–1698
Janaki Ramayya P, Vinukonda VP, Singh UM, Alam S, Venkateshwarlu C, Vipparla AK, Dixit S, Yadav S, Abbai R, Badri J (2021) Marker-assisted forward and backcross breeding for improvement of elite Indian rice variety Naveen for multiple biotic and abiotic stress tolerance. PLoS One 16:e0256721
Kaw R, Khush G (1986) Combining ability for low-temperature tolerance in rice. In: Rice genetics I: (in 2 parts). World Scientific, pp 593–612
Khataar M, Mohammadi MH, Shabani F (2018) Soil salinity and matric potential interaction on water use, water use efficiency and yield response factor of bean and wheat. Sci Rep 8:2679
Kim SM, Kim CS, Jeong JU, Reinke RF, Jeong JM (2019) Marker-assisted breeding for improvement of anaerobic germination in japonica rice (Oryza sativa). Plant Breed 138:810–819
Kondo M, Murty MV, Aragones DV (2000) Characteristics of root growth and water uptake from soil in upland rice and maize under water stress. Soil Sci Plant Nutr 46:721–732
Korres N, Norsworthy J, Burgos N, Oosterhuis D (2017) Temperature and drought impacts on rice production: an agronomic perspective regarding short-and long-term adaptation measures. Water Resources Rural Dev 9:12–27
Kumar C (2014) Correlation and path coefficient analysis of yield components in aerobic rice (Oryza sativa L.). Bioscan 9:907–913
Kumar V, Ladha JK (2011) Direct seeding of rice: recent developments and future research needs. Adv Agron 111:297–413
Kumar A, Bernier J, Verulkar S, Lafitte H, Atlin G (2008) Breeding for drought tolerance: direct selection for yield, response to selection and use of drought-tolerant donors in upland and lowland-adapted populations. Field Crop Res 107:221–231
Kumar P, Singh G, Singh PD, Singh T, Singh A, Lakra K (2021) Effect of different crop establishment methods and nitrogen levels on the performance of Kharif season rice (Oryza sativa) in the indo-Gangatic plains of eastern Uttar Pradesh. Crop Res 56:1–7
Kumar V, Mahajan G, Sheng Q, Chauhan BS (2023) Weed management in wet direct-seeded rice (Oryza sativa L.): issues and opportunities. Adv Agron 179:91–133
Linh LH, Linh TH, Xuan TD, Ham LH, Ismail AM, Khanh TD (2012) Molecular breeding to improve salt tolerance of rice (Oryza sativa L.) in the red River Delta of Vietnam. Int J Plant Genomics 2012:1
Liu C-T, Wang W, Mao B-G, Chu C-C (2018) Cold stress tolerance in rice: physiological changes, molecular mechanism, and future prospects. Yi Chuan = Hereditas 40:171–185
Liu W, Pan X, Xiong H, Huang F, Duan Y, Li Y, Min J, Liu S, Liu L, Wei X (2020) QTL mapping and validation of a segment responsible for early-seedling cold tolerance in rice. Euphytica 216:1–9
Lobell DB, Burke MB, Tebaldi C, Mastrandrea MD, Falcon WP, Naylor RL (2008) Prioritizing climate change adaptation needs for food security in 2030. Science 319:607–610

Lutts S, Kinet J, Bouharmont J (1995) Changes in plant response to NaCl during development of rice (Oryza sativa L.) varieties differing in salinity resistance. J Exp Bot 46:1843–1852
Lv Y, Hussain MA, Luo D, Tang N (2019) Current understanding of genetic and molecular basis of cold tolerance in rice. Mol Breed 39:1–18
Maclean JL (2002) Rice almanac: source book for the most important economic activity on earth. International Rice Research Institute
Majhi PK, Singh SK, Anandan A, Khaire AR, Korada M, Habde SV, Singh DK, Bhagvan AP, Jayasudha S, Singh A (2022) Parental polymorphism survey for evaluation and selection of contrasting parents for drought tolerance in rice (Oryza sativa L.) by using SSR markers. Int J Environ Clim 12:2507–2519
Mb M (1999) Seed deterioration: physiology, repair and assessment. Seed Sci Technol 27:177–237
Mishra M, Wungrampha S, Kumar G, Singla-Pareek SL, Pareek A (2021) How do rice seedlings of landrace Pokkali survive in saline fields after transplantation? Physiology, biochemistry, and photosynthesis. Photosynth Res 150:117–135
Mohammed AR, Tarpley L (2011) Effects of night temperature, spikelet position and salicylic acid on yield and yield-related parameters of rice (Oryza sativa L.) plants. J Agron Crop Sci 197:40–49
Moraes de Freitas GP, Basu S, Ramegowda V, Thomas J, Benitez LC, Braga EB, Pereira A (2019) Physiological and transcriptional responses to low-temperature stress in rice genotypes at the reproductive stage. Plant Signal Behav 14:e1581557
Muthu V, Abbai R, Nallathambi J, Rahman H, Ramasam S, Kambale R, Thulasinathan T, Ayyenar B, Muthurajan R (2020) Pyramiding QTLs controlling tolerance against drought, salinity, and submergence in rice through marker assisted breeding. PLoS One 15:e0227421
Najafi E, Pal I, Khanbilvardi R (2020) Larger-scale ocean-atmospheric patterns drive synergistic variability and worldwide volatility of wheat yields. Sci Rep 10:5193
Nakashima K, Tran LSP, Van Nguyen D, Fujita M, Maruyama K, Todaka D, Ito Y, Hayashi N, Shinozaki K, Yamaguchi-Shinozaki K (2007) Functional analysis of a NAC-type transcription factor OsNAC6 involved in abiotic and biotic stress-responsive gene expression in rice. Plant J 51:617–630
Nations U (2019) 2019 revision of world population prospects
Neeraja CN, Maghirang-Rodriguez R, Pamplona A, Heuer S, Collard BC, Septiningsih EM, Vergara G, Sanchez D, Xu K, Ismail AM (2007) A marker-assisted backcross approach for developing submergence-tolerant rice cultivars. Theor Appl Genet 115:767–776
O'Toole JC (2004) Rice and water: the final frontier. Rockefeller Foundation Bangkok, Thailand
Oladosu Y, Rafii MY, Samuel C, Fatai A, Magaji U, Kareem I, Kamarudin ZS, Ii M, Kolapo K (2019) Drought resistance in rice from conventional to molecular breeding: a review. Int J Mol Sci 20:3519
Oladosu Y, Rafii MY, Arolu F, Chukwu SC, Muhammad I, Kareem I, Salisu MA, Arolu IW (2020) Submergence tolerance in rice: review of mechanism, breeding and, future prospects. Sustain For 12:1632
Organization FaA (2008) Land and plant nutrition management service. Available online: https://www.fao.org/documents/card/en/c/cb9929en/. Accessed on 22 January 2023
Ozga JA, Kaur H, Savada RP, Reinecke DM (2017) Hormonal regulation of reproductive growth under normal and heat-stress conditions in legume and other model crop species. J Exp Bot 68:1885–1894
Pandit E, Pawar S, Barik SR, Mohanty SP, Meher J, Pradhan SK (2021) Marker-assisted backcross breeding for improvement of submergence tolerance and grain yield in the popular rice variety 'Maudamani'. Agronomy 11:1263
Pareek A, Dhankher OP, Foyer CH (2020) Mitigating the impact of climate change on plant productivity and ecosystem sustainability. J Exp Bot 71:451–456
Pendergrass AG, Meehl GA, Pulwarty R, Hobbins M, Hoell A, AghaKouchak A, Bonfils CJ, Gallant AJ, Hoerling M, Hoffmann D (2020) Flash droughts present a new challenge for subseasonal-to-seasonal prediction. Nat Clim Chang 10:191–199

Pradhan S, Pandit E, Pawar S, Bharati B, Chatopadhyay K, Singh S, Dash P, Reddy J (2019) Association mapping reveals multiple QTLs for grain protein content in rice useful for biofortification. Mol Gen Genomics 294:963–983

Prasad P, Boote K, Allen L Jr, Sheehy J, Thomas J (2006) Species, ecotype and cultivar differences in spikelet fertility and harvest index of rice in response to high temperature stress. Field Crop Res 95:398–411

Quilloy FA, Labaco B, Casal C, Dixit S (2021) Crop establishment in direct-seeded rice: traits, physiology, and genetics. Rice Improvement:171–202

Rahman S, Haque T, Rahman MS, Seraj Z (2008) Salt tolerant BR11 and salt tolerant BR28 through marker assisted backcrossing (MAB). In: Proceedings of the International scientific conference on, p 50

Rathna Priya TS, Eliazer Nelson ARL, Ravichandran K, Antony U (2019) Nutritional and functional properties of coloured rice varieties of South India: a review. J Ethnic Foods 6:11. https://doi.org/10.1186/s42779-019-0017-3

Rehman Arif M, Nagel M, Neumann K, Kobiljski B, Lohwasser U, Börner A (2012) Genetic studies of seed longevity in hexaploid wheat using segregation and association mapping approaches. Euphytica 186:1–13

Rivero RM, Mittler R, Blumwald E, Zandalinas SI (2022) Developing climate-resilient crops: improving plant tolerance to stress combination. Plant J 109:373–389

Sahebi M, Hanafi MM, Rafii M, Mahmud T, Azizi P, Osman M, Abiri R, Taheri S, Kalhori N, Shabanimofrad M (2018) Improvement of drought tolerance in rice (Oryza sativa L.): genetics, genomic tools, and the WRKY gene family. Biomed Res Int 2018:1

Sales E, Viruel J, Domingo C, Marqués L (2017) Genome wide association analysis of cold tolerance at germination in temperate japonica rice (Oryza sativa L.) varieties. PLoS One 12:e0183416

Sandhu N, Dixit S, Swamy B, Raman A, Kumar S, Singh S, Yadaw R, Singh O, Reddy J, Anandan A (2019) Marker assisted breeding to develop multiple stress tolerant varieties for flood and drought prone areas. Rice 12:1–16

Sandhu N, Sagare DB, Singh VK, Yadav S, Kumar A (2021) Environment-friendly direct seeding rice technology to foster sustainable rice production. In: Scaling-up solutions for farmers: technology, partnerships and convergence, pp 279–305

Sarkar R, Reddy J, Sharma S, Ismail AM (2006) Physiological basis of submergence tolerance in rice and implications for crop improvement. Curr Sci:899–906

Sarkar R, Panda D, Reddy J, Patnaik S, Mackill DJ, Ismail AM (2009) Performance of submergence tolerant rice (Oryza sativa) genotypes carrying the Sub1 quantitative trait locus under stressed and non-stressed natural field conditions. Indian J Agric Sci 79(11):876–883

Senguttuvel P, Jaldhani V, Raju NS, Balakrishnan D, Beulah P, Bhadana VP, Mangrauthia SK, Neeraja CN, Subrahmanyam D, Rao PR (2022) Breeding rice for heat tolerance and climate change scenario, possibilities and way forward. A review. Arch Agron Soil Sci 68:115–132

Setter T, Ellis M, Laureles E, Ella E, Senadhira D, Mishra S, Sarkarung S, Datta S (1997) Physiology and genetics of submergence tolerance in rice. Ann Bot 79:67–77

Sharma N, Khanna R (2019) Rice grain quality: current developments and future prospects. Recent Adv Grain Crops Res 5772:89367

Shekhawat K, Rathore SS, Chauhan BS (2020) Weed management in dry direct-seeded rice: a review on challenges and opportunities for sustainable rice production. Agronomy 10:1264

Shimono H, Abe A, Aoki N, Koumoto T, Sato M, Yokoi S, Kuroda E, Endo T, Ki S, Nagano K (2016) Combining mapping of physiological quantitative trait loci and transcriptome for cold tolerance for counteracting male sterility induced by low temperatures during reproductive stage in rice. Physiol Plant 157:175–192

Siddiq E, Vemireddy LR (2021) Advances in genetics and breeding of rice: an overview. In: Rice improvement: physiological, molecular breeding and genetic perspectives, pp 1–29

Singh AK, Ansari MW, Pareek A, Singla-Pareek SL (2008) Raising salinity tolerant rice: recent progress and future perspectives. Physiol Mol Biol Plants 14:137–154

Singh S, Mackill DJ, Ismail AM (2009) Responses of SUB1 rice introgression lines to submergence in the field: yield and grain quality. Field Crop Res 113:12–23

Singh A, Gopalakrishnan S, Singh V, Prabhu K, Mohapatra T, Singh N, Sharma T, Nagarajan M, Vinod K, Singh D (2011) Marker assisted selection: a paradigm shift in basmati breeding. Indian J Genet Plant Breed 71:120

Singh R, Singh Y, Xalaxo S, Verulkar S, Yadav N, Singh S, Singh N, Prasad K, Kondayya K, Rao PR (2016) From QTL to variety-harnessing the benefits of QTLs for drought, flood and salt tolerance in mega rice varieties of India through a multi-institutional network. Plant Sci 242:278–287

Singh SK, Majhi PK, Anandan A, Korada M, Habde SV, Khaire AR, Singh DK, Jayasudha S, Singh A, Bhagvan AP (2022) Microsatellites based parental polymorphism survey for moisture stress in Rice (Oryza sativa L.) between the parental genotypes HUR-1309 and CR Dhan 801. Emergent Life Sci Res 8:248–258

Singh A, Rami E, Upadhyay P, Gangawane AK (2023) The impact of climate change on crop production and combat strategies. In: Climate change and sustainable development. CRC Press, pp 103–115

Sridevi V, Chellamuthu V (2015) Impact of weather on rice—a review. Int J Appl Res 1:825–831

Suh J, Jeung J, Lee J, Choi Y, Yea J, Virk P, Mackill D, Jena K (2010) Identification and analysis of QTLs controlling cold tolerance at the reproductive stage and validation of effective QTLs in cold-tolerant genotypes of rice (Oryza sativa L.). Theor Appl Genet 120:985–995

Susanti Z, Snell P, Fukai S, Mitchell JH (2019) Importance of anther dehiscence for low-temperature tolerance in rice at the young microspore and flowering stages. Crop Pasture Sci 70:113–120

Swamy BM, Kumar A (2013) Genomics-based precision breeding approaches to improve drought tolerance in rice. Biotechnol Adv 31:1308–1318

Thomson MJ, de Ocampo M, Egdane J, Rahman MA, Sajise AG, Adorada DL, Tumimbang-Raiz E, Blumwald E, Seraj ZI, Singh RK (2010) Characterizing the Saltol quantitative trait locus for salinity tolerance in rice. Rice 3:148–160

Trenberth KE (2011) Changes in precipitation with climate change. Clim Res 47:123–138

Van Houtan KS, Tanaka KR, Gagné TO, Becker SL (2021) The geographic disparity of historical greenhouse emissions and projected climate change. Sci Adv 7:eabe4342

Von der Gathen P, Kivi R, Wohltmann I, Salawitch RJ, Rex M (2021) Climate change favours large seasonal loss of Arctic ozone. Nat Commun 12:3886

Wing RA, Purugganan MD, Zhang Q (2018) The rice genome revolution: from an ancient grain to green super Rice. Nat Rev Genet 19:505–517

Withanawasam DM, Kommana M, Pulindala S, Eragam A, Moode VN, Kolimigundla A, Puram RV, Palagiri S, Balam R, Vemireddy LR (2022) Improvement of grain yield under moisture and heat stress conditions through marker-assisted pedigree breeding in rice (Oryza sativa L.). Crop Pasture Sci 73:356–369

Xie G, Kato H, Sasaki K, Imai R (2009) A cold-induced thioredoxin h of rice, OsTrx23, negatively regulates kinase activities of OsMPK3 and OsMPK6 in vitro. FEBS Lett 583:2734–2738

Yadav S, Sandhu N, Dixit S, Singh VK, Catolos M, Mazumder RR, Rahman MA, Kumar A (2021) Genomics-assisted breeding for successful development of multiple-stress-tolerant, climate-smart rice for southern and southeastern Asia. Plant Genome 14:e20074

Yang Y, Yu J, Qian Q, Shang L (2022) Enhancement of heat and drought stress tolerance in rice by genetic manipulation: a systematic review. Rice 15:67

Ye C, Li X, Redoña E, Ishimaru T, Jagadish K (2021) Genetics and breeding of heat tolerance in rice. In: Rice improvement: physiological, molecular breeding and genetic perspectives. Springer International Publishing, Cham, pp 203–220

Yu S, Ali J, Zhang C, Li Z, Zhang Q (2020) Genomic breeding of green super rice varieties and their deployment in Asia and Africa. Theor Appl Genet 133:1427–1442

Yuan S, Linquist BA, Wilson LT, Cassman KG, Stuart AM, Pede V, Miro B, Saito K, Agustiani N, Aristya VE (2021) Sustainable intensification for a larger global rice bowl. Nat Commun 12:7163

Zafar SA, Arif MH, Uzair M, Rashid U, Naeem MK, Rehman OU, Rehman N, Zaid IU, Farooq MS, Zahra N (2022) Agronomic and physiological indices for reproductive stage heat stress tolerance in green super rice. Agronomy 12:1907

Zaman M, Shahid SA, Heng L, Shahid SA, Zaman M, Heng L (2018) Soil salinity: historical perspectives and a world overview of the problem. In: Guideline for salinity assessment, mitigation and adaptation using nuclear and related techniques, pp 43–53

Zhang Q, Chen Q, Wang S, Hong Y, Wang Z (2014) Rice and cold stress: methods for its evaluation and summary of cold tolerance-related quantitative trait loci. Rice 7:1–12

Zhang C, Yu S, Zhang Q (2018) Recent advances in green super rice development. Chin Bull Life Sci 30:1083–1089

Zhang C, Li M, Rey JD, Feng T, Lafitte R, Zheng T, Lv Y, Wu F, Fu B, Xu J (2023) Simultaneous improvement and genetic dissection of drought and submergence tolerances in rice (Oryza sativa L.) by selective introgression. Front Plant Sci 14:1134450

Zhao C, Liu B, Piao S, Wang X, Lobell DB, Huang Y, Huang M, Yao Y, Bassu S, Ciais P (2017) Temperature increase reduces global yields of major crops in four independent estimates. Proc Natl Acad Sci 114(35):9326–9331

Zhen F, Wang W, Wang H, Zhou J, Liu B, Zhu Y, Liu L, Cao W, Tang L (2019) Effects of short-term heat stress at booting stage on rice-grain quality. Crop Pasture Sci 70:486–498

Zhou L, Zeng Y, Zheng W, Tang B, Yang S, Zhang H, Li J, Li Z (2010) Fine mapping a QTL qCTB7 for cold tolerance at the booting stage on rice chromosome 7 using a near-isogenic line. Theor Appl Genet 121:895–905

Zhu J-K (2016) Abiotic stress signaling and responses in plants. Cell 167:313–324

Zhu J, Dong C-H, Zhu J-K (2007) Interplay between cold-responsive gene regulation, metabolism and RNA processing during plant cold acclimation. Curr Opin Plant Biol 10:290–295

Untapped Rice Genetic Resources and Pre-breeding in Genomics Era

2

Jitendra Kumar Tiwari and Ashish Kumar Tiwari

Abstract

Rice production faces mounting challenges due to factors such as population growth, shifting dietary preferences, and climate variability. To ensure sustainable rice cultivation, it's imperative to tap into the diverse genetic resources of rice. This chapter explores the significance of untapped genetic resources in enhancing rice productivity and resilience, explain the concept of pre-breeding and its role in modern rice breeding programs and discuss the advancements in genomic tools and their application in pre-breeding for untapped potential of wild and traditional rice varieties and their integration into modern breeding strategies through pre-breeding in the genomics era. These rice genetic resources possess a diverse range of traits, including resistance to multiple diseases (like blast, bacterial blight, sheath blight and tungro virus), pests (such as stem borers, gall midges and plant hoppers) and various abiotic stresses (including drought, submergence, heat, cold, salinity and soil toxicities). Moreover, they provide nutritional improvements and the ability to adapt to evolving environmental conditions. Despite their potential, the utilization of these resources has been constrained by challenges like inadequate characterization and limited availability. Nonetheless, unlocking these resources holds significant promise for preserving biodiversity, bolstering stress resilience, and addressing nutritional deficiencies. In the genomics era, pre-breeding has undergone a transformative evolution, enabled by the integration of high-capacity sequencing, advanced phenotyping

J. K. Tiwari (✉)
Department of Genetics and Plant Breeding, Rani Lakshmi Bai Central Agricultural University, Jhansi, Uttar Pradesh, India

A. K. Tiwari
Department of Genetics and Plant Breeding, College of Agriculture, IGKV, Raipur, Chhattisgarh, India

A. Singh et al. (eds.), *Climate-Smart Rice Breeding*,
https://doi.org/10.1007/978-981-97-7098-4_2

techniques, and bioinformatics tools. This revolutionizes the identification and integration of favorable traits from untapped resources to elite cultivars. Key steps in the pre-breeding process include trait identification, genomic analysis, gene introgression, and the utilization of molecular markers for precise trait transfer. However, challenges persist, including ethical and regulatory considerations regarding genetic modification, intellectual property rights, and the necessity to balance introduced traits with agronomic characteristics. To fully unlock the potential of untapped rice genetic resources through pre-breeding, collaborative efforts involving researchers, policymakers, and farmers are paramount.

Keywords

Rice · Trait identification · Genomic analysis · Molecular markers · Genetic resources · Pre-breeding

2.1 Introduction

Rice is a crucial staple crop worldwide, supplying essential nutrients to more than half of the global population. Despite significant breeding efforts, numerous genetic resources within rice remain underutilized, representing a valuable reservoir of traits that could improve yield, disease resistance, and environmental adaptability. The significance of these untapped genetic assets lies in their potential to tackle future food security challenges arising from climate change, population growth, and shifting agricultural demands (McCouch et al. 2016; Xie et al. 2018). Exploration and utilization of these untapped resources are crucial for the continued development of superior rice cultivars. Wild relatives and landraces of rice harbor a wealth of genetic diversity that modern breeding programs can exploit. This genetic diversity is essential for introducing new traits into cultivated varieties, such as resilience to abiotic stresses (salinity, drought, and extreme temperatures) and resistance to biotic stresses (insects and diseases) (Brar and Khush 2018). Pre-breeding, also known as germplasm enhancement, involves the identification and transfer of desirable traits from wild or unadapted germplasm into breeding lines that are more easily used in commercial breeding programs. This process often includes the initial stages of hybridization, selection, and evaluation to develop lines that possess target traits and are compatible with elite breeding materials (Pérez-de-Castro et al. 2012). Rice pre-breeding plays a central role in addressing the world's escalating food security concerns. In response to climate change and an ever-increasing global population, it is essential to develop rice cultivars that can adapt to new challenges while ensuring sustainable agriculture. The primary drivers of pre-breeding are the creation and introduction of genetic diversity and the utilization of wild relatives to elevate the genetic pool of cultivated rice.In the genomic era, pre-breeding has gained significant importance due to advancements in molecular biology and genomics tools. These tools facilitate the precise detection and alteration of genes linked to desirable traits. Approaches including marker-assisted selection (MAS), gene editing (CRISPR/Cas9), and genomic selection (GS) have revolutionized pre-breeding by enabling the rapid and accurate transfer of beneficial traits (Varshney et al. 2019).

The application of genomics in pre-breeding enhances the efficiency of developing new rice varieties with improved yield, nutritional quality, and stress tolerance. By integrating high-throughput phenotyping and genotyping platforms, researchers can more effectively screen large populations and identify promising candidates for further breeding (Li et al. 2018).

Genomics has been a game-changer in rice pre-breeding. The complete sequencing of the rice genome, combined with advanced sequencing technologies, has provided an intricate understanding of the genetic makeup of rice. This knowledge has paved the way for the identification of key genes and markers associated with desirable traits, allowing breeders to make informed selections during the breeding process. Furthermore, comparative genomics has opened doors to leverage genetic information from wild relatives of rice, broadening the genetic diversity available for breeding.Untapped rice genetic resources, along with the cutting-edge tools of genomics, offer promising avenues to achieve these goals.

Research on the utilization of untapped rice genetic resources and pre-breeding has yielded promising results. For example (Xu et al. 2016), demonstrated the successful introgression of disease resistance genes from wild rice into elite cultivars using MAS. Similarly (Swamy et al. 2017), reported the identification of novel QTLs for yield under drought conditions from wild rice accessions, which were then introduced into high-yielding varieties.

Recent studies have also highlighted the potential of genomic selection in pre-breeding programs. A study by Spindel et al. (2015) showed that genomic selection could significantly accelerate the breeding cycle and improve the accuracy of selecting desirable traits. Furthermore, the use of CRISPR/Cas9 has enabled precise gene editing, allowing for the targeted improvement of specific traits such as disease resistance and grain quality (Li et al. 2020).

Additionally, pre-breeding processes are essential for identifying favorable traits or genes from unadapted materials that are not immediately suitable for breeding populations. These traits are then transferred to an intermediate set of materials that breeders can use to develop new varieties for farmers. Pre-breeding serves as a crucial initial step in harnessing the diversity found in wild relatives and other unimproved materials.

Genetic diversity is the cornerstone for developing sustainable solutions to the current and future challenges posed by various biotic and abiotic pressor. Pre-breeding offers a unique opportunity to utilize this genetic diversity by transferring desirable traits or innovative techniques enable the transfer of genes from wild crop relatives to cultivated varieties. Germplasm enhancement involves all preliminary plant breeding research prior to cultivar development, testing, and release. Several factors drive the need for pre-breeding: the limited genetic diversity of current crops, the extensive distribution of genetically uniform contemporary varieties, the impact of climate change necessitating better-adapted genes, and the evolving populations of pests and pathogens requiring new resistance genes. Pre-breeding activities include characterizing landraces/accessions, establishing new base populations, inducing polyploidy, employing novel breeding techniques, conducting wide crosses, and generating double haploids. These activities transform germplasm for immediate use in conventional breeding programs. Despite facing various challenges, pre-breeding presents numerous opportunities to leverage its advantages for future agricultural resilience and productivity.

The chief objective of this book chapter is to furnish a comprehensive overview of the current status, challenges, and opportunities in the utilization of untapped rice genetic resources through pre-breeding.

2.2 Rice Germplasm

The genetic diversity in rice species is fundamental for global economies, as rice is an essential cereal crop that provides sustenance to over half of the global population. The genus *Oryza*, part of the extensive Poaceae family, comprises more than 20 grass species, including the two cultivated rice species: Asian rice (*Oryza sativa*) and African rice (*Oryza glaberrima*). *Oryza sativa*, originating in South and Southeast Asia, is now grown globally, while *Oryza glaberrima* is cultivated in Africa. Over thousands of years, farmers have selectively bred rice to meet their diverse needs, resulting in the extensive genetic diversity seen in the multitude of rice varieties today. *Oryza sativa* includes both primitive varieties (landraces) and improved ones, estimated to number over 140,000. This vast diversity forms a vital component of the primary rice gene pool, essential for further advancements in rice cultivars. According to Harlan and de Wet (1971), the primary gene pool includes races that typically belong to the same biological species as the cultivated varieties and can produce viable hybrids when crossed with them, allowing for the introduction of new genetic variations through sexual hybridization. Research into the genetic structure of the *Oryza* genus (AA genome) has been extensive, involving morphological, cytogenetic, interspecific hybridization, biochemical, and molecular marker studies. Most experts agree that *O. glaberrima* originated in West Africa from *O. barthii*, also known as *O. breviligulata*. Various hypotheses exist regarding the domestication of *O. sativa*. Some suggest it was domesticated from the perennial type of common wild rice *O. rufipogon* or the annual common wild rice *O. nivara*. However, the most consistent hypothesis, supported by isozyme and DNA marker data, suggests a diphyletic origin of *O. sativa*. This theory proposes two independent domestications on either side of the Himalayan barrier from *O. nivara* populations, leading to the two main subspecies of *O. sativa*, indica and japonica. Continuous hybridization processes have contributed to the current structure of *O. sativa*. The most practical categorization of *O. sativa* up to date has been achieved through isozyme studies. Glaszmann 1987 successfully grouped *O. sativa* cultivar into six clusters. The primary groups, I and VI, correspond to the indica and japonica cultivars, respectively. This classification, which aligns with other biological criteria, has gained widespread acceptance among plant breeders. However, it is important to note that the isozyme-defined groups should not be seen as completely distinct. Although there are discriminant marker loci between these groups, recent research has shown that indica and japonica varieties are not "pure" but rather combinations of both indica and japonica alleles. This hybridity results from the ongoing hybridization process between indica and japonica varieties. In addition to the cultivated rice species and their immediate predecessors, there are over a dozen other wild species within the *Oryza* genus, distributed across tropical regions.

These species are distributed across various regions, including Asia, Africa, South and Central America, and Australia (Table 2.1).The *O. sativa* complex includes species like *O. glaberrima, O. barthii, O. longistaminata, O. glumaepatula, O. nivara, O. meridionalis* and *O. rufipogon*. These species are notable for their resistance to various pests, such as green leafhopper, grassy stunt virus, bacterial blight, blast, and stem rot. They also possess traits like drought avoidance, elongation ability, and cytoplasmic male sterility, which are critical for developing rice varieties capable of thriving under adverse environmental conditions. For instance, *O. longistaminata* from Africa is known for its resistance to bacterial blight and drought tolerance, making it a valuable resource for breeding programs aimed at improving stress resistance in rice.

Similarly, the *O. officinalis* complex includes species such as *O. australiensis, O. alta, O. eichingeri, O. grandiglumis, O. latifolia, O. minuta, O. officinalis, O. punctata*, and *O. rhizomatis*. These species are distributed across diverse regions, including tropical Asia, Africa, and the Americas, and offer resistance to a wide range of pests like brown planthopper, yellow mottle virus, thrips, and sheath blight. They also exhibit traits such as high biomass production and drought tolerance. For example, *O. australiensis* from tropical Australia is valued for its drought tolerance and resistance to brown planthopper, traits that are increasingly important as climate change exacerbates water scarcity and pest pressures.

The *O. ridleyi* and *O. meyeriana* complexes, including species like *O. ridleyi, O. meyeriana, O. granulata*, and *O. longiglumis*, contribute additional traits such as resistance to pests like stem borer, whorl maggot, and blast, and shade tolerance, adaptation to aerobic soil. These traits are particularly beneficial for expanding rice cultivation into non-traditional areas and improving sustainability in rice farming systems.

The wild species within the *Oryza* genus possess distinct genomes and exhibit varying bio-systematic relationships with cultivated rice. They can be classified into either the secondary gene pool, which includes races, or species that can cross with cultivated species but have restricted gene transfer capabilities, or the tertiary gene pool, which comprises races or species that are difficult to cross with cultivated varieties and require special genetic techniques for gene transfer. The classification of species into these categories depends on their genetic proximity to cultivated rice and the level of difficulty associated with transferring genes from these species to cultivars (Fig. 2.1). Additionally, there are several species in related genera within the *Oryzeae* tribe. These species are distantly related to cultivated rice and are considered part of the tertiary gene pool for rice (Table 2.1).

Table 2.1 Rice germplasm and their source institutions

Germplasm	Species	Genome	Source institution	Useful traits
IR64	*O. Sativa*	AA	International Rice research institute (IRRI)	High yield, disease resistance
NERICA (new Rice for Africa)	*O. Glaberrima x O. Sativa*	AA	Africa Rice Center (AfricaRice)	Drought tolerance, early maturity
Kasalath	*O. Sativa*	AA	National Institute of Agrobiological sciences (NIAS), Japan	Phosphorus uptake efficiency
Pokkali	*O. Sativa*	AA	Kerala agricultural university, India	Salt tolerance
Khao Dawk Mali 105	*O. Sativa*	AA	Rice Department, Ministry of Agriculture and cooperatives, Thailand	Aromatic rice, drought resistance
DJ123	*O. Sativa*	AA	International Rice research institute, Philippines	Submergence tolerance
Azucena	*O. Sativa*	AA	International Rice research institute, Philippines	Root architecture traits
O. Rufipogon	*O. Rufipogon*	AA	International Rice research institute, Philippines	Elongation ability, disease resistance
O. Barthii	*O. Barthii*	AA	Africa Rice Center (AfricaRice)	Resistance to green leafhopper, drought avoidance
O. Longistaminata	*O. Longistaminata*	AA	Africa Rice Center (AfricaRice)	Resistance to bacterial blight, drought tolerance
O. Glumaepatula	*O. Glumaepatula*	AA	International Center for Tropical Agriculture (CIAT)	Elongation ability, cytoplasmic male sterility
O. Nivara	*O. Nivara*	AA	International Rice research institute, Philippines	Drought avoidance, resistance to grassy stunt virus
O. Officinalis	*O. Officinalis*	CC	International Rice research institute, Philippines	Resistance to green leafhopper, brown planthopper

(continued)

Table 2.1 (continued)

Germplasm	Species	Genome	Source institution	Useful traits
O. Australiensis	*O. Australiensis*	EE	Australian tropical grains germplasm Centre	Resistance to brown planthopper, drought tolerance
O. Punctata	*O. Punctata*	BB, BBCC	Africa Rice Center (AfricaRice)	Resistance to zigzag leafhopper, brown planthopper
O. Brachyantha	*O. Brachyantha*	FF	Africa rice center (AfricaRice)	Resistance to yellow stem borer, tolerance to lateritic soil

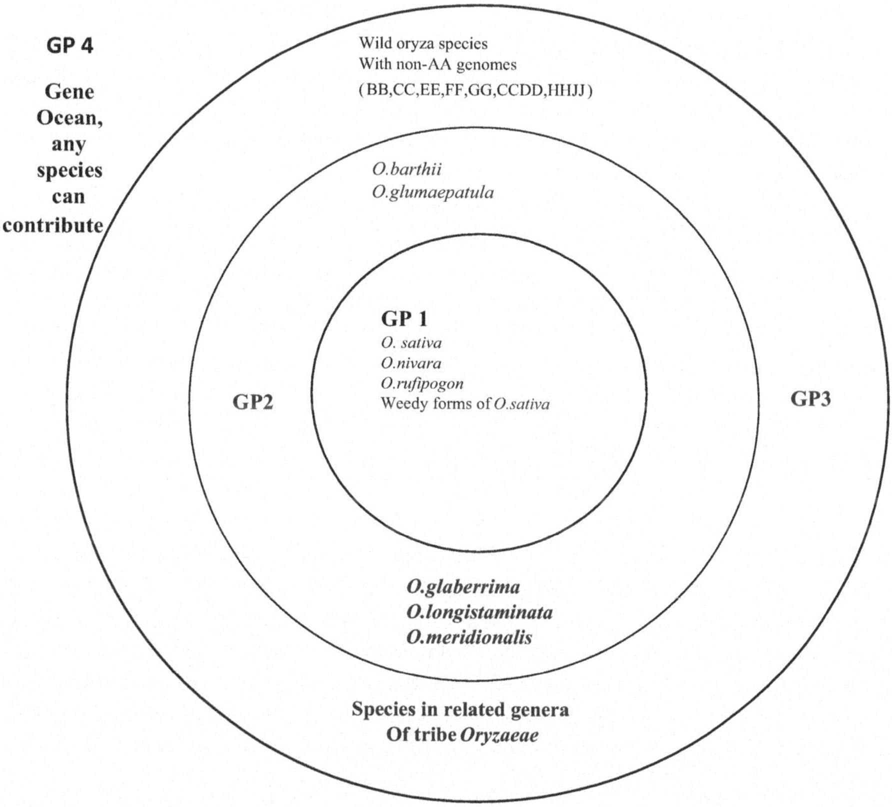

Fig. 2.1 Gene pool in rice pre-breeding genetic enhancement

2.3 The Importance of Biodiversity

Biodiversity in crop genetics serves as the foundation of our food supply, making it a cornerstone of our survival. This holds true whether we are discussing a society based on agricultural subsistence or one integrated into markets with advanced technology. Genetic diversity is the key that empowers farmers and researchers to tailor crops to diverse and ever-changing environments, equipping them with resilience against pests and diseases. Planting vast expanses with just one variety or a few closely related cultivars exposes them to heightened vulnerability when it comes to pests, diseases, and extreme weather conditions (Wilkes 1989).

Rice, together with maize and wheat, stands among the world's top three staple food crops. It assumes a crucial role as the primary crop grown in humid and subhumid regions of Asia, where it serves as a dietary staple. Rice serves as a vital economic driver in this region, contributing 30–50% to agricultural earnings and supplying 50–80% of the population's daily calories. Forecasts for this region predict that rice production must increase to more than 800 million tons in the next three decades, from the current 480 million tons (Hossain 1997). In the last 30 years, rice production has increased by an annual average of 2.4%, primarily driven by the widespread adoption of modern varieties (MVs) in irrigated areas and significant public and private investments in converting rainfed regions to irrigated ones (Hossain 1997).

The development of new cultivars hinges on the sustained availability of genetic resources (Chang 1984; Jackson and Huggan 1993). Traditional varieties, which have been cultivated and selectively bred by rice farmers for generations, serve as the primary source of this diversity. All modern varieties can trace their lineage back to these traditional landraces. For example, the lineage of IRRI varieties up to 1994 can be attributed to 40 landraces originating from 12 different countries (Leon and Carpena 1995). Similarly, wild rice species provide rich diversity, especially in their capacity to resist pests and diseases (Jackson 1995). They have played a vital role in enhancing rice (Savenko et al. 2023). For instance, a strain of *O. nivara* gathered from Uttar Pradesh, India, has been the exclusive reservoir of resistance against grassy stunt virus, significantly reducing its incidence on millions of hectares and similarly many more land races contributed biotic, abiotic stress tolerance and quality including nutritional quality traits (Sahoo et al. 2023). Emerging biotechnology tools hold the promise of increasing the value of genes from wild relatives for rice improvement (Khush and Aquino 1994). Furthermore, wild rice species contribute significantly to fundamental research. For instance, the first saturated restriction fragment length polymorphism (RFLP) genetic map of rice was developed through a cross between *O. sativa* and *O. longistaminata*, a wild species native to Africa (Causse et al. 1994).

Although modern varieties are widely adopted, especially in irrigated areas, traditional varieties remain pivotal for the livelihoods of numerous rice-growing farmers. Small-scale rice farmers in Asia and Africa cultivate a wide array of varieties, each chosen for traits like aroma, cooking quality, or suitability to specific agroecological conditions.

Preserving genetic diversity on farms is critical, benefiting both traditional small-scale farmers and technologically advanced ones alike. A wide range of genetic diversity, including both genic and cytoplasmic factors, is essential for tackling the challenges posed by intensive cultivation and continuous monocropping. This diversity is indispensable in retarding genetic shifts in major pathogens or insect pests, preventing minor pests from evolving into major threats, mitigating yield losses resulting from unusual weather patterns, countering the risk of epidemics associated with continuous monoculture in tropical and subtropical regions, and creating the potential for further genetic advancements (Manohar et al. 2022).

2.4 Untapped Rice Genetic Resources

2.4.1 Wild Rice Species

Oryza rufipogon and *Oryza nivara*, among other wild rice species, are crucial genetic reservoirs for enhancing cultivated rice varieties. They possess a range of beneficial traits including disease resistance, drought and salinity tolerance, and unique grain characteristics. These traits can be incorporated into cultivated rice through breeding efforts. Utilizing these wild relatives in breeding programs has resulted in the creation of improved rice varieties with increased resilience and higher yield potential.

2.4.2 Landraces

Landraces are traditional rice varieties adapted to specific agroecological niches. They possess valuable gene for resistance to local insects and diseases, adaptation to specific growing conditions, special nutritional and medicinal values, and unique culinary qualities. Preserving and studying landraces can provide essential genetic diversity for breeding programs and help conserve cultural heritage.

2.4.3 Farmer Varieties

Farmers' varieties, often overlooked in formal breeding programs, have adapted to local conditions over generations. These varieties may contain unique genetic traits that can be valuable for addressing specific challenges faced by farmers, such as changing climate conditions.

2.4.4 Management of Untapped Rice Genetic Resources

Managing untapped rice genetic resources is essential for conserving, characterizing, and utilizing these valuable genetic materials effectively. Strategic management ensures their availability for current and future breeding programs, facilitating the development of rice varieties with enhanced traits such as yield, stress resistance, and nutritional quality. Different Drought tolerant genotypes of rice are presented in Fig. 2.2. Conservation of these resources in rice involves two primary methods: In situ and Ex situ conservation. In situ conservation preserves rice species in their natural habitats, maintaining their evolutionary processes and natural adaptation.

Fig. 2.2 Different Drought tolerant genotypes of rice (**a**) at Flowering stage (**b**) at 50% Flowering stage (**c**) at grain filling stage (**d**) at maturity stage (**e**) grains after harvesting

Protected areas, natural reserves, and farmers' fields where traditional varieties are cultivated play a crucial role in In situ conservation. Collaborations with local communities and stakeholders are vital to safeguard these areas and ensure sustainable use of genetic resources. Ex situ conservation involves storing genetic materials outside their natural habitats, typically in gene banks. Major institutions such as the International Rice Research Institute (IRRI), Africa Rice Center (Africa Rice), and national agricultural research organizations like the Indian Institute of Rice Research (IIRR, Hyderabad), National Rice Research Institute (NRRI, Cuttack), and Indira Gandhi Agriculture University (IGAU, Raipur) maintain extensive rice germplasm collections. These gene banks store seeds under optimal conditions to preserve their longevity and genetic integrity. Cryopreservation and tissue culture techniques are also employed to conserve species that do not produce viable seeds. Terminologies used to study pre-breeding in genomic era are presented in Table 2.2.

Table 2.2 Terminologies used to study Pre-breeding in genomics era

Aspect	Definition/description	Importance	References
Genomics era	Refers to the period marked by advancedgenomic technologies and their application in plant breeding	Enables comprehensive understanding of plant genomes, aiding targeted improvements	–
Pre-breeding	The process of enhancing genetic diversity in crops by introducing wild or un-adapted germplasm	Provides a diverse pool of traits for use in breeding programs	Sharma et al. (2013)
Germplasm	Genetic material, such as seeds or plant tissues, used for breeding purposes	Acts as the foundation for breeding by introducing novel traits	Wilkes and Williams (1983)
Target traits	Specific characteristics or traits that breeders aim to improve in a crop	Guides the selection of germplasm for introduction in pre-breeding	Passioura (2012)
Marker-assisted selection (MAS)	Utilizes DNA markers to identify and select individuals with desired traits	Accelerates breeding by enabling more efficient and precise trait selection	Hasan et al. (2021)
Genetic diversity	The diversity of genetic material, including genes and alleles, within a population	Crucial for adapting and maintaining resilience amidst changing environments	Govindaraj et al. (2015)
Adaptation	The process of becoming better suited to the environment	Facilitates crop resilience to various environmental challenges	–
Breeding programs	Systematic plans for developing new crop varieties	Integrates pre-breeding efforts into larger breeding strategies	–
Public-private partnerships	Collaboration between public and private sectors in pre-breeding initiatives	Enhances resource sharing and knowledge exchange for more effective outcomes	Smyth et al. (2021)

(continued)

Table 2.2 (continued)

Aspect	Definition/description	Importance	References
Data integration	Incorporation of diverse data sources, including genomics, phenomics, and environmental data	Optimizes decision-making by utilizing comprehensive information	–
Ethical considerations	Addressing ethical issues related to genetic modification and breeding practices	Ensures responsible and sustainable use of emerging technologies	–
DNA sequencing	Determines the nucleotide sequence of DNA strands	Genome-wide analysis, identification of genes, and marker development	Adhikari et al. (2017)
QTL mapping	Identifies regions of the genome influencing specific traits	Pinpoints genomic regions linked to complex traits such as yield or stress tolerance	Kadirvel et al. (2015)
Genomic selection	Predicts the genetic value of individuals based on genomic information	Enhances breeding accuracy by considering multiple markers across the genome	Heffner et al. (2009)
Genome editing	CRISPR/Cas9 and other techniques for targeted gene editing	Enables precise modification of genes, facilitating trait improvement	Arora and Narula (2017)
Transcriptomics	Studies the expression of genes by analyzing RNA transcripts	Provides insights into gene function, especially during different developmental stages or stress conditions	Evans (2015)
Proteomics	Studies the structure and function of proteins within the rice genome	Offers insights into protein composition, aiding in understanding physiological processes and stress responses	Komatsu (2008)
Metabolomics	Analyzes the complete set of small-molecule metabolites within the plant	Provides information on metabolic pathways and helps understand biochemical changes	Hall (2006)
Bioinformatics	Utilizes computational tools to analyze and interpret genomic data	Facilitates data management, analysis, and interpretation for large-scale genomics	Rathore et al. (2018)
High-throughput phenotyping	Enables the rapid screening of a large number of plants for specific traits	Accelerates trait discovery and validation by assessing plant performance on a large scale	Mir et al. (2019)
Comparative genomics	Compares genomic sequences between different species to understand evolutionary relationships	Identifies conserved regions, facilitating the transfer of traits between species	Paterson et al. (2010)

2.5 Pre-Breeding in the Genomics Era

The necessity for genomics-assisted management of Plant Genetic Resources (PGR) and pre-breeding stems from the field of genomics, which includes the study of genomes, covering their structure, function, evolution, mapping, and genome editing. Incorporating modern genomics techniques like next-generation sequencing (NGS), cost-effective high-throughput genotyping, and high-throughput phenotyping (phenomics), and harnessing bioinformatics and statistical decision support tools, facilitates substantial acceleration of genetic improvements over time (Varshney et al. 2014).

Plant breeders widely acknowledge the significance of germplasm resources in enhancing crop quality. However, an excessive reliance on some elite parents with similar genetic backgrounds has resulted in an unsustainable narrowing of the genetic base in many breeding materials. To address this issue, previously overlooked reservoirs of heritable genetic variations must be tapped into to develop new and improved crop varieties. Pre-breeding involves the identification of desirable traits and the creation of materials that can be utilized by breeders. This process, reliant on close collaboration between genebank curators and plant breeders, serves as the crucial initial step in harnessing diversity from wild relatives and other underutilized materials.

Genomics plays a vital role in the management of PGR and pre-breeding by providing various benefits. These include the ability to uniquely identify individual accessions, prevent the intrusion of insect-pests with quarantine significance, assess the genetic worth of germplasm through trait-specific characterization, and ultimately ensure access to germplasm with specific traits for breeders.

The pre-breeding process is complex and requires substantial time, historically hindered by difficulties in phenotypic evaluation due to factors such as concealed environmental impacts, the polygenic complexity of key traits, crossing barriers, linkage drag, and negative trait correlations (Prohens et al. 2011). Breaking down the barriers to favorable allele transfer often requires multiple backcrosses, making the process time-intensive. Genomics-assisted breeding (GAB) can help circumvent the introduction of linkage drags and expedite the process. Moreover, for complex traits, identifying desirable allelic variants and genetic combinations can pose challenges, yet, genomics methodologies assist in the identification and selection of superior haplotypes and alleles for integration into pre-breeding efforts and their subsequent transfer to modern cultivars. Genomics-assisted pre-breeding strategies are enhancing the efficiency of developing climate-resilient crops (Varshney et al. 2018).

Molecular markers, association mapping, and QTL mapping have played a pivotal role in utilizing PGR and progressing pre-breeding efforts (Riar et al. 2012; Neelam et al. 2016; Li et al. 2014). In the era of genomics, the accessibility of high-quality reference genomes, high-throughput sequencing and re-sequencing platforms, and automated, cost-effective high-throughput genotyping technologies has significantly enhanced the productivity and efficiency of utilizing PGR and pre-breeding efforts (Kim and Rush 2016; Zhou et al. 2017). Whole-genome information provides the necessary data for genomics-assisted pre-breeding, including the identification of optimal haplotypes or allelic combinations, specific genomic regions, and gene networks (Xu et al. 2012). Key features of genomics-assisted pre-breeding include shorter breeding cycles, high accuracy in selection, and improved efficiency, allowing for direct enhancements (Varshney et al. 2018). It's worth noting that pre-breeding strategies for minor crops or non-crop plants may require different approaches due to distinct genomic challenges. When traditional resources are lacking, crops can benefit from the utilization of a genome annotation approach derived from translational genomics. This approach is invaluable for investigating phenotypic expressions and identifying specific genetic markers for the purpose of conducting marker-assisted breeding and genome selection (Kang et al. 2016).

Utilizing genomics approaches to enhance the utilization of Plant Genetic Resources (PGR) and empower pre-breeding is imperative in contemporary agriculture. Genomics offers a multitude of technologies, such as sequencing and resequencing platforms, the availability of reference genome sequences, SNP arrays, genome editing tools, and high-throughput genotyping platforms. These technologies are Genome Sequencing and Resequencing, Pangenomics, Molecular Markers, QTL, and Genomic Maps in Enhancing PGR Utilization and Pre-breeding, Genome-Wide Association Studies (GWAS), SNP Chips Approach or SNP Marker Arrays, Genotyping by Sequencing (GBS) Approach, Genomic Selection, and Genomic Selection for Trait Improvement.

2.6 Discovery and Accumulation of Yield-Related Genes Concealed in Wild and Weedy Species

The gene pools of wild and weedy varieties serve as abundant repositories of genetic resources for breeders aiming to enhance the traits of crop plants, with rice being a notable example. In addition to identifying and utilizing various Mendelian genes that predominantly confer resistance to biotic stresses, the exploration for genes/quantitative trait loci responsible for polygenically controlled yield and its primary components began with the assumption that many of these genetic elements might have eluded capture in modern rice varieties during the evolutionary process. It was believed that these genes/QTLs could still exist within the untamed gene pool (Xiao et al. 1996).The prospect of identifying these elusive genes through QTL mapping and integrating theminto current high-yielding varieties via marker-assisted breeding raised the possibility of further enhancing genetic yield. This initiative gained momentum following the successful mapping and validation of two yield QTLs, yld 1.1 and yld 2.1, in *O. rufpogan* using the marker-associated QTL approach by Cornell University, USA (Rath et al. 2021). This achievement sparked global interest among rice geneticists, prompting an extensive search for additional yield genes/QTLs in wild and weedy relatives as well as primitive cultivars. This endeavor led to the identification of numerous promising yield QTLs in *O. rufpogan*, *O. nivara*, and landraces (Marri et al. 2005; Kaladhar et al. 2008; Swamy and Sarla 2008).The groundbreaking work culminated in the development of higher-yielding varieties such as Jefferson, incorporating yld2.1 from *O. rufpogan* (Imai et al. 2013) DRR Dhan 40 featuring yield genes/QTLs derived from *O. nivara* (Haritha et al. 2017) and the high-yielding salinity-tolerant IET21943, based on a yield QTL from *O. rufpogon* (Ganeshan et al. 2016).

2.7 Challenges and Future Perspectives

Pre-breeding, while immensely promising, is not without its challenges. Regulatory and ethical considerations, issues related to the utilization of wild relatives, and the need for robust interdisciplinary collaborations are among the hurdles that must be addressed. However, these challenges are met with exciting opportunities in the form of cutting-edge technologies and collaborative efforts that promise to further enhance rice pre-breeding.

The issues surrounding food security hold global importance, exacerbated by the unprecedented growth in the world population, leading to the overexploitation of natural resources and, consequently, the depletion of plant genetic diversity. Recently, the genetic diversity present in landraces, wild, and weedy cultivars has been noted for its role in mitigating diseases, pests, and environmental fluctuations affecting both animal and plant populations. However, these valuable resources are rapidly diminishing due to human activities. While numerous varieties of rice are preserved in the gene bank, there's incomplete information about them. Germplasm lacking characterization and evaluation data cannot effectively contribute to crop improvement efforts. Therefore, a methodical process of characterizing, evaluating, and documenting crucial traits for each germplasm is necessary to enhance their utilization by breeders.

While the genetic diversity preserved in gene banks harbors valuable genes and alleles that could potentially address various constraints in crop production, selecting the most suitable set of accessions remains challenging without adequate information such as phenotypic and genotypic data (Lee et al. 2020). Consequently, researchers and breeders often utilize only a fraction of the genetic diversity available in their research programs, highlighting a significant gap between gene bank collections and breeding outcomes (Ata-Ul-Karim et al. 2022). The implementation of high-throughput phenotyping as part of gene bank routine operations yields valuable insights, aiding seed users in conserving resources for germplasm screening and subsequent discovery phases. The establishment of a digital rice gene bank, housing genomic data for all conserved accessions, is now feasible, albeit requiring substantial support and specialized teams (He et al. 2022). These extensive resources, combined with comprehensive multi-omics analyses encompassing genomics, proteomics, and metabolomics on selected materials, facilitate the discovery of genotypes, novel alleles, and haplotypes for diverse traits. Harnessing these resources is expected to accelerate the breeding process and optimize the utilization of the diverse conserved germplasm.

2.8 Conclusion

The combination of untapped rice genetic resources and genomics tools presents a promising path for addressing the complex challenges facing rice production in the twenty-firstcentury. Pre-breeding efforts that leverage genomics can unlock the full potential of these resources, leading to the development of rice varieties that are more productive, resilient, and sustainable, ultimately benefiting both farmers and consumers worldwide. However, it is crucial to navigate the ethical and regulatory aspects of these advancements to ensure that they are harnessed responsibly for the greater good of global food security.

The genomics era has introduced a variety of groundbreaking technologies, including sequencing and re-sequencing platforms, high-throughput markers associated with traits, affordable genotyping platforms, and genome editing. These innovations hold the potential to revolutionize the management of Plant Genetic Resources (PGR) by improving their efficient utilization and optimizing pre-breeding endeavors. Despite the undeniable progress facilitated by genomics tools in managing PGR and advancing pre-breeding, significant challenges persist in fully exploiting the potential of these genomic advancements. Particularly, the limited availability of high-throughput phenotyping platforms remains a critical bottleneck.

References

Adhikari S, Saha S, Biswas A, Rana TS, Bandyopadhyay TK, Ghosh P (2017) Application of molecular markers in plant genome analysis: a review. Nucleus 60:283–297. https://doi.org/10.1007/s13237-017-0214-7

Arora L, Narula A (2017) Gene editing and crop improvement using CRISPR-Cas9 system. Front Plant Sci 8:296675. https://doi.org/10.3389/fpls.2017.01932

Ata-Ul-Karim ST, Begum H, Lopena V, Borromeo T, Virk P, Hernandez JE et al (2022) Genotypic variation of yield-related traits in an irrigated rice breeding program for tropical Asia. Crop Environ 1(3):173–181

Brar DS, Khush GS (2018) Wild relatives of rice: a valuable genetic resource for genomics and breeding research. Rice 11(1):17

Causse MA, Fulton TM, Cho YG, Ahn SN, Chunwongse J, Wu K, Xiao J, Yu Z, Ronald PC, Harrington SE (1994) Saturated molecular map of the rice genome based on an interspecific backcross population. Genetics 138(4):1251–1274. https://doi.org/10.1093/genetics/138.4.1251

Chang TT (1984) The ethnobotany of rice in Island Southeast Asia. Asian Perspect 26(1):69–76. http://hdl.handle.net/10125/16919

Evans TG (2015) Considerations for the use of transcriptomics in identifying the 'genes that matter' for environmental adaptation. J Exp Biol 218(12):1925–1935. https://doi.org/10.1242/jeb.114306

Ganeshan P, Jain A, Parmar B, Rao AR, Sreenu K, Mishra P et al (2016) Identification of salt tolerant rice lines among interspecific BILs developed by crossing *'Oryza sativa*× *O. rufipogon*and*O. Sativa*× *O. nivara'*. Aust J Crop Sci 10(2):220–228. https://doi.org/10.3316/informit.031022341370500

Glaszmann JC (1987) Isozymes and classification of Asian rice varieties. Theor Appl Genet 74(1):21–30. https://doi.org/10.1007/BF00290078

Govindaraj M, Vetriventhan M, Srinivasan M (2015) Importance of genetic diversity assessment in crop plants and its recent advances: an overview of its analytical perspectives. Genet Res Int 2015:1. https://doi.org/10.1155/2015/431487

Hall RD (2006) Plant metabolomics: from holistic hope, to hype, to hot topic. New Phytol 169(3):453–468. https://doi.org/10.1111/j.1469-8137.2005.01632.x

Haritha G, Vishnukiran T, Yugandhar P, Sarla N, Subrahmanyam D (2017) Introgressions from *Oryza rufipogon* increase photosynthetic efficiency of KMR3 rice lines. Rice Sci 24(2):85–96. https://doi.org/10.1016/j.rsci.2016.07.006

Hasan N, Choudhary S, Naaz N, Sharma N, Laskar RA (2021) Recent advancements in molecular marker-assisted selection and applications in plant breeding programmes. J Genet Eng Biotechnol 19(1):128. https://doi.org/10.1186/s43141-021-00231-1

He S, Liu H, Zhan J, Meng Y, Wang Y, Wang F, Ye G (2022) Genomic prediction using composite training sets is an effective method for exploiting germplasm conserved in rice gene banks. Crop J 10:1073–1082

Heffner EL, Sorrells ME, Jannink JL (2009) Genomic selection for crop improvement. Crop Sci 49(1):1–12. https://doi.org/10.2135/cropsci2008.08.0512

Hossain M (1997) Rice supply and demand in Asia: a socioeconomic and biophysical analysis. In: Applications of systems approaches at the farm and regional levels volume 1: proceedings of the second international symposium on systems approaches for agricultural development, held at IRRI, Los Baños, Philippines, 6–8 December 1995. Springer, pp 263–279. https://doi.org/10.1007/978-94-011-5416-1_20

Imai I, Kimball JA, Conway B, Yeater KM, McCouch SR, McClung A (2013) Validation of yield-enhancing quantitative trait loci from a low-yielding wild ancestor of rice. Mol Breed 32:101–120. https://doi.org/10.1007/s11032-013-9855-7

Jackson MT (1995) Protecting the heritage of rice biodiversity. GeoJournal 35:267–274. https://doi.org/10.1007/BF00989134

Jackson MT, Huggan R (1993) Sharing the diversity of rice to feed the world. Diversity 9(3):22–25. https://doi.org/10.5555/19941601353

Kadirvel P, Senthilvel S, Geethanjali S, Sujatha M, Varaprasad KS (2015) Genetic markers, trait mapping and marker-assisted selection in plant breeding. Plant Biol Biotechnol Vol II: Plant Genomics Biotechnol:65–88. https://doi.org/10.1007/978-81-322-2283-5_4

Kaladhar K, Swamy BPM, Babu AP, Reddy CS, Sarla N (2008) Mapping quantitative trait loci for yield traits in BC2F2 population derived from Swarna x *O. Nivara* cross. Rice Genet Newslett 24:34–36

Kang L, Settlage R, McMahon W, Michalak K, Tae H, Garner HR et al (2016) Genomic signatures of speciation in sympatric and allopatric Hawaiian picture-winged drosophila. Genome Biol Evol 8(5):1482–1488. https://doi.org/10.1093/gbe/evw095

Khush GS, Aquino RC (1994) Breeding tropical japonicas for hybrid rice production. In: Hybrid rice technology: new developments and future prospects. International Rice Research Institute, Manila, Philippines, pp 33–36

Kim Y, Rush AM (2016) Sequence-level knowledge distillation. Comput Lang. arXiv preprint arXiv:1606.07947. https://doi.org/10.48550/arXiv.1606.07947

Komatsu S (2008) Research on the rice proteome: the contribution of proteomics technology in the creation of abiotic stress-tolerant plants. Rice 1(2):154–165. https://doi.org/10.1007/s12284-008-9013-8

Lee JS, Chebotarov D, Platten JD, McNally K, Kohli A (2020) Advanced strategic research to promote the use of rice genetic resources. Agronomy 10(11):1629. https://doi.org/10.3390/agronomy10111629

Leon JC, Carpena AL (1995) Morphology-based diversity analysis of improved irrigated lowland rice (*Oryza sativa* L.) varieties in the Philippines. Philipp J Crop Sci 20(2):113–121

Li JY, Wang J, Zeigler RS (2014) The 3000 rice genomes project: new opportunities and challenges for future rice research. Gigascience 3(1):2047–217X

Li Z et al (2018) Advancements in high-throughput phenotyping and genomic selection for rice breeding. Theor Appl Genet 131(1):39–51

Li C et al (2020) CRISPR/Cas9-mediated gene editing in rice. Curr Protoc Plant Biol 5(2):e20119
Manohar RV, Nivethitha T, Jadhav BN, Raveendran M, Sritharan N, Pushpam R, Joel AJ (2022) Utilising genetic variability and diversity analysis as a tool to identify drought tolerant pre-breeding genetic materials in rice (Oryza sativa L.). Pharma Innov J 11(8):1374–1381
Marri PR, Sarla N, Reddy LV, Siddiq EA (2005) Identification and mapping of yield and yield related QTLs from an Indian accession of *Oryza rufipogon*. BMC Genet 6:1–14. https://doi.org/10.1186/1471-2156-6-33
McCouch SR et al (2016) Open access resources for genome-wide association mapping in rice. Nat Commun 4:1050
Mir RR, Reynolds M, Pinto F, Khan MA, Bhat MA (2019) High-throughput phenotyping for crop improvement in the genomics era. Plant Sci 282:60–72. https://doi.org/10.1016/j.plantsci.2019.01.007
Neelam K, Kumar K, Dhaliwal HS, Singh K (2016) Introgression and exploitation of QTL for yield and yield components from related wild species in rice cultivars. Mol Breed Sustain Crop Improv 2:171–202. https://doi.org/10.1007/978-3-319-27090-6_8
Passioura JB (2012) Phenotyping for drought tolerance in grain crops: when is it useful to breeders. Funct Plant Biol 39(11):851–859. https://doi.org/10.1071/FP12079
Paterson AH, Freeling M, Tang H, Wang X (2010) Insights from the comparison of plant genome sequences. Ann Rev Plant Biol 61:349–372. https://doi.org/10.1146/annurev-arplant-042809-112235
Pérez-de-Castro AM et al (2012) Application of genomic tools in plant breeding. Curr Genomics 13(3):179–195
Prohens J, Fita AM, Rodríguez-Burruezo A, Raigón MD, Plazas M, Vilanova S (2011) Breeding for the present and the future: achievements and constraints of conventional plant breeding and contributions of genomics to a new green revolution. Bullet Univ Agric Sci Vet Med Cluj-Napoca Hortic 68(1) http://journals.usamvcj.ro/horticulture
Rath PC, Bose LK, Jambhulkar NN, Subudhi HN, Kar MK (2021) Broadening genetic base of rice and identification of pre-breeding lines for resistance to brown plant hopper, Nilaparvatalugens (Stål.). J Exp Agric Int 43(8):76–80
Rathore A, Singh VK, Pandey SK, Rao CS, Thakur V, Pandey MK, Anil Kumar V, Das RR (2018) Current status and future prospects of next-generation data management and analytical decision support tools for enhancing genetic gains in crops. Plant Genet Mol Biol:277–292. https://doi.org/10.1007/10_2017_56
Riar AK, Kaur S, Dhaliwal HS, Singh K, Chhuneja P (2012) Introgression of a leaf rust resistance gene from *Aegilops caudata* to bread wheat. J Genet 91:155–161. https://doi.org/10.1007/s12041-012-0161-7
Sahoo SK, Deo I, Prashad H, Kumar A, Kumawat S (2023) Evaluation of rice (*Oryza sativa* L.) pre-breeding genotypes for resistance to bacterial leaf blight (*Xanthomonas oryzaepv. Oryzae*) disease under field condition. Pharma Innov 12(5):1093–1095
Savenko EG et al (2023) Development of rice pre-breeding resources with antioxidant grain properties basing on hybridization and experimental haploidy methods. In: AIP conference proceedings, vol 2777. No. 1. AIP Publishing
Sharma S, Upadhyaya HD, Varshney RK, Gowda CLL (2013) Pre-breeding for diversification of primary gene pool and genetic enhancement of grain legumes. Front Plant Sci 4:57969. https://doi.org/10.3389/fpls.2013.00309
Smyth SJ, Webb SR, Phillips PW (2021) The role of public-private partnerships in improving global food security. Glob Food Sec 31:100588. https://doi.org/10.1016/j.gfs.2021.100588
Spindel J et al (2015) Genomic selection and association mapping in rice. J Exp Bot 66(8):1913–1926
Swamy BM, Sarla N (2008) Yield-enhancing quantitative trait loci (QTLs) from wild species. Biotechnol Adv 26(1):106–120. https://doi.org/10.1016/j.biotechadv.2007.09.005
Swamy BPM et al (2017) Identification of genomic regions associated with root traits under drought stress in rice. BMC Genet 18(1):66

Varshney RK, Terauchi R, McCouch SR (2014) Harvesting the promising fruits of genomics: applying genome sequencing technologies to crop breeding. PLoS Biol 12(6):e1001883. https://doi.org/10.1371/journal.pbio.1001883

Varshney RK, Singh VK, Kumar A, Powell W, Sorrells ME (2018) Can genomics deliver climate-change ready crops? Curr Opin Plant Biol 45:205–211. https://doi.org/10.1016/j.pbi.2018.03.007

Varshney RK et al (2019) Genomic selection for crop improvement: status and prospects. Trends Plant Sci 24(7):531–537

Wilkes G (1989) Germplasm preservation: objectives and needs. In: Biotic diversity and germplasm preservation, global imperatives, pp 13–41

Wilkes G, Williams JT (1983) Current status of crop plant germplasm. Crit Rev Plant Sci 1(2):133–181

Xiao J, Li J, Yuan L, Tanksley SD (1996) Identification of QTLs affecting traits of agronomic importance in a recombinant inbred population derived from a subspecific rice cross. Theor Appl Genet 92:230–244

Xie W et al (2018) Genomic variation in 3,010 diverse accessions of Asian cultivated rice. Nature 557(7703):43–49

Xu Y, An D, Liu D, Zhang A, Xu H, Li B (2012) Molecular mapping of QTLs for grain zinc, iron and protein concentration of wheat across two environments. Field Crop Res 138:57–62. https://doi.org/10.1016/j.fcr.2012.09.017

Xu K et al (2016) Development of a marker-assisted selection system for bacterial blight resistance in rice. Mol Breed 36(11):144

Zhou J, Xiong Y, Xing Z, Deng J, Du K (2017) Characterizing and sourcing ambient PM2.5 over key emission regions in China II: organic molecular markers andCMB modeling. Atmos Environ 163:57–64. https://doi.org/10.1016/j.atmosenv.2017.05.033

Conservation and Development of Rice Germplasm for Natural Farming

3

Anubhav Thakur, Akashdeep Singh, and Narender Kumar Bharat

Abstract

Natural farming practices have gathered colossal attention due to their long-term sustainability and economic viability. The non-dependence on chemical fertilizers in the new farming practices makes preserving the vast diversity of genetic resources such as landraces, wild relatives, and traditional varieties essential because they store the genetic traits that might be useful for adapting rice to new farming practices. The wide germplasm offers opportunities to the scientific community to explore genetic traits to develop varieties/cultivars tailored to specific agro ecological zones, leading to enhanced crop yields and reduced economic burden. Besides, climate change poses a massive threat to agriculture production systems across the globe. Temperature aberration, erratic precipitation, frequency of extreme weather events, etc., can impact rice production. The conservation of germplasm emerges as a stalwart defense against this menace. It equips the agricultural community with a potent arsenal for combating climate change by providing access to genetic resources that harbor resilience-enhancing traits critical for sustaining rice production amidst the unpredictable challenges posed by a changing climate. Meanwhile, the continuous release of modern rice varieties (hybrids), as a replacement to traditional varieties to boost rice productivity, has inadvertently led to increased pests and diseases under intensive cultivation. As natural farming forms an unformidable pillar of sustainable agriculture,

A. Thakur · N. K. Bharat
Department of Seed Science and Technology, Dr Yashwant Singh Parmar University of Horticulture and Forestry, Nauni, Solan, Himachal Pradesh, India

A. Singh (✉)
Department of Agronomy, Chaudhary Sarwan Kumar Himachal Pradesh Krishi Vishavydyalaya, Palampur, Himachal Pradesh, India

Faculty of Agriculture, School of Agricultural Sciences, Nirwan University, Jaipur, Rajasthan, India

A. Singh et al. (eds.), *Climate-Smart Rice Breeding*,
https://doi.org/10.1007/978-981-97-7098-4_3

the conservation of rice germplasm to preserve genetic diversity becomes essential. It provides a safety net to guard against the unforeseen effects of modernization.

Keywords

Germplasm · Natural farming · Rice · Variety

3.1 Introduction

Rice (*Oryza sativa* L.) is the staple food for almost 50% of the world's population and is cultivated in all continents except Antarctica. Continents like Asia produce and consume 90% of world rice (FAO 2021), contributing to 80% daily calorie consumption and 50–80% protein content (Bierschenk et al. 2020; Sahebi et al. 2018). The world's most populated nation, China, consumed 154.9 million metric tons of rice in 2021–2022, the highest during the same year. India, on the other hand, consumed 103.5 million metric tons in the same period, placing it second after China (Noor et al. 2022). The area of rice in India is 43.86 million hectares (Anonymous 2022). The genus *Oryza* has 25 wild progenitors, some of whom are *Oryza nivara* and *Oryza rufipogon* (Ricepedia 2021). Only two species are under cultivation, *Oryza sativa* and *Oryza glaberrima*. Three major genetically divergent rice cultivars are *indica*, *japonica* and *javanica* (Prasad et al. 2017).

Indian rice biodiversity is unparalleled by any other country in the world, with over 2 lakh (200,000) varieties, as supported by *vedic* evidence (Deb 2005). Due to its wide acreage and huge germplasm diversity, it is a crop significantly related to food security (Husaini and Sofi 2018a). In rice, over the last few decades, the enormous mosaic of conventional crop varieties has rapidly been replaced by high-yielding varieties, which include crop hybrids created by national and international agricultural organizations (Husaini and Sofi 2018a). When paired with complementary management techniques like irrigation, weedicide, and fertilizer treatment, these new rice varieties had a higher harvest index and yielded noticeably greater yields. High-yielding cultivars are unquestionably necessary to meet the food demands of growing human populations, but doing so comes at a cost in terms of genetic erosion or loss of genetic variability (Wissuwa et al. 1998). The main way to address the genetic depletion of crop germplasm is to preserve genetic variation and use it in breeding by utilizing the natural population of cultivated plants, wild ancestors and landraces (Gomez et al. 2022). Local landraces have become increasingly important, either for direct exploitation or indirect use, as a result of new, rising concerns, including climate change, population boom, limited land and water resources, and demand for organic products (Gomez et al. 2022). The results of a study are shown here to highlight how crucial it is to preserve and investigate the genetic value of landraces. *Pup1*, a significant QTL for phosphorus-deficiency tolerance, was discovered in the traditional *aus*-type Indian rice variety *Kasalath* about a decade ago (Husaini and Sofi 2018a; Wissuwa et al. 1998). This QTL was sequenced & *Pup1*-specific protein kinase gene was located and renamed as

PSTOL1 gene (phosphorous starvation tolerance 1). It is crucial to stress that neither the reference genome for rice nor any other phosphorus-starved contemporary variety contains this gene. In phosphorus-deficient soil, the over-expression of *PSTOL1* in these cultivars greatly increases early root growth and grain output by allowing plants to absorb more phosphorus and other minerals (Wissuwa and Ae 2001). It is clear from this example how important ancient rice germplasms are because contemporary rice types lack genes like *PSTOL1*. Likewise, many more examples of the important gene transferred from landraces to rice cultivars viz., qDTY for drought stress and Sub1 for submergence tolerance, etc.

Natural farming is such a farming practice in which the conservation of landraces and crop wild relatives is promoted (Bana et al. 2022). Natural Farming is an environment-friendly, chemical-free, multifarious farming method that combines animals, trees, and crops to enable functional biodiversity (Ghasal et al. 2023). It is an on-farm conservation practice where farmers are involved in the conservation of landraces and wild relatives of fields, vegetables, and other crops. Farmers can choose and develop rice varieties that are more suited to the environmental conditions in their area, such as tolerance to pests, diseases, and drought, by maintaining a diversity of genetic stock. Sustaining crop diversity in farmers' fields is a key goal of in situ on-farm conservation and natural farming, as it can lead to beneficial adaptations that can support the resilience of food and agriculture systems to changing conditions. Rice germplasm conservation and development for natural farming are crucial to supporting sustainable agriculture. Therefore, in this chapter, we will discuss how conserving traditional rice germplasm and using it with natural farming practices can help create a sustainable food system.

3.2 Sustainable Farming System

Sustainable farming systems are methods of raising crops to produce food and other agricultural products in a socially, economically, and environmentally responsible way. These programs prioritize protecting the environment, preserving natural resources, and advancing the welfare of farmers and rural communities. Key principles of a sustainable farming system are conservation of natural resources, soil health management, IPM, and agro ecology. Natural farming is such an example of a sustainable farming system that we discuss in this book chapter.

3.2.1 Importance of Natural Farming

Natural farming is gaining momentum as a transformative approach in agriculture, addressing the limitations of conventional farming practices. This shift is supported by a dedication to environmental sustainability, including reducing dependency on chemical inputs and promoting soil health (Hoekstra and Hung 2003). Unlike conventional monoculture, natural farming emphasizes diversified cultivation practices, contributing to the preservation of biodiversity and creating resilient ecosystems

(Nicolopoulou-Stamati et al. 2016). The strategy is especially important in rice-growing places, which normally require a lot of water, and it aligns with global concerns about freshwater scarcity (FAO 2020). A fundamental aspect of natural farming is its prioritization of human health and safety. By minimizing pesticide applications, the method directly addresses health concerns associated with chemical exposure in conventional farming (Damalas and Koutroubas 2016). The method also addresses soil health issues by incorporating strategies such as minimal tillage, cover cropping, and organic matter incorporation to improve fertility and structure (Brown et al. 2021). Natural farming forms part of a larger shift toward agroecology, which involves incorporating ecological concepts into agricultural systems (Tiwari et al. 2008). This transformation encourages local innovation, community empowerment, and economic resilience while decreasing reliance on external inputs and industrial methods (Knorr and Watkins 1984). Natural farming emerges as a revolutionary alternative amid agricultural crises, ensuring long-term food security and environmental well-being (Liu et al. 2014). In essence, natural farming emerges as a transformative model with implications far beyond its local origins, embodying a vision for a more sustainable and harmonious future in agriculture.

3.2.2 Reduced Reliance on Chemical Fertilizers

Over 2.6 million tonnes of pesticides are used annually, with a market value of over USD 25 billion (Abd-Aziz et al. 2022). Reducing the heavy reliance on chemical fertilizers & pesticides is crucial for achieving sustainable development in agriculture. The prolonged use of chemical-based technology has compromised soil quality, negatively impacting its physical, biological, and chemical properties. This not only impedes healthy plant growth but also renders crop production unsustainable. The environmental risks and pollution associated with depleting nutrients and relying on exhaustible fossil fuels for fertilizer production further underscore the need for a shift in agricultural practices. In India, the subsidy-driven pricing policy has led to opaque and indiscriminate fertilizer use. Returning to traditional, organic farming practices rooted in indigenous knowledge emerges as a viable alternative (Ghosh 2004). This perspective critiques India's extensively promoted fertilizer-intensive agricultural technology, advocating a transition toward a more balanced approach. Balla and Goswami (2022) study delves into the factors influencing the adoption of natural farming (NF) among rice-growing farmers in Andhra Pradesh, India. Critical constraints to NF adoption include higher labor costs, limited agro-ecological experience, poor soil fertility, and increased pest incidence compared to chemical fertilizers. However, the study highlights the potential of promoting crop diversification and staggered planting and providing detailed knowledge, local input shops, soil testing, and treatment to boost farmer confidence in transitioning to natural farming. Natural farming, emphasizing organic practices, offers a sustainable alternative to chemical fertilizers, promoting enhanced soil fertility and reduced environmental impact. This shift towards natural farming aligns with the imperative for eco-friendly and resilient agricultural practices. Educating farmers, providing government support, and promoting market initiatives are crucial to ensure successful adoption.

3.2.3 Preserving Genetic Diversity

Any variation within and between crop cultivars, including their genotypic and phenotypic characteristics, is referred to as genetic diversity (Joshi et al. 2023). Agricultural genetic diversity in the fields is the most crucial component of secure and sustainable agriculture. Genetic diversity makes agricultural systems more stable and allows farmers to use different microclimates and genetic variation between species for a range of nutritional and other uses (Nonić and Šijačić-Nikolić 2021; Singh 2021). From fields and institutes, agricultural researchers gather germplasm (basket of genetic diversity), which includes both native and foreign species as well as wild relatives. Additionally, processes like genetic engineering, hybridization, and mutation may be used to produce and preserve diversity. The majority of the diversity has been preserved within structures like gene banks and other reservoirs, and only a small number of specifically selected lines have been cultivated because of their excellent yields. As a result, the genetic diversity of crops in the field has decreased (Joshi 2017; Joshi et al. 2020). More crucially, because most research is done inside research stations, genotypes in research fields interact with environmental elements far less than in conventional breeding systems. Therefore, farming practices like natural farming can help preserve the genetic diversity of crop because it only uses Indigenous and local landraces for cultivation and involves farmers' traditional knowledge. Farmers' traditional knowledge and practices favor creating and maintaining both intra and inter-varietal diversity. Most farmers let nature choose the genotypes for the crops they will grow the following season. Therefore, compared to crops harvested from a formal seed system, there is greater variation observed in crops maintained by farmers, from seeding to storing seeds independently.

3.3 Genetic Resources in Agriculture (Landraces, Wild Relatives, and Traditional Varieties)

Landraces exhibit adaptation to specific agro-climatic conditions, preserving substantial diversity within and between populations (Brown et al. 2021). This diversity serves as a genetic reservoir with implications for future breeding endeavors and the innovation of agricultural systems and products. In rice Nona Bokra and Pokkali landraces were identified as sources of salt tolerance for rainfed and irrigated agroecosystems. The discovery of the submergence-tolerant landrace FR13A led to identifying the SUBMERGENCE 1 (SUB1) locus on chromosome 9, coding for the ethylene response factor resulting in submergence tolerance in rice (Xu and Mackill 1996). Therefore, it is essential to explore genetic diversity and conserve landraces for the benefit of future generations. Upland rice, a major global production system, faces challenges due to drought stress. Landraces such as Aus 257, Aus *Bak Tulsi*, Azucena, Basmati 370, *Dular*, Kalia, *Kali* Aus, *Lal* Aus, and N22 have been pivotal for grain yield under drought stress, leading to the development and release of 17 high-yielding drought-tolerant rice cultivars in Asia and Africa. Additionally, landraces contributed 14 significant QTLs associated with drought adaptation. The pyramiding of these large-effect QTLs has improved drought adaptation in widely grown rice cultivars, including Swarna, IR64, Sabitri, TDK1, and BR11. Moreover, agronomic evaluations are underway for introgressed offspring carrying SUB1 and drought-tolerant QTL alleles on various rice backgrounds. The genetic resources of traditional rice in India encompass a wide range of genetic variations, offering medicinal properties and aromatic qualities. Table 3.1 shows some examples of traditional rice varieties and landraces cultivated in different parts of India. These

Table 3.1 Some examples of traditional rice varieties and landraces cultivated in different parts of India

Type	Common name	Feature	Major cultivated area	References
Landrace	*Kamad, Mushk budji*	Aromatic	Anantnag district, Kashmir	Husaini and Sofi (2018b)
	Khuch, Khuch niver, Mir sagi	Black pericarp, high anthocyanin content	Kashmir	
Traditional rice varieties	*Adukkan, Cheattuveliyan Kannichennellu, Kuruva Kumkumashali, Marathondi Onnamottan, Paalthondimatta*	High se, Zn, Cr content	Kerala	Veena et al. (2023)
Landrace	*Sigappukuruvikar, Sennel*	Red-colored grain	Tamil Nadu	Amudha et al. (2023)
	Thattan samba, Panamara samba	Heat tolerant	Tamil Nadu	Megala et al. (2022)
	Norungan, Mattaikar	Suitable for grown under sodic soils	Tamil Nadu	Geetha et al. (2017)
	Kavuni	Fifty times higher luetin than traditional white rice	Tamil Nadu	Valarmathi et al. (2015)
Landrace	Lal Dhan	Used in fever	Himachal Pradesh	Ghosh and Dasgupta (2023)
Landrace	Lusaibiroin, Sona birion-2 Ranga bora1	Very low amylose content	Assam	Gautam et al. (2023)
Landrace	*Karanga*	Cure dysentry	Bihar	Das et al. (2023)
Landrace	*Chak-Hao amub*	Black rice	Manipur	Saikia et al. (2012)

traditional varieties serve as viable sources of agricultural attributes and bioactive nutrients, rich in vitamin D, calcium, thiamine, riboflavin, glutamic acids, and fiber. They are low in fat, sugar, and gluten, containing oryzanol, which reduces fat production and makes them suitable for diabetics and hypertensive individuals. Medicinal rice varieties, including *Kattuyanam*, *Mapillai samba*, *Navara*, *Karunguruvai*, *Kavuni*, *Kichadi samba*, *Illupaipoo samba*, *Kalanamak*, *Karudan samba*, and *Seeraga samba*, exhibit antioxidant potential and physiological benefits. Unpolished traditional rice offers more health benefits than polished rice, with its bran layers containing high bioactive components like polyphenols, phytochemicals, antioxidants, vitamins, and minerals. These components can help control various biological diseases such as carcinoma, cardiovascular disease, neurological disorders, nephrological disorders, and diabetes. Despite their weedy and inferior morphological traits, wild rice species serve as crucial genetic resources for breeding and genomics research. They house valuable genes/QTLs to tolerate major biotic and abiotic stresses, yield-related traits, cytoplasmic male sterility, and other attributes contributing to improving rice. For example, the landmark rice variety IR 36 incorporated resistance to grassy stunt virus (GSV) derived from the wild rice *Oryza nivara*, which remains the sole source of GSV resistance in rice. Various rice varieties also host the *Xa21* gene for bacterial blight resistance derived from the wild perennial rice *Oryza longistaminata*. Over 20 viruses infect rice, with *O. nivara* identified as the sole source of resistance to rice grassy stunt disease due to the *Gs* gene. The transfer of GSV resistance from *O. nivara* to cultivated *O. sativa* marked the successful introduction of a useful agronomic gene from a wild to cultivated rice species. Rice tungro disease, a significant viral disease transmitted by the green leafhopper *Nephotettix virescens*, is addressed by the identification of tolerance in wild species like *O. longistaminata* and *O. rufipogon*, leading to the development of tungro-tolerant lines. The increasing challenges of diverse biotic and abiotic stresses, climate change, and the rising demand for food production call for innovative solutions in global agriculture. Superior cultivars developed by harnessing the untapped genetic diversity in crop wild relatives, landraces, and traditional varieties using modern molecular and genomic tools have the potential to enhance the food production process. Millennia of plant domestication focused primarily on yield improvement and edibility, resulting in genetic bottlenecks and reduced genetic variation. In contrast, crop wild relatives, landraces, and traditional varieties, residing in their natural environments, have consistently faced challenges, maintaining higher levels of genetic diversity. Leveraging this untapped genetic diversity for crop improvement through modern technologies presents an attractive option to enhance global food production. The use of advanced technologies for identifying and understanding the molecular, genetic, and genomic bases of traits in crop wild relatives, landraces, and traditional varieties can further expedite this process.

3.4 Enhancing Crop Yields and Economic Viability

With the current climate problems and the need to strengthen sustainability in agricultural production systems, finding alternatives to enhance crop yields becomes imperative (Mrabet et al. 2023). It is indisputable that resource constraints and climate variability have put traditional methods under severe strain. An integrated and comprehensive strategy is essential to addressing these issues. The adoption of precision agriculture, a technique based on the prudent application of technology to maximize resource utilization, is one viable path. Precision agriculture enables real-time monitoring of soil health, crop performance, and environmental variables by utilizing advanced sensors, satellite imaging, and machine learning algorithms (Mohammed et al. 2021; Sharma et al. 2023a).

With the help of this data-driven strategy, inputs like pesticides, fertilizers, and water can be applied precisely and strategically, reducing waste and having a minimal negative impact on the environment (Abdelraouf et al. 2020; Argento et al. 2021). The impact of real-time evapotranspiration data-based irrigation application was observed (Ayankojo et al. 2018), and maximum tomato yields were observed with 100% evapo-transpiration-based irrigation. The evapo-transpiration data and the irrigation recommendations were generated using the Smart Irrigation Vegetable App (SI). Incorporating agroecological concepts can also promote resilient farming systems. Crop diversification and the use of cover crops improve soil fertility and structure and aid in the control of pests and diseases (Mrabet et al. 2023). Agroforestry, a method of incorporating trees into agricultural landscapes, has several advantages, including the ability to improve microclimates, create new revenue sources, and sequester carbon (Kay et al. 2019). However, such impacts as in carbon sequestration are highly variable, especially on a global scale, varying from 0.29–15.21 Mg/ha/yr. for different agro-forestry systems (Patel et al. 2023). Adopting these creative approaches supports the overall objective of increasing crop yields while simultaneously addressing the demands of sustainability and climate resilience in agriculture.

3.4.1 Tailoring Varieties to Agroecological Zones

Determining geographic zones by thoroughly examining soil, landform, and climate features is known as agroecological zoning, or AEZ. Using particular criteria, this technique considers the climatic and edaphic requirements of crops and the related management practices used in their production (Gajbhiye and Mandal 2000). Each zone is defined to capture a harmonious combination of land use potentials and restrictions. After being described in this way, these zones become the center of specific recommendations meant to improve the current land-use dynamics. These ideas are centered on two main objectives: raising production levels and reducing land degradation. There is no denying that the adoption of high-yield crop types and the extensive use of artificial fertilizers have increased crop productivity worldwide and improved food security (Sharma et al. 2023b). But this progress has come at a price; the environment experiences excessive reactive nitrogen and phosphate releases. The nitrogen and phosphorus cycles have been upset by an over-reliance on inorganic fertilizers, which has had a negative effect on the atmosphere, human health, and biodiversity (John and Babu 2021). Despite stabilizing yields and curbing the need for expanding crop areas, nitrogenous fertilizers contribute significantly to greenhouse gas emissions, exacerbating climate change (Malhi et al. 2021). The future agricultural landscape will have difficulty providing consumer needs for wholesome food, fiber, and animal feed in the face of a highly unpredictable climate while minimizing environmental effects (Owino et al. 2022). A key tactic in tackling this intricate problem is to develop a thorough understanding of genetic diversity and the molecular mechanisms by which plants react to infections and environmental stimuli. The objective is to create novel crop varieties with genetic features that mitigate crop losses from various environmental and insect limitations experienced during the crop's life cycle. The optimization of photosynthesis, water, and nutrient use efficiency is a developing goal that includes promoting advantageous interactions between microbes and plants to improve nutrient uptake (Bailey-Serres et al. 2019).

3.4.2 Reducing Economic Burden

Tailored crop varieties play a pivotal role in improving the economies of stakeholders within the agricultural sector. Recent studies have underscored the significant impact of customized crop varieties on both smallholder farmers and larger agribusinesses. The meta-analysis included 147 original research from field trials and worldwide farm surveys (Klümper and Qaim 2014). To meet the inclusion requirements, primary data about the effects of genetically modified (GM) soybean, maize, or cotton on crop yields, pesticide use, and farmer profits must be used. Results show that the use of chemical pesticides was reduced by an average of 37%, crop yields increased by 22%, and farmer earnings significantly increased by 68% as a result of the adoption of GM technology. Interestingly, compared to their herbicide-tolerant counterparts, insect-resistant crops show more notable increases in yield and decreases in pesticide use (Gao et al. 2017). Furthermore, developing countries see larger yield and profit improvements compared to developed countries. Martey et al. (2020) explored the pressing problem of drought and high temperatures in Africa, which seriously threaten the continent's agricultural sustainability and the livelihoods of its largely agrarian populace. The research targets the introduction of drought-tolerant maize (DTM) varieties as a solution to this problem. The paper finds major adoption drivers, such as labor availability, geographical location, accessibility of seeds, and extension services, using empirical analysis involving 200 farm households in Northern Ghana. Adoption of DTM technology is noteworthy for showing significant benefits, especially in yield and commercialization intensity, with a notable 150% increase in yield (936 kg/ha). The thorough examination conducted (Martey et al. 2020) indicated that DTM has a significant role in maintaining yield stability, especially in moderate-to-extreme drought periods. It is possible for farmers to recover losses and increase productivity to higher levels when DTM is used alongside crop insurance. The DTM cultivar is remarkably more effective than conventional maize cultivars at maintaining average yields.

3.5 Impact of Climate Change on Agriculture Production

Within the context of global agricultural challenges, climate change is a ubiquitous threat to the global paradigm of food and nutritional security (Mrabet et al. 2023). As greenhouse gas emissions increase and become more concentrated in the earth's atmosphere, the greenhouse effect gets stronger and the average world temperature rises steadily. Forecasts suggest that temperatures will increase by 2 °C by 2100, which might have substantial effects on the world economy (Fernandez-Cornejo et al. 2014) (Fig. 3.1). Carbon dioxide (CO_2) is one of the main constituents of these greenhouse gases, and its concentration is rising at an alarming rate (Verma et al. 2023). By increasing photosynthesis, this spike has ironically led to increased growth and plant output. Nonetheless, the simultaneous rise in temperature acts as a counteracting factor, causing crop respiration and evapo-transpiration rates to increase. In addition, higher temperatures cause pest infestation to grow, weed flora to change, and crop duration to decrease (Chauhan et al. 2023). Changes in temperature, precipitation patterns, and the frequency of extreme weather events are all signs of climate change, and they all have a big impact on rice crop growth, development, and yield (Saud et al. 2022).

Temperature increases, a defining feature of climate change, have a double effect on rice production. High temperatures have the potential to promote photosynthesis and, in turn, plant growth in specific stages, but they also hasten respiration and evapotranspiration. The crop may develop more slowly as a result of this imbalance,

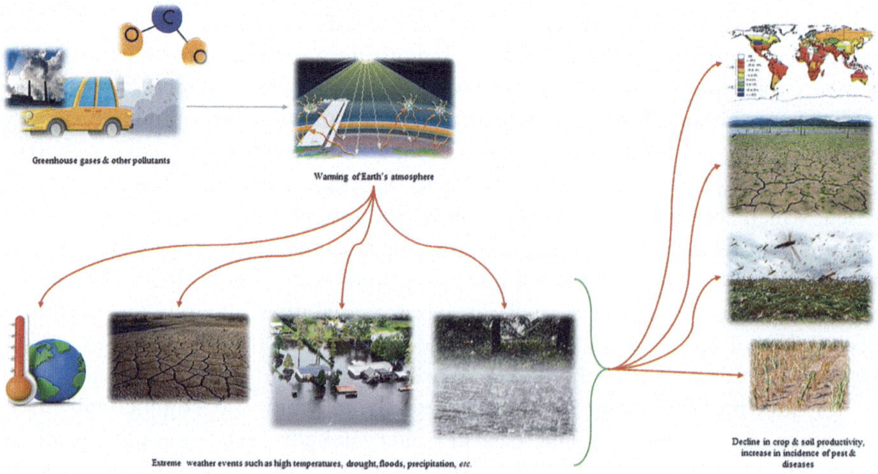

Fig. 3.1 Impact of climate change on agriculture

which would lower total productivity (Wasaya et al. 2022). Another aspect of climate change is changes in precipitation patterns, which bring with them fluctuation in water availability. This is important for rice production, which is frequently done in flooded conditions. Both rainfed and irrigated rice crops may experience variations in water supply as a result of altered precipitation. Yields can be adversely affected by drought stress during critical growth phases, and water-logging from excessive rainfall can result in related problems, including nutrient leaching and increased susceptibility to disease (Nguyen 2002). Furthermore, it is important to take into account the complex relationship between atmospheric CO_2 levels and climate change. In the beginning, increased CO_2 concentrations can promote photosynthesis and accelerate plant growth (Nguyen 2002). However, the difficulties associated with temperature stated earlier and modifications to the dynamics of the agroecosystem as a whole can counteract this beneficial effect. Rice adaptation to high temperatures has been the subject of numerous research, most of which have concentrated on critical growth stages such as meiosis, blooming, and grain filling (Ren et al. 2023). Selecting rice varieties that can withstand extreme temperatures, variation in sowing time, and utilizing plant growth regulators are all effective strategies. The physiology of rice is greatly impacted by high temperatures, which suppress hormone levels, metabolic processes, and photosynthetic and respiratory rates. Key enzymes (SOD, POD, APX, CAT, GR) and non-enzymatic antioxidants (ascorbate, phenolics, glutathione, alkaloids, α-tocopherols, carotenoids, nonprotein amino acids) are involved in molecular responses under stress (Zhang et al. 2022). The complex biological reaction involves the production of genes and proteins, including heat-tolerance-related genes like SBPase and FAD7, heat shock transcription factors (HSFs), and heat stress proteins (HSPs) (Zhang et al. 2022). The investigations on the effect of global warming on rice production, as demonstrated by field warming tests conducted in key rice-growing districts of China, highlight the complexity of rising temperatures affecting rice phenology (Saud et al. 2022). The single-cropping rice experiment was conducted in Harbin between 2016 and 2020, and the medium-cropping rice warming experiment was conducted in Nanjing during the night in 2008 amid flooding and drought conditions. As the temperature rises by 1.5 °C, the rice growing period is significantly shortened, especially from sowing to heading and flowering. Curiously, the time following blooming is essentially unaffected, despite a reduction in the reproductive period preceding flowering. A thorough analysis using the RCP 8.5 scenario on the Keduang subbasin scale in Central Java predicted a significant 11.77% decrease in rice harvests in the second dry season of the 2050s. Notably, it affected planting timing in addition to just lowering output. Climate change in West Java caused an earlier planting schedule corresponding to the need for the wet season and more accessible groundwater. Studies showed that between 2021 and 2030, East and West Java saw a lengthening of the planting pattern in comparison to earlier and later decades (Ikhwali et al. 2022).

3.6 Natural Farming and Sustainability

Natural Farming (NF) stands out as a distinctive, chemical-free agricultural methodology deeply rooted in agroecology. Based on a farming methodology developed by Japanese farmer Masanobu Fukuoka, it has been promoted by Sh. Subash Palekar under the banner of 'Zero Budget Natural Farming (ZBNF)'. ZBNF has been primarily based on liquid bio-stimulant mixtures such as *Jeevamritha* and *Beejamritha* (Sharma et al. 2021; Kumar et al. 2023). Such bio-stimulant mixtures represent a liquid blend crafted from cow dung, cow urine, jaggery, pulses flour, and bund soil amalgamated with water, teeming with a diverse consortium of beneficial microbes (Laishram et al. 2022). Such an application of ZBNF-based biostimulators has been advocated to enhance soil physical, chemical, and biological properties. Enhanced water holding capacity, earthworm population and crop yields were advocated under field experiments conducted (Duddigan et al. 2023) across farmers' fields in southern India. These beneficial microbes play a pivotal role in colonizing germinating seed roots and leaves, bolstering the robust growth of plants. Saharan et al. (2023) advocated a significant increase in primary nutrients such as nitrogen, phosphorus, potassium, and micronutrients such as iron, zinc, copper, and manganese under natural farming-based practices. Researchers further observed an elevated level of bacterial and fungal populations. A similar positive impact on primary nutrients under the influence of natural farming-based bio-stimulant formulations was also observed (Sharma et al. 2023b). Bio-stimulant applications can further boost soil organic matter levels, thus providing a favorable environment for the proliferation of soil microbes. Integral components such as *Achhadana* (bio-mulching), intercropping practices, and utilization of indigenous seeds along with bio-stimulant applications promote self-sustenance of soil ecological systems with positive impacts over biological nutrient cycling. ZBNF-based plant protection measures include variable biopesticide formulations such as *neemastra, agniastra* and *bramhastra* (Kumar and Kumari 2020). Such formulations have been advocated to manage plant pests such as sucking pests, mealy bugs, pod borers, leaf rollers and fruit and stem borers below threshold levels, thus avoiding economic crop damage. *Neemastra* has been advocated to effectively control the population of *Empoasca flavescens* in castor whole, preserving the populations of *Microplitis* cocoons, coccinellid and spiders under field conditions (Kumar and Sarada 2020). Further, *agniastra*-based application @ 800 mL/10 L water was found to be responsible for reducing the root-knot nematode infestation as well as the root-knot index (Chaudhari and Maru 2023).

3.7 The Necessity of Genetic Diversity Preservation

Genetic diversity serves as the foundation for advancing crop enhancement. Diverse plant genetic resources (PGRs) empower plant breeders to develop or refine crop varieties with desirable attributes. When creating new cultivars, careful attention must be given to farmers' preferences, encompassing traits such as high yield, quality, and resistance to diseases and pests. In ancient times, humans selected favorable genotypes based on the natural genetic variability within populations (Govindaraj et al. 2015). The inclination toward developing new crop varieties evolves in response to environmental changes. Plant breeders focus on creating climate-resilient varieties with a comprehensive set of desirable traits, including resistance to various biotic and abiotic stresses. Genetic diversity, represented by mutant lines, wild species, and breeding stocks, is harnessed for the improvement and modernization of crop varieties (Bhandari et al. 2017). To develop climate-resilient varieties, it is imperative to preserve novel genes conferring tolerance to environmental stresses for future integration into breeding programs. Similarly, preserving plant genotypes housing genes for quality traits and aesthetic properties is vital within the available germplasm. The genetic diversity within and between plant species empowers plant breeders to identify superior genotypes for establishing genetic stock in hybridization programs or releasing new crop varieties (Bhandari et al. 2017). This diversity enables PGRs to adapt to diverse climatic conditions (Wernberg et al. 2018), mitigating the negative impact of inbreeding in populations. Elevated levels of genetic diversity in PGRs bolster resilience against adverse environmental changes, ensuring the integrity of community structures and ecosystem functions. Additionally, it allows plant breeders to incorporate genetically diverse parents into breeding programs, ultimately enhancing the productivity of agriculture and horticulture crops.

3.8 Conclusion

This chapter clarifies how genetic stock and natural farming methods play a critical role in sustainable rice production. In recent decades, there has been a radical change in rice cultivation, with high-yielding cultivars replacing traditional crops. However, there are now worries about genetic erosion and the loss of important features due to this change. By encouraging the preservation of landraces and crop wild relatives, natural farming gives farmers the freedom to choose and develop rice varieties specific to their area's environmental conditions. This technique addresses issues related to conventional farming methods, including chemical reliance, deteriorating soil health, and biodiversity loss. Plant breeders can develop climate-resilient varieties by preserving various plant genetic resources and incorporating qualities, high yields, and resilience to a range of stresses. To maintain ecosystem integrity, increase agricultural output, and strengthen resistance to environmental change, genetic diversity is essential.

References

Abd-Aziz S, Gozan M, Ibrahim MF, Phang LY (2022) Demand and sustainability of palm oil plantation. In: Biorefinery of oil producing plants for value-added products. Wiley, pp 11–28

Abdelraouf RE, Ghanem HG, Bukhari NA, El-Zaidy M (2020) Field and modeling study on manual and automatic irrigation scheduling under deficit irrigation of greenhouse cucumber. Sustain For 12(23):1–20

Amudha K, Geetha S, Manimekalai M, Ganesamurthy K (2023) Rice landraces of Tamil Nadu—a review. Indian J Tradit Knowl 22(1):17–29

Anonymous (2022). https://pib.gov.in/PressReleseDetail.aspx?PRID=1899193

Argento F, Anken T, Abt F, Vogelsanger E, Walter A, Liebisch F (2021) Site-specific nitrogen management in winter wheat supported by low-altitude remote sensing and soil data. Precis Agric 22:364–386

Ayankojo IT, Morgan KT, Ozores-Hampton M, Migliaccio KW (2018) Effects of real-time location-specific drip irrigation scheduling on water use, plant growth, nutrient accumulation, and yield of Florida fresh-market tomato. Hort Sci 53(9):1372–1378

Bailey-Serres J, Parker JE, Ainsworth EA, Oldroyd GE, Schroeder JI (2019) Genetic strategies for improving crop yields. Nature 575(7781):109–118

Balla J, Goswami K (2022) Understanding the constraints and reasons to adopt natural farming-a study on rice growing farmers of Andhra Pradesh, India. Int J Agric Sustain 20(6):1209–1224

Bana RS, Dawar R, Haldhar SM, Godara S, Singh A, Bamboriya SD, Choudhary M (2022) Natural farming: is it safe to march ahead? J Agric Ecol 14:1–11

Bhandari HR, Bhanu AN, Srivastava K, Singh MN, Shreya HA (2017) Assessment of genetic diversity in crop plants—an overview. Adv Plants Agric Res 7(3):279–286

Bierschenk B, Tagele MT, Ali B, Ashrafuzzaman MD, Wu LB, Becker M, Frei M (2020) Evaluation of rice wild relatives as a source of traits for adaptation to iron toxicity and enhanced grain quality. PLoS One 15:e0223086

Brown B, Samaddar A, Singh K, Leipzig A, Kumar A, Kumar P, Singh DK, Malik R, Craufurd P, Kumar V, McDonald A (2021) Understanding decision processes in becoming a fee-for-hire service provider: a case study on direct seeded rice in Bihar, India. J Rural Stud 87:254–266. https://doi.org/10.1016/j.jrurstud.2021.09.02

Chaudhari DR, Maru AK (2023) Effect of natural inputs against root-knot nematode (*M. Incognita*) on potato in pot conditions. Int J Pharm Res Appl 5:354–363

Chauhan G, Kumar A, Kumar S, Sharma A, Sharma T (2023) Impact of climate change on agricultural systems and their adaptations. In: Impact of climate change on agriculture. SR Scientific Publications, pp 1–20

Damalas CA, Koutroubas SD (2016) Farmers' exposure to pesticides: toxicity types and ways of prevention. Toxics 4(1):1

Das K, Roy P, Tiwari RKS (2023) Biofortification of rice, an impactful strategy for nutritional security: current perspectives and future prospect. In: Plant-based diet. Intechopen

Deb D (2005) Seeds of tradition, seeds of future, folk rice varieties of eastern India. Research Foundation for Science Technology & Ecology, New Delhi, p 135

Duddigan S, Shaw LJ, Sizmur T, Gogu D, Hussain Z, Jirra K, Collins CD (2023) Natural farming improves crop yield in SE India when compared to conventional or organic systems by enhancing soil quality. Agron Sustain Dev 43(2):31

FAO (2020) UN Food and Agriculture Organization, statistical year book. World Food and Agriculture. Accessed on November 23, 2021, at http://www.fao.org/3/cb1329en/online/cb1329en.html

FAO (2021) FAOSTAT Database. https://www.fao.org/faostat/en/#data/QCL/visualize

Fernandez-Cornejo J, Wechsler S, Livington M, Mitchell L (2014) Genetically engineered crops in the United States. Economic Research Report Number 162, United States, Department of Agriculture

Gajbhiye KS, Mandal C (2000) Agroecological zones, their soil resource and cropping systems. In: Status of farm mechanization in India, cropping systems, status of farm mechanization in India, pp 1–32

Gao Y, Gao X, Zhang X (2017) The 2 °C global temperature target and the evolution of the long-term goal of addressing climate change—from the United Nations framework convention on

climate change to the Paris agreement. Engineering 3(2):272–278. https://doi.org/10.1016/J.ENG.2017.01.022
Gautam RK, Langyan S, Devi SV, Singh R, Semwal DP, Ali S, Singh K (2023) Genetic resources of sticky rice in India: status and prospects. Genet Resour Crop Evol 70(1):95–106
Geetha S, Vasuki A, Selvam PJ, Saraswathi R, Krishnamurthy SL, Dhasarathan M, Baskar M (2017) Development of sodicity tolerant rice varieties through marker assisted backcross breeding. Electron J Plant Breed 8(4):1013–1031
Ghasal PC, Mishra RP, Choudhary J, Dutta D, Bhanu C, Meena AL, Panwar AS (2023) Evaluation of integrated crop management, organic management and natural farming in basmati rice-wheat system under upper indo-Gangetic plains. J Plant Nutr 12:1–11
Ghosh N (2004) Reducing dependence on chemical fertilizers and its financial implications for farmers in India. Ecol Econ 49(2):149–162
Ghosh, SC, Dasgupta T (2023) Medicinal health benefits of traditional rice (*Oryza sativa* L), SATSA Mukhapatra, p. 1–10
Gomez RJA, Tad-awan BA, Fagyan AW, Paing JN, Sagalla EJD (2022) Native/heirloom rice in the cordilleras: status, conservation, and utilization. In: Plant genetic resources, inventory, collection and conservation. Singapore, Springer Nature Singapore, pp 121–134
Govindaraj M, Vetriventhan M, Srinivasan M (2015) Importance of genetic diversity assessment in crop plants and its recent advances: an overview of its analytical perspectives. Genet Reses Int 2015:1–14
Hoekstra AY, Hung PQ (2003) Virtual water trade. Proceedings of the international expert meeting on virtual water trade 12:1–244
Husaini AM, Sofi N (2018a) Rice biodiversity in cold hill zones of Kashmir Himalayas and conservation of its landraces. In: Grillo O (ed) Rediscovery of landraces as a resource for the future. Intech Open, London, pp 39–58
Husaini AM, Sofi NR (2018b) Rice biodiversity in cold hill zones of Kashmir Himalayas and conservation of its landraces. In: Rediscovery of landraces as a resource for the future. Intechopen, London, p 39
Ikhwali MF, Nur S, Darmansyah D, Hamdan AM, Ersa NS, Aida N, Satria A (2022) A review of climate change studies on paddy agriculture in Indonesia. In: IOP conference series: earth and environmental science, vol 1116, No. 1. IOP Publishing, p 012052
John DA, Babu GR (2021) Lessons from the aftermaths of green revolution on food system and health. Front Sustain Food Syst 5:644559
Joshi BK (2017) Plant breeding in Nepal: past, present and future. J Agric For Univ 1:1–33
Joshi BK, Ayer DK, Gauchan D, Jarvis D (2020) Concept and rationale of evolutionary plant breeding and its status in Nepal. J Agri For Univ 4:1–11
Joshi BK, Ghimire KH, Neupane SP, Gauchan D, Mengistu DK (2023) Approaches and advantages of increased crop genetic diversity in the fields. Diversity 15:603. https://doi.org/10.3390/d15050603
Kay S, Rega C, Moreno G, den Herder M, Palma JH, Borek R, Herzog F (2019) Agroforestry creates carbon sinks whilst enhancing the environment in agricultural landscapes in Europe. Land Use Policy 83:581–585
Klümper W, Qaim M (2014) A meta-analysis of the impacts of genetically modified crops. PLoS One 9(11):e111629
Knorr D, Watkins TR (1984) Alternation in food production. Van Nostrand Reinhold, Print
Kumar A, Kumari S (2020) A review on zero budget natural farming: a path towards sustainable agriculture. Pharma Innov 9(4):236–239
Kumar GS, Sarada O (2020) Evaluation of cow based fermented organic products for non-insecticidal pest management in castor. Int J Curr Microbiol Appl Sci 9(10):292–300
Kumar R, Kumar S, Yashavanth BS, Venu N, Meena PC, Dhandapani A, Kumar A (2023) Natural farming practices for chemical-free agriculture: implications for crop yield and profitability. Agriculture 13(3):647

Laishram C, Vashishat RK, Sharma S, Rajkumari B, Mishra N, Barwal P, Sharma N (2022) Impact of natural farming cropping system on rural households—evidence from Solan district of Himachal Pradesh. India Front Sustain Food Syst 6:878015

Liu W, Liu J, Triplett L, Leach JE, Wang GL (2014) Novel insights into rice innate immunity against bacterial and fungal pathogens. Annu Rev Phytopathol 52(1):213–241. https://doi.org/10.1146/annurev-phyto-102313-045926

Malhi GS, Kaur M, Kaushik P (2021) Impact of climate change on agriculture and its mitigation strategies: a review. Sustain For 13(3):1318

Martey E, Etwire PM, Kuwornu JK (2020) Economic impacts of smallholder farmers' adoption of drought-tolerant maize varieties. Land Use Policy 94:104524

Megala R, Kalarani MK, Jeyakumar P, Senthil N, Pushpam R (2022) Exogenous melatonin application mitigates the adverse impact of drought stress at panicle initiation and anthesis stages in rice (*Oryza sativa*). Res Crops 23(4):711–718

Mohammed M, Riad K, Alqahtani N (2021) Efficient iot-based control for a smart subsurface irrigation system to enhance irrigation management of date palm. Sensors 21(12):3942

Mrabet R, Singh A, Sharma T (2023) Conservation agriculture as sustainable and smart soil management: when food systems meet sustainability. In: Agroecological approaches for sustainable soil management. Wiley, New York, pp 339–356

Nguyen NV (2002) Global climate changes and rice food security. FAO, Rome

Nicolopoulou-Stamati P, Maipas S, Kotampasi C, Stamatis P, Hens L (2016) Chemical pesticides and human health: the urgent need for a new concept in agriculture. Front Public Health 4:148

Nonić M, Šijačić-Nikolić M (2021) Genetic diversity: sources, threats, and conservation. In: Life on land. Springer International Publishing, pp 421–435

Noor W et al (2022) Callus induction and regeneration in high altitude Himalayan rice genotype SR4 via seed explant. Biotechnol Rep 36:e00762

Owino V, Kumwenda C, Ekesa B, Parker ME, Ewoldt L, Roos N, Tome D (2022) The impact of climate change on food systems, diet quality, nutrition, and health outcomes: a narrative review. Front Clim 4

Patel R, Patel B, Jaison M, Dash B, Mukherjee S (2023) Impact of agroforestry ecosystem on carbon sequestration potential and climate change. In: Agricultural soil sustainability and carbon management. Academic Press, pp 269–297

Prasad R, Shivay YS, Kumar D (2017) Current status, challenges, and opportunities in rice production. In: Chauhan B, Jabran K, Mahajan G (eds) Rice production worldwide. Springer, pp 1–3

Ren H, Bao J, Gao Z, Sun D, Zheng S, Bai J (2023) How rice adapts to high temperatures. Front Plant Sci 14:1137923

Ricepedia (2021) The online authority on rice. http://ricepedia.org/. Accessed 27 November 2023

Saharan BS, Tyagi S, Kumar R, Vijay OH, Mandal BS, Duhan JS (2023) Application of Jeevamrit improves soil properties in zero budget natural farming fields. Agriculture 13(1):196

Sahebi M, Hanafi MM, Rafii MY, Mahmud TM, Azizi P, Osman M, Abiri R, Taheri S, Kalhori N, Shabanimofrad M, Miah G (2018) Improvement of drought tolerance in rice (Oryza sativa L.): genetics, genomic tools, and the WRKY gene family. Biomed Res Int 2018:3158474

Saikia S, Dutta H, Saikia D, Mahanta CL (2012) Quality characterisation and estimation of phytochemicals content and antioxidant capacity of aromatic pigmented and non-pigmented rice varieties. Food Res Int 46(1):334–340

Saud S, Wang D, Fahad S, Alharby HF, Bamagoos AA, Mjrashi A, Hassan S (2022) Comprehensive impacts of climate change on rice production and adaptive strategies in China. Front Microbiol 13:926059

Sharma T, Singh J, Kaur N (2021) Effect of organic sources of nutrients on growth and growth indices of cowpea (*Vigna unguiculata*) under mid hill conditions of Himachal Pradesh. HJAR 46(2):201–204

Sharma T, Singh A, Kumar N, Chauhan G (2023a) Digital technologies and tools for the ensuing digital era. In: Applying Drone technologies and robotics for agricultural sustainability. IGI Global, pp 1–11

Sharma T, Singh J, Singh A, Sharma R, Chauhan G (2023b) Effect of organic nutrient sources on the yield, nutrient uptake and nodulation in cowpea (*Vigna unguiculata*) under mid-hill conditions of Western Himalayas. Environ Conserv J 24(2):250–256

Singh RP (2021) Genetically distinct cultivar hybrids for the treatment of insect pests and increased agricultural productivity. Acad Int Multidiscip Res J 11:755–764

Tiwari KR, Ingrid LP, Nyborg BKS, Giridhari SP (2008) Analysis of the sustainability of upland farming systems in the middle mountains region of Nepal. Int J Agric Sustain 6(4):289–306. https://doi.org/10.3763/ijas.2008.0390

Valarmathi R, Raveendran M, Robin S, Senthil N (2015) Unraveling the nutritional and therapeutic properties of 'Kavuni'a traditional rice variety of Tamil Nadu. J Plant Biochem Biotechnol 24:305–315

Veena M, Puthur JT, Stępień P, Kalaji HM (2023) Minerals profile and nutraceutical factors in landraces and hybrid varieties of rice: a comparison. Food Biosci 53:102779

Verma B, Porwal M, Jha AK, Sharma T (2023) Agriculture practices to reduce in-field greenhouse gas emissions. In: Climate smart agriculture: principles and practices. KD Publishers, pp 166–180

Wasaya A, Yasir TA, Sarwar N, Mubeen K, Rajendran K, Hadifa A, Sabagh AE (2022) Climate change and global rice security. In: Modern techniques of rice crop production. Singapore, Springer Singapore, pp 13–26

Wernberg T, Coleman MA, Bennett S, Thomsen MS, Tuya F, Kelaher BP (2018) Genetic diversity and kelp forest vulnerability to climatic stress. Sci Rep 8:1851

Wissuwa M, Ae N (2001) Genotypic variation for tolerance to phosphorus deficiency in rice and the potential for its exploitation in rice improvement. Plant Breed 120:43–48

Wissuwa M, Yano M, Ae N (1998) Mapping of QTLs for phosphorus-deficiency tolerance in rice (*Oryza sativa* L.). Theor Appl Genet 97:777–783

Xu K, Mackill DJ (1996) A major locus for submergence tolerance mapped on rice chromosome 9. Mol Breed 2:219–224

Zhang M, Feng K, Dong X, Zhou Y, Chen J, Teng B, Wu W (2022) Screening of heat-tolerant Indica rice varieties in the middle and lower Yangtze River. Agronomy 12(10):2462

Doubled Haploid Rice Breeding

4

M. Tamilzharasi, D. Umamaheswari, J. Umabalan, T. Anandhan, and V. Vengadessan

Abstract

Rice (*Oryza sativa* L.) is a vital component of global food security, sustaining one-third of the world's population. The *Oryza* genus encompasses 24 species, with *Oryza sativa* and *Oryza glaberrima* being cultivated. Due to the challenges posed by climate change and the depletion of natural resources, there is an urgent need for extensive research to enhance rice yield and quality. Traditional breeding methods, involving continuous selfing over generations, are labor-intensive and time-consuming, prompting the adoption of Doubled Haploid (DH) technology. DH technology rapidly fixes desired genes in the immediate generation, offering a solution to challenges faced by rice breeders. The process involves producing haploids from hybrids through anther, pollen, or ovule culture, doubling the chromosomes, and resulting in stable and fertile plants. Various methods contribute to haploid induction in rice, including in vitro-based approaches (anther culture, microspore culture, and ovary culture) and in vivo methods (distant hybridization, CENH3-mediated, and HIS). The in vitro-based approach involves challenges such as early anther necrosis and poor callus response in *indica* rice varieties. Microspore culture refines anther culture by isolating microspores, while ovary culture explores inducing haploids from unpollinated ovaries. In vivo methods present promising alternatives, with genome editing technologies enhancing their applicability. Diploidization of haploids can be achieved through spontaneous or artificial doubling agents. Characterizing DH lines involves analyzing ploidy, phenotype, pollen determination, cytology, isozyme markers, and DNA markers. DH technology finds diverse applications, including accelerating breeding cycles, mapping genes for qualitative and quan-

M. Tamilzharasi (✉) · D. Umamaheswari · J. Umabalan · T. Anandhan · V. Vengadessan
Department of Plant Breeding and Genetics, Pandit Jawaharlal Nehru College of Agriculture and Research Institute, Karaikal, Puducherry (U.T), India
e-mail: tamil94gene@pajancoa.ac.in

A. Singh et al. (eds.), *Climate-Smart Rice Breeding*,
https://doi.org/10.1007/978-981-97-7098-4_4

titative traits, studying genetic modifications, conserving germplasm, and developing biotic and abiotic stress-tolerant lines. The versatility of DH technology significantly contributes to advancing rice breeding, ensuring timely delivery of improved cultivars to meet the increasing global demand for rice sustainably.

4.1 Introduction

Rice (*Oryza sativa* L.) is a major crop of global concern, and maintaining sustainable food production is very important to meet the demand of Asian people who depend on rice for their daily intake (Sarao and Gosal 2018). There are 24 species in the genus *Oryza, of which only 2 are cultivated species, namely, Oryza sativa* and *Oryza glaberrima.* It is a very efficient crop on which huge amount of research has been done for increasing yield as well as quality (Wędzony et al. 2009).

In the financial year 2023, rice cultivation globally reached approximately 513.54 million metric tons, making it the staple food crop for many countries (Statista 2024). However, rice research has faced challenges in recent years, compounded by the imperative to meet the growing global demand amidst escalating competition for agricultural land, water, and labor, particularly in developing nations. The COVID-19 pandemic further exacerbated these challenges, not only causing a decline in rice production but also reversing previous gains (Sankof et al. 2020). This downturn in production has significantly impacted populations, with more than half of those in Asia and one-third in Africa experiencing heightened poverty levels (FAO, IFAD, UNICEF, WFP, WHO 2021). The versatility in climatic conditions and the scarcity of natural resources present substantial hurdles for scientists striving to ensure sustainable food production. In response, recent trends in rice research have shifted towards mega projects and the development of stress-tolerant rice varieties (AfricaRice 2022). Given its vital role as a staple food, addressing these challenges and advancing research in rice cultivation and breeding are crucial for global food security and poverty alleviation efforts.

This underscores the need for timely delivery of improved rice cultivars, a requirement presently addressed through the augmentation of rice varieties and hybrids utilizing both traditional and marker-assisted breeding methods (Osman et al. 2020; Palanisamy et al. 2019). The traditional approaches for developing improved rice varieties involves stabilizing a population obtained from crossing selected parents and through repeated selfing for many generations. Nonetheless, these techniques are typically labor-intensive, expensive, and time-consuming, hindering progress in meeting the escalating demand for food (Yan et al. 2017).

To overcome these problems, doubled haploid (DH) technology has become a powerful driving method in plant breeding. It helps in producing an elite line with reduced breeding cycles by fixing a desired gene in the immediate generation by various approaches like androgenesis (Anther culture and Microspore culture) and gynogenesis (Ovary/Ovule culture) especially under in vitro conditions (Central Rice Research Institute (CRRI) 2007). Production of DH can be achieved by doubling the haploid through various in vitro (anther/pollen culture, microspore culture and ovary culture) and in vivo methods (Distant hybridization, CENH3 (Centromeric specific variant of histone H3)) mediated technique, and Haploid Inducer Stock (HIS) (Mohd Kyum et al. 2023). To raise DH, a protocol such as spontaneous or colchicine treatment of chromosomal doubling is used (Sarao and Gosal 2018).

Double haploid (DH) technology revolutionizes rice breeding by rapidly producing homozygous lines, taking several generations with traditional methods. This acceleration is vital for addressing urgent challenges like climate change and biotic and abiotic stresses, ensuring the timely adaptation of rice crops (Samantaray et al. 2021). DH technology enhances precision by eliminating repeated selfing, reducing the risk of unwanted genetic variability. It facilitates in-depth genetic studies, identifying key genes associated with yield and stress resistance traits. Moreover, generating uniform homozygous lines enhances predictability, enabling breeders to develop rice cultivars with consistent and desirable characteristics efficiently (Mishra et al. 2015). It can also facilitate hybrid and speed breeding, ensuring a more rapid and efficient response to the increasing global demand for rice (Gosal et al. 2020).

4.2 History of DH

The history of DHs began in 1921 when A.D. Bergner observed haploid production in *Datura stramonium* L. (Jimson weed) (Fig. 4.1). Guha and Maheshwari (1964) later achieved artificial haploids through anther culture in *Datura innoxia*. It is further recorded in many crops such as tobacco (*Nicotiana tabacum*) (Zarsky et al. 1995), wheat (*Triticum compactum*) (Barnabás 2003) followed by chilli (Barany et al. 2001), barley (Davies 2003), maize (Barnabás et al. 1999), rice (Miah et al.

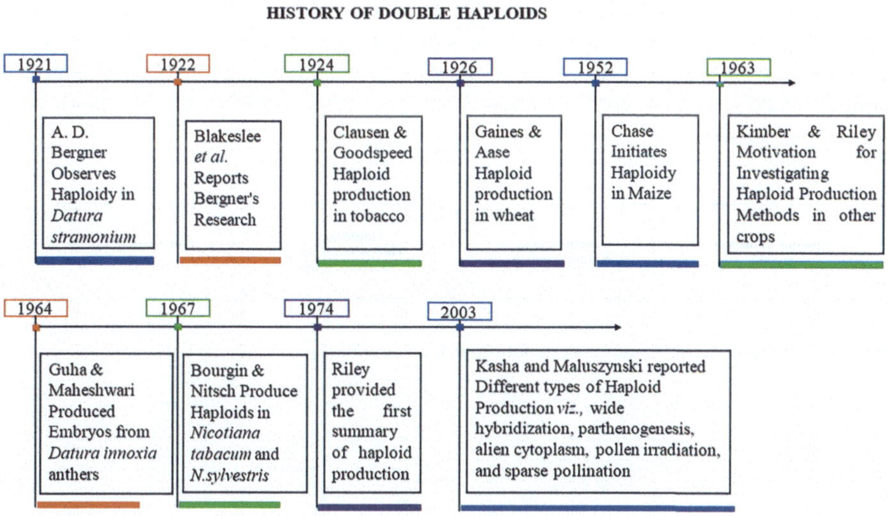

Fig. 4.1 History of double haploid production

1985), triticale (Tuvesson et al. 2003), rye (Immonen 1999), rapeseed (Ferrie et al. 1995) and cabbage (Dias 2001), having a long history of utilizing haploid embryogenesis in plant breeding. Though the study of haploids in maize was started in the early nineteenth century, the identification of mutant gene *igI* in 1969 increased the frequency of androgenic haploids (Chang and Coe Jr 2009). The recognition of the breeding efficacy of doubled haploids prompted a shift in research focus toward identifying methods, problems and strategies to enhance the percentage of diploid plantlets.

Notable reviews discuss haploid development and application of many cereals, including rice and other crops (Khush and Virmani 1996; Sood and Dwivedi 2015; Touraev et al. 2009). Initial attempts in rice by Niizeki and Oono (1968) involved MS medium with NAA, followed by extensive studies on factors affecting haploid yield (Mishra and Rao 2016). The journey from the early observations in *Datura stramonium* to the sophisticated methods applied in rice underscores the continuous evolution and refinement of doubled haploid breeding techniques, making significant contributions to crop improvement and global food security.

4.3 Principles of DH Technology

DH is the technology involved in the production of haploids from varieties or any hybrids by crossing distinct parents and doubling the chromosome content, providing a complement pair to univalent chromosomes (Fig. 4.2). This process aids in the separation of F_1 gametes during meiosis, ultimately leading to the production of stable and fertile plants (Sood and Dwivedi 2015).

In contrast with the conventional breeding technology, DH can produce the complete homozygous elite lines from the F_1 in fewer generations. Explants from the F_1 of a distinct cross are developed into haploid plants through anther, pollen, or ovule culture (Samantaray et al. 2021). Using doubling agents like colchicine, haploids with a single set of genomes are doubled to produce stable, homozygous doubled haploid plants, which are considered an immortal population (Bosland and Barchenger 2024).

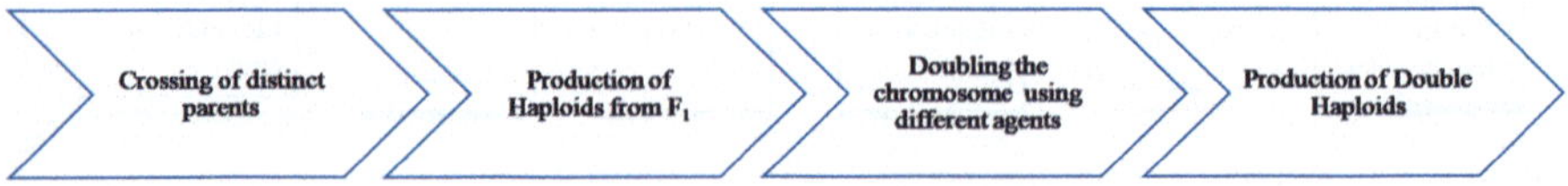

Fig. 4.2 Simple diagrammatic representation of DH technology

4.4 Methods to Induce Haploids

Driven by the advantages of the Double haploids, two methods of haploid induction (HI) in rice are in vitro and in vivo-based techniques (Mohd Kyum et al. 2023). In vitro methods involve culturing gametophytic cells in modified media, while in vivo methods employ Distant hybridization, CENH3 mediation (Fig. 4.3), and HIS to eliminate the uniparental chromosome set during early zygotic development (Comai and Tan 2019; Hu et al. 2016).

4.4.1 In Vitro-Based (Androgenesis and Gynogenesis) Approach

Androgenesis: "*Andro*" denotes male and "*genesis*" means production. As a whole, androgenesis is the production of an embryo or zygote from pollen grains or cells from anther tissues.

Gynogenesis: "*Gynogenesis*" refers to the embryo developed from the female gamete (the egg). It induces the sporophyte development in the unfertilized egg cell. Among the various haploid production methods, androgenesis is the most compatible method to produce DH in rice (Węzony et al. 2009; Seguí-Simarro 2021).

4.4.1.1 Anther Culture

Anther culture is one of the methods of androgenesis where the in vitro culture of the male gametophyte (anther) in the suitable media to produce haploids either through direct embryogenesis (without callus formation) or indirect embryogenesis (via callus) (Fig. 4.4.). Approximately 100 varieties with improved traits have been released through anther culture in various crops (Sasmita 2011). Genetic analysis shows most DH lines are fertile and genetically uniform, facilitating early selection

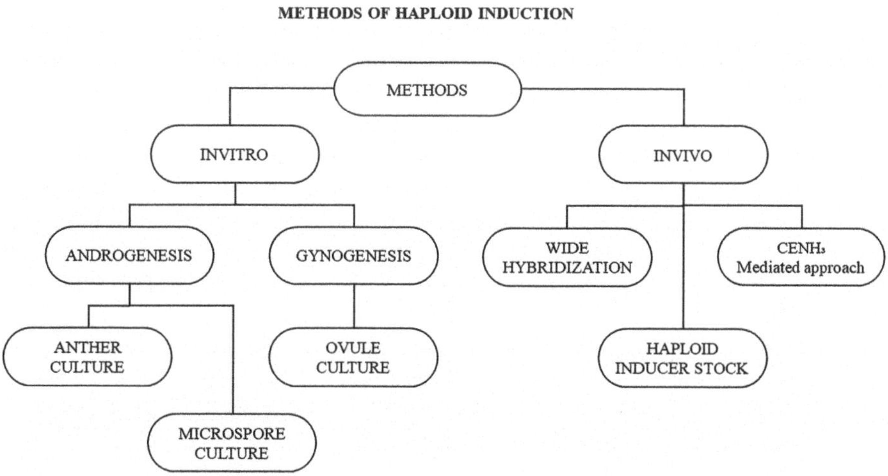

Fig. 4.3 Overview of haploid production methods in rice

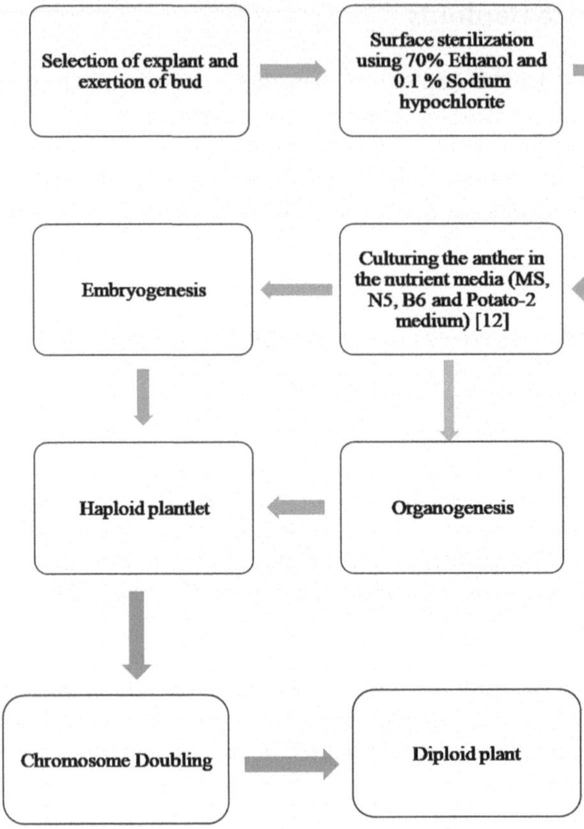

Fig. 4.4 The illustrative depiction of the process-generating doubled haploids (DH)

(Patra et al. 2019). Anther culture-mediated double haploid (DH) breeding overcomes the sterility in F_1 hybrids from interspecific crosses in *indica* rice (Enriquez et al. 2000).

Anther culture responses vary with species and genotype—*O. glaberrima* surpasses *O. sativa* in callusing and regeneration (Sarao and Gosal 2018; Dash et al. 2014; Gueye and Ndir 2010); *Japonica* exhibits greater androgenic responsiveness than *indica* due to the recalcitrant nature and other attributed traits are as follows

(i) Early anther necrosis: Premature death of anthers can hinder the success of the process.
(ii) Poor callus response and proliferation: The formation and multiplication of callus tissue may be insufficient or ineffective.
(iii) Albinism in *indica* rice: The development of albino plants is a significant challenge in *indica* rice, which has an adverse impact on anther culture.

Addressing these obstacles is essential for the broad implementation of anther culture in *indica* rice varieties.

4.4.1.1.1 Factors Affecting the Anther Culture

The efficacy of anther culture is contingent upon multiple factors, including the developmental stage of microspores (Chen 1976), the duration of cold pretreatment of anthers (Zapata et al. 1982), the growth conditions of donor plants (Lee et al. 2003), the composition of the culture medium (Mandal and Gupta 1997), the orientation of plated anthers (Mercy and Zapata 1987), and the genotype of the donor plant (Shen et al. 1983). Attempts have been undertaken to enhance media specifications for androgenesis and the regeneration of green plants in *indica* rice, aiming to achieve a comparable and improved response in conjunction with *japonica* varieties (Guiderdoni et al. 1992; Sathish et al. 1995).

4.4.1.1.1.1 The Developmental Stage

The explant serves as the raw material for tissue culture, and the success of the program hinges on the stage at which it is collected. In anther culture, the developmental stage of the mother plant is crucial. The plant's growth conditions influence the release of microspores, affecting further divisions and regeneration. Optimal pollen collection occurs during the first pollen meiotic division, typically between vacuolate microspores and early (mid-bicellular pollen) (Germana 2011). The anther culture response depends on the donor plant's age and other in vitro conditions. Physiological traits of the donor plant, such as thinner exine structure, weak acetocarmine staining, vacuole presence, and starch grain absence, connect with phenomena like nitrogen starvation (Heberle-Bors 1984) and negatively influence the anther culture.

4.4.1.1.1.2 Pretreatment

Pre-culture treatment, a crucial factor for the induction of diploid plants from the female gametophyte (Baldursson et al. 1993), involves physical or chemical methods applied to spikelet, inflorescence, or anthers. Popular pretreatments, including lower and higher temperatures, water stress, relative humidity variations, anoxia treatment, centrifugation, change in carbon source of media, alcohols, gamma radiation, and increased media pH, all influence genotype-dependent outcomes.

4.4.1.1.1.3 Cold Shock

It is widely applied technique in anther culture of rice because it delays the senescence of anther wall, increases the equational division of pollen and increases the protein breakdown. The efficacy of subjecting *indica* and *japonica* rice varieties and inter-subspecific hybrids to a 14-day cold pretreatment at 8 °C has been demonstrated (Herath et al. 2009).

4.4.1.1.1.4 Temperature Shock

Among the treatments, temperature shock is the preferred technique to develop sporophytic plant from pollen. In rice, callus induction was achieved by pretreating anther for 5 min at 35 °C, which was succeeded by exposing it to 10 °C for 7 days (Reddy et al. 1985). Increasing the duration of heat treatment by immersing panicles at 32 °C for either 2.5 or 5 h improves the callus initiation (Sathish et al. 1995; Rout et al. 2016). Additionally, a two-day pre-incubation cycle at 10 °C impacts the effectiveness of androgenesis in an *indica* rice hybrid (Rout et al. 2016; Raina and Irfan 1998).

In addition to that, *indica* lines respond well in producing androgenic haploids when it is pretreated by sugar starvation (microspore for 3 days) (Raina and Irfan 1998) and gamma irradiation (20 Gy) (Chen et al. 2001). The induction of microspore division and gametic embryogenesis in various species is facilitated by colchicine-induced chromosome doubling (Levan 1945).

4.4.1.1.1.5 Culture Media

The culture medium is crucial in providing nutrients to the growing explants and supports anchorage to the plant. So, the composition of culture media is very important for any in vitro culture. The list of media used in the development of different DH rice hybrids is presented in Table 4.1.

4.4.1.1.1.6 Basal Media Requirement

Basal Media Requirement varies with diverse genotypes. The widely used mediums for anther cultures are N6, B5, Murashige and skoog medium, potato dextrose medium and Nitsch and Nitsch medium (Nitsch and Nitsch 1969). The best basal medium for inducing androgenesis in cereals is N6, and shoot generation is half-strength MS (Murashige and Skoog medium) (Min et al. 2016). The liquid basal medium is preferred over solid media due to two advantages: it provides nutrients in an easily absorbable form for quick uptake and disperses toxic elements produced by dead anthers, thereby reducing the infection rate.

A vital factor for haploid anther embryo production is the carbohydrate source. Sucrose, commonly used as the carbon source, requires higher levels (6–17%) for tricellular pollen in the poaceae family, while lower levels suffice for bicellular pollen.

Nitrogen is the essential element for the development of plants from the anther by encouraging metabolism. It is supplied in both inorganic (Nitrate and Ammonia) and organic form (amino acids). Media with lower levels of NH_4^+ and high NO_3^+ is usually beneficial for the induction of callus during embryogenesis. Casein hydrolysate (CH) is the organic nitrogen supplement of Ca, along with micronutrients, vitamins and amino acids added to the medium. In addition, glutamine, alanine and some organic extracts (coconut water and yeast extracts) helps to enhance callus formation and subsequent plant regeneration. Ethylene inhibitors like $AgNO_3$ and Putrescine are crucial in reducing anther-produced ethylene accumulation during early senescence. Activated charcoal additionally aids in removing inhibitory substances from the culture medium. Since anther culture is a tedious and time-consuming method, microspore culture is used to develop DH.

4.4.1.2 Microspore Culture

Isolated microspore culture, a preferred technique for generating double haploids (DH), especially in *O. sativa* (Seguí-Simarro 2021; Gueye and Ndir 2010), involves isolating microspores from the anther. It is a refined version of anther culture, allowing for easy nutrient uptake and enhanced nourishment to the embryo. Although widely adopted in cereal crops, there is no specified protocol for this technique (Ferrie and Caswell 2011). However, it facilitates the study of embryo maturation and development.

Table 4.1 List of rice double haploid lines (DHLs) produced using different media

SI. No.	Name of double haploid lines (DHLs)	Subspecies	Media	References
1.	CRHR32	*Indica*	MS	Rout et al. (2016)
2.	Janka	*Japonica*	N6	Pauk et al. (2009)
3.	CRHR7	*Indica*	MS	Zarsky et al. (1995)

4.4.1.2.1 Procedure for Microspore Culture

4.4.1.2.1.1 Explant Source

The donor plants should be grown in an appropriate growth chamber to produce green plants essential for regeneration. Microspore culture is influenced by the development stage and pretreatment. Temperature pretreatments typically range from 4 to 10 °C. While subjecting anthers to a shorter duration of heat shock conditions at 33 °C for 48–72 h proved to be efficacious (Liu et al. 2002).

4.4.1.2.1.2 Removal of Floral Parts

Buds are excised at a precise developmental stage from the donor plant. During this process, the microspores should be at the stage between mid or late uni-nucleate (developed to some extent) to the early phase of bi-nucleate using stains like aceto carmine and DAPI (4′, 6-diamidino-2-pheny-lindole) (Ferrie and Caswell 2011; Shariatpanahi et al. 2006).

4.4.1.2.1.3 Microspore Isolation

In the microspore culture technique, sterilized buds (treated with 70% ethanol and 0.1% sodium hypochlorite) are utilized. The process involves two methods: crushing anthers using a pestle and mortar to release microspores or allowing anthers to dehisce in a liquid medium.

4.4.1.2.1.4 Induction of Microspore Culture

Isolated microspores are induced to produce a haploid plant using a nutrient-rich medium, such as Hunter (FHG) media in barley microspore culture (Hunter 1988), comprising 6.2% maltose with a pH of 5.8 (Kasha et al. 2001). Similar to in vivo culture media, it includes a carbohydrate source, nitrogen source, vitamins, and antibiotics. Ethylene inhibitors are crucial to minimize contamination by toxins from dead explants. Additives like growth regulators, amino acids, and other components modify nutrient formulation, enhancing culture efficiency.

4.4.1.2.1.5 Induce Embryogenesis

Stress serves as a useful tool to induce haploid regeneration from microspores, as normal conditions may lead to the formation of pollen grains (Rout et al. 2016).

4.4.1.3 Ovary Culture

Though anther culture and some extent of microspore isolated culture are effective in rice haploid production, several scientists explored ovary culture research. It is reported that haploids of rice can be induced from the culture of unpollinated ovaries from the 1980s. Female cells are induced to produce embryos by using exogenous hormones like MCPA and Picloram (Segui-Simarro and Nuez 2008). However, the efficiency is still low in ovary culture comparatively; hence, it is untapped.

4.4.1.4 Megaspore Cell Culture

In the 1980s, attempts were made to produce haploid plants from megaspores, considered a viable option to address issues like androgenic recalcitrance and albino plant production in rice cultivars (Bhat and Murthy 2007; Chen and Lin 1981; Chen et al. 2011). Success in mega sporogenesis depends on the stage of the ovary (late uni-nucleate to early bi-nucleate), especially with unhusked flowers. The regeneration efficiency of haploid plants is lower in ovary culture because of the presence of one embryo sac per ovary compared to anther where several microspores are present (Gosal et al. 1997; Zhou and Yang 1980). To enhance anther culture efficiency in *indica* rice improvement, focus should be on improving callus induction, addressing early anther necrosis, and understanding the genetic basis of albino traits (Jacquier et al. 2020; Yao et al. 2018).

4.4.2 In Vivo Methods

4.4.2.1 Distant Hybridization

Distant hybridization is a method that involves crossing two genetically diverse genomes, subsequently eliminating the genomes of selected species, leading to the production of haploid (n) embryos (Dwivedi et al. 2015; Watts et al. 2018). The "bulbosum" method was the pioneering approach, where crossing *Hordeum vulgare* with *H. bulbosum* resulted in the elimination of *H. bulbosum* chromosomes during embryo development, producing seedlings with haploid chromosomes of *H. vulgare* (Kasha and Kao 1970). Similarly, maize chromosomes were rapidly lost in intergeneric cross between wheat and maize, generating wheat haploid embryos (Laurie and Bennett 1988). In wheat, distant hybridization with various species, such as eastern gamagrass, teosinte, pearl millet, and sorghum, was reported for haploid induction (Ishii et al. 2016; Chaudhary et al. 2015).

Despite attempts to explore haploid induction in rice through various intergeneric crosses of rice, maize pearl millet poses challenges such as pre-fertilization barriers, hybrid viability, and sterility, which have hindered its application. The *Oryza* genus, comprising 22 wild species, presents opportunities for distant hybridization, but concerted research is essential to unravel the genes present in wild species for haploid induction (Kumari 2019).

4.4.2.2 Centromeric-Specific Variant of Histone H3 (CENH3)-Mediated Approach

Discovered the variant associated with centromere (CENH3) led to the generation of haploid plants (Ravi and Chan 2010). CENH3, crucial for faithful chromosome segregation, consists of N-terminal (variable) and C-terminal histone fold domain (evolutionarily conserved). Researchers focused on altering the N-terminal domain, employing genome editing techniques to induce haploids.

Two approaches for CENH3-mediated haploid plant production are generating CENH3 null mutants and complementing them with transgenic constructs, while the second identifies CENH3 mutants without deleterious effects. In rice, elucidation of the gene encoding CENH3 and its characterization aid researchers in understanding centromere functionality. The CENH3 nucleosomes in rice consist of CentO satellite repeats and the centromeric retrotransposon of rice (CRR). Rice CENH3 genes exhibited strong conservation from diploid to allotetraploid species unaffected by allopolyploidization.

Rice, a model system for genetic studies, was explored for HI using CENH3 mutations similar to those in *Arabidopsis*. Although initial results showed haploid production only in *Arabidopsis*, three inducer lines with different amino acid substitutions in rice were recently identified, yielding haploids at around 1%. The precise mechanism of CENH3-mediated HI on rice and its characterization on crop plants beyond *Arabidopsis* require further exploration, especially with the advancements in genome editing technology (Kalinowska et al. 2019; Wang et al. 2019a).

4.4.2.3 Haploid Induction Stock (HIS) in Rice

The Haploid Inducer Stocks (HIS) is another simple technology to develop haploids in rice preferred over anther culture. In the 1950s, Chase introduced the concept of HIS in Maize. Haploid embryos are produced by crossing HIS with a recurrent parent (Jackson 2017). "Stock 6," an inbred line identified by Coe in 1959, became a pivotal product of the hybrid seed business in maize (Jacquier et al. 2020; Chaikam et al. 2015).

Advances in high throughput techniques such as sequencing and genome editing have enabled researchers to elucidate the pathway of the HIS-mediated haploid embryo in *Zea mays*. The QTLs such as qhir1 (ggi1) and qhir8, have been detected in connection with the inheritance and response of maternal haploids (Dong et al. 2013).

Recent studies by three independent groups have identified the key player, ZmMTL (Matrilineal)/PLP1 (Phospholipase1)/NLD (Not like dad), in the qhir1 region, unraveling the genetic basis of high haploid induction rates (HIR) in corn. The discovery of a 4-bp insertion at the C-terminal of the phospholipase gene explained the mechanism of haploid induction. Aligning the ZmMTL gene's amino acid sequence with publicly available genome databases helps to reveal its conservation and gamete-specific expression patterns in various crops, including rice, wheat, barley, sorghum, foxtail millet, and Arabidopsis (Kelliher et al. 2017).

CRISPR/Cas9-mediated genome editing was used to target the ZmMTL (Matrilineal)/PLP1 (Phospholipase1)/NLD (Not Like Dad) orthologue in rice, generating knockout mutants that achieved 2–6% HIR with a reduction in seed number. The rice matl gene, designated as OsMATL, exhibited similar effects as that of maize, demonstrating the potential of MATL in improving HIS in rice (Yao et al. 2018).

The utilization of CRISPR/Cas9 to knock out the OsMATL gene in rice, based on the breakthrough work done by various researchers (Yao et al. 2018; Wang et al.

2019a), marks a pivotal step in developing a novel breeding approach for rice. The haploid frequency achieved through OsMATL knockout in rice resembles that of the original Stock 6 in maize, suggesting the potential for similar modifiers in the genetic background to influence haploid induction (Yao et al. 2018). The diverse behaviors of OsMATL in different genetic backgrounds could enhance haploid induction rates (HIR), necessitating the introgressing of the osmatl gene into varied genotypes, potentially resulting in increased HIR with a variable seed set.

The small change in ZmDMP, a membrane protein-encoding gene, can significantly increase HIR in maize when combined with ZmMTL. The identification of these genes (ZmMTL and ZmDMP) highlights their potential in elevating HIR in maize, prompting the exploration of their orthologues in rice (OsDMP) for potential roles in HIR (Zhong et al. 2019).

Additionally, screening the global sorghum population led to the discovery of two in vivo haploid inducer sorghum lines (SMHI01 and SMHI02) with a mean rate of 1–2% HIR (Hussain and Franks 2019). Given the shared family between sorghum and rice, this novel finding in sorghum holds promise for potential applications in rice breeding to enhance haploid induction.

4.5 Integration of a Robust Screening Trait for Efficient Haploid Selection

The development of doubled haploid (DH) breeding in maize has seen success through a systematic and effective haploid screening trait at either the seed or seedling level (Geiger 2009; Meng et al. 2021). Morphological marker (R1-Navajo) based haploid identification is widely used in maize based on the presence or absence of purple color in embryo and endosperm (Meng et al. 2021). Identifying such similar phenotypic traits in rice is the major challenge due to small seed size and complex biochemical pathways (Dong et al. 2018).

Alternative methods for identifying haploids in rice include unique molecular markers, microscopic analysis, chromosome counting, and seedling-based identification. As seen in maize, fluorescence-based approaches offer promising accuracy and specificity in rice (Dong et al. 2018; Hua et al. 2019). Spontaneous development of DHs varies from 30% to 40% in rice with genotypic specific effects (Rukmini et al. 2013). Antimitotic agents like colchicine induce chromosome duplication, with colchicine at 500 and 625 mg/L proving most effective in rice (Hooghvorst et al. 2020). Colchicine application can be done in vivo or in vitro, providing flexibility in DH plant production. Standardization of chromosome doubling methods, such as complete immersion in colchicine solution, can enhance chromosome doubling rates in rice while ensuring efficient haploid screening trait incorporation (Chaikam et al. 2020).

The incorporation of a Highly Inducible System (HIS) creates significant opportunities for precise and flexible enhancement of agronomic traits in rice, particularly those controlled by complex genetic mechanisms that traditionally limit plant selection using classical and marker-assisted breeding methods (Kelliher et al. 2019).

Recently, revolutionary mechanisms like Haploid Induction-Edits (HI-Edit) and Inducer Mediated Genome Editing (IMGE) in maize have transformed agricultural sciences (Kelliher et al. 2019; Wang et al. 2019b). This dual-target feature in rice could produce DNA-free Doubled Haploids (DHs) with specific genetic modifications in fewer generations with efficient resources (Dong et al. 2018).

The autogamous nature of rice presents challenges for hybrid breeding, but the involvement of HIS exploits the basis of heterosis, offering a potential solution to the burdensome tasks of large-scale crossing and maintenance. Additionally, HIS can contribute to overcoming challenges associated with maintaining genetic vigor in subsequent generations through synthetic apomixis (Mieulet et al. 2016; Wang 2020). Recent advancements in clonal seed production, achieved through the editing of MiMe genes (Mitosis instead of Meiosis) and the Baby Boom 1 (BBM1) gene, demonstrate the potential for practical outcomes in the near future (Wang et al. 2019a; Conner et al. 2017; Khanday et al. 2019).

4.6 Diploidization of the Produced Haploids

Diploidization of the produced haploids in any crops is a crucial step in the development of stable and fertile plants through the Doubled Haploid (DH) breeding technique (Fig. 4.4). The process involves doubling the chromosome content of haploid cells to restore them to a diploid state. This diploidization is achieved by applying antimitotic agents, commonly colchicine, which induces chromosome duplication.

The incorporation of diploidization in rice breeding follows the production of haploids, typically derived from in vitro or in vivo induction methods. Once haploid plants are obtained, they undergo treatment with diploidizing agents to promote the doubling of chromosomes. The efficiency of diploidization can vary among different genotypes, and the success of the process relies on various factors, as discussed.

It is reported that a high percentage of 72% diploid plants can be achieved in rice through spontaneous doubling (Sarao and Gosal 2018). Several agents used for doubling are enlisted in Table 4.2. The two major classifications of doubling agents are listed below,

Table 4.2 List of doubling agents used in the production of DH rice

Name of the DH	Subspecies	Doubling agent used	Concentration	References
BS6444G (rice DH)	*Indica*	Spontaneous	–	Samantaray et al. (2021)
TM7-5	*Indica*	Gamma rays	20 Gy	Chen et al. (2001)
Risabell	*Japonica*	Colchicine	0.1%	Pauk et al. (2009)
Janka	*Japonica*	Colchicine	0.1%	Pauk et al. (2009)
Abel	*Japonica*	Colchicine	0.1%	Pauk et al. (2009)
CR5-10, CR5-49, CR5-61	*Indica*	BAP and kinetin	0.75 and 0.25 mg, respectively	Shariatpanahi et al. (2006)

4.6.1 Artificial Doubling

The most commonly used agent is *Colchicine*, an antimicrotubule agent extracted from *Colchicum autumnale*. Though it is not carcinogenic, it has higher toxic levels. So, some lower toxic agents like amiprophos-methyl (APM), oryzalin, pronamide, trifuralin, and nitrous oxide gas can be used. Unlike colchicine, it cannot be useful in a wide range of species. For example, N_2O has been widely used in the chromosome doubling of monocots, mostly cereals; however, it has lower efficiency than colchicine in dicots (Palmer et al. 1996).

4.6.2 Spontaneous Doubling

Spontaneous doubling of the haploid genome by using the natural mechanism called *Endomitosis* (repeated division of parental cells with non-separation of daughter cells). Endomitosis is induced easily by growing the haploids in the medium with higher levels of auxin and cytokinin, which divert the unstable haploid to double the genome and then differentiate into shoots (Samantaray et al. 2021). In this case, cereals have a higher success rate, especially rice Spontaneous, with reported frequencies of 50–60% (Segui-Simarro and Nuez 2008). Notably, the hybrid BS6444G exhibits a remarkably high rate of 90–99% doubling (Naik et al. 2017). Doubling rates show significant fluctuations among different genotypes (Kleiber et al. 2012).

Another concept is that haploid embryos can double their genome spontaneously, which is called "spontaneous haploid genome duplication," an inherent trait tendency through which haploids balance their genome naturally for survival (Naik et al. 2017; Hofinger et al. 2013).

4.7 Analysis of DH

Once the genome is doubled, it is worthwhile to identify and confirm the doubled haploid lines through various methods. The characterization of double haploid lines (DHL) using different methods is presented in the Table. 4.3.

Table 4.3 Characterization of DHL through various methods

<table>
<tr><th>S. no.</th><th colspan="4">Description</th><th>Reference</th></tr>
<tr><td>1</td><td>Ploidy</td><td colspan="3">Successful regenerants of DH need to distinguish diploid homozygous plants from redundant heterozygous plants. These plants may arise from somatic tissue of inoculated plant organs, viz., anther wall cells, filaments, and tapetum
The ploidy level can be studied by counting chromosome numbers from the root tip</td><td>Gueye and Ndir (2010)</td></tr>
<tr><td rowspan="5">2</td><td rowspan="5">Phenotype</td><td>Ploid type</td><td>Morphological characteristics</td><td>Grain fertility</td><td rowspan="5">Gueye and Ndir (2010)</td></tr>
<tr><td>Fertile diploids</td><td>Normal morphological appearance</td><td>70–80%</td></tr>
<tr><td>Haploids</td><td>Short stature, no spikelet fertility</td><td>0%</td></tr>
<tr><td>Polyploids</td><td>Tall, large, broad thick leaves</td><td><1%</td></tr>
<tr><td>Mixploids</td><td>Grain-type discrimination after maturity</td><td>Variable</td></tr>
<tr><td>3</td><td>Pollen size determination</td><td colspan="3">Study of the size of the pollen</td><td rowspan="7">Samantaray et al. (2021), Hofinger et al. (2013)</td></tr>
<tr><td>4</td><td>Cytology</td><td colspan="3">The root tip sample is collected and fixed to count the chromosome during the metaphase stage
Flow cytometry is widely utilized for plant-ploidy analysis due to its reliability in assessing nuclear DNA content, which closely correlates with ploidy levels</td></tr>
<tr><td>5</td><td>Isozyme markers</td><td colspan="3">The enzymes produced by androgenetic and somatic tissue are different in identifying heterozygote diploids from the DH regenerants</td></tr>
<tr><td rowspan="4">6</td><td rowspan="4">DNA markers viz., SSR and SNP</td><td>Source of origin</td><td colspan="2">Markers employed for discerning the origin, whether from microspore mother cells or diploid somatic tissue embryogenesis</td></tr>
<tr><td>Parental contribution</td><td colspan="2">Utilization of markers to understand parental contribution, focusing on allelic frequencies</td></tr>
<tr><td>Homozygosity in plant derivatives</td><td colspan="2">Examination of homozygosity in plant derivatives through the use of these markers</td></tr>
<tr><td>Application in rice hybrid identification</td><td colspan="2">Successful utilization of SSRs for identifying homozygous spontaneous-doubled haploids in rice hybrids</td></tr>
</table>

4.8 Application of Doubled Haploids

4.8.1 Mapping a Gene Using DH

Mapping a gene or gene block in the genomic region is an essential step for locating a desirable gene in the chromosome of a plant species. Further it is used in introgression from the donor (wild species) to a recipient through the technique called marker-assisted selection (MAS) (Samantaray et al. 2021). Most of the agronomically important characters are governed by the block of genes as QTL. Mapping a QTL is generally done using an immortal mapping population. DH has its own significance in mapping a gene since it is highly immortal and produced in fewer generations compared to other populations like NILs and RILs (Ochatt et al. 2009). Gene identified from DH are listed in the Table 4.4.

4.8.2 Development of Improved Varieties in Fewer Generations

DH technology is unique and helps to develop improved varieties in fewer generations. Though the procedure is tedious, it can fix all the segregating loci in a lesser generation (especially one) compared to conventional breeding techniques. Most research ideas are to develop improved varieties in context to nutritional quality, resistance and higher yield. In India, promising double haploids, namely CR Dhan 10 (CRAC2221–43/Satyakrishna) (Central Rice Research Institute (CRRI) 2007), Risabell, Abel (Pauk et al. 2009), CR Dhan 801 (CRAC2224–1041/Phalguni) (Central Rice Research Institute (CRRI) 2009), Parag 401 (Patil et al. 1997) were released for resistant to various biotic, and abiotic stresses and quality traits.

4.8.3 Double Haploids Derived from Hybrids

Doubled Haploids (DHs) obtained from hybrid rice signify a crucial element in contemporary agriculture, showcasing the efficiency and precision of breeding methodologies (Rout et al. 2016). DHs offer a streamlined path to genetic uniformity and stability, playing a key role in accelerating the breeding process for desirable traits in hybrid rice. Table 4.5 mentions doubled haploid Varieties derived from hybrids. The evaluation of DH performance encompasses various agronomic and physiological parameters, influencing factors such as yield potential, plant height, panicle characteristics, and resistance to biotic and abiotic stresses. Several of the DHs demonstrated superior grain quality and grain yield comparable to that of the parent rice hybrids. Utilizing the DH technique addressed challenges linked to *indica* rice hybrids, including the high cost of seeds, which limited their accessibility to marginal Indian farmers over consecutive years, and the uncertainties of environmental conditions and synchronized flowering (Naik et al. 2017).

Table 4.4 List of genes/QTLs identified for different traits in double haploids

Name of the cross (DH parental lines)	Genes/QTLs	Trait of gene	Reference
A subset of 151 accessions from a USDA rice core collection	SSR markers RM1359 and RM273 on chromosome 4; RM1359 is physically close to the pi21gene	Blast resistant genes	Wang et al. (2015)
Zhai ye Qing 8 × Jin Xi17	R8 cDNA	Blast resistant genes	Zheng et al. (2004)
Japonica (Maybelle) × *Indica* (Baiyeqiu)	qShB1	Sheath blight resistance	Xu et al. (2011)
CJ06 × TN1	8 QTLs	Sheath blight resistance	Zeng et al. (2015)
IR64 and *Azucena*	6 QTLs	Brown plant hopper (BPH) resistance	Soundararajan et al. (2004)
IR64 and *Azucena*	Chromosome 1 (W1-pMRF1), 2 (XLRfrI7-RG157) and 7 (RG711-CDO418)	White-backed plant hopper (WBPH) resistance	Geethanjali et al. (2009)
Zhai ye Qing 8 (ZYQ8) and Jing xi 17 (JX17)	qSBR-11	Sheath blight resistance	Kunihiro et al. (2002)
	qSBR-7		
	qSBR-3		
	qSBR-2		
IR64 and Azucena	qNBL-10	Blast resistance	Hittalmani et al. (2000)
	qNBL-9		
	qNBL-5		
Savitri and Pokkali	LOC_Os01g09550 and LOC_Os01g09560	Salt tolerance at the establishment stage	
	LOC_Os12g06560 and LOC_Os12g06570		

Table 4.5 Doubled haploid varieties derived from hybrids

Name of the parental cross	Released DH varieties	Methodology	References
HB71 × KRH2	Satyakrishna and Phalguni	Standardized technology	Rout et al. (2016), Kleiber et al. (2012)
CRHR32 × BS6444G	–	Androgenic protocols were followed and it produced a substantial number of DHs	

4.8.4 Reverse Breeding

Reverse breeding is a sophisticated genetic technique closely related to doubled haploid breeding. Unlike traditional breeding methods, which involve selecting desirable phenotypes and crossing them to generate offspring with desired traits, reverse breeding reverses this process (Samantaray et al. 2021). It starts with a desired phenotype and aims to produce parental lines that give rise to that phenotype when crossed.

In the context of doubled haploid breeding, reverse breeding can be a valuable tool for rapidly generating pure-breeding parental lines. By starting with the desired trait and working backward to identify the genetic factors responsible, reverse breeding facilitates the creation of homozygous lines in a single generation (Gosal et al. 2020). These homozygous lines can then be used to produce doubled haploid offspring with the desired trait consistently expressed.

4.8.5 Biofortification

Rice, a vital food crop, is a primary source of energy, protein, and essential nutrients. However, micronutrient deficiencies, particularly in iron and zinc, contribute to "hidden hunger," impacting global nutrition (Calayugan et al. 2020). Addressing this, biofortification aims to enhance rice varieties by increasing protein, iron, and zinc content while reducing phytate levels (Jeong et al. 2020). This approach, crucial for overcoming micronutrient malnutrition, utilizes genetic variation through breeding or genetic engineering. The doubled-haploid method emerges as a promising strategy, providing faster development of fixed genetic materials with reduced background noise (Lee et al. 2020). Recent evaluations of doubled-haploid populations have effectively identified QTLs for grain micronutrients, highlighting their promise in breeding initiatives and multi-location trials for improved crop performance, as summarized in Table 4.6.

Table 4.6 List of identified QTLs for biofortification in DH rice lines

S. no.	Name of the parents	Methodology	QTL identification	References
1.	IR05F102 × IR69428	6 K SNP genotyping inclusive composite interval mapping	2 QTLs for Zn on chromosomes 5 and 12 Identification of QTLs with candidate genes, e.g., OsZIP6 on QTL qZn5.1	Calayugan et al. (2020)
2.	Derived from an intra-japonica cross between 'Hwaseonchal' × 'Goami 2'	6 K SNP genotyping composite interval mapping (CIM) with QGene Genome annotation using the Rice genome annotation project (RGAP) browser	Major-effect QTL, qFe7 and qZn7 Eleven iron- and zinc-related candidate genes co located with qFe7, qZn7, and the chromosome 7 region	Jeong et al. (2020)
3.	93–11 × Milyang 352	Inclusive composite interval mapping	Co-localized QTLs, qFe3-1 and qZn3-1, identified on chromosome 3	Lee et al. (2020)
4.	PSBRc82 × Joryeongbyeo and PSBRc82 × IR69428	6 K SNP genotyping inclusive composite interval mapping	Identified candidate genes near QTLs for grain Zn (OsNRAMP, OsNAS, OsZIP, OsYSL, OsFER, and OsZIFL family) Eight epistatic interactions were detected for cu, mg, Na, and Zn	Swamy et al. (2018)

4.9 Conclusion

Doubled haploids (DHs) represent a crucial tool in plant breeding, particularly in crops like rice, where achieving homozygosity quickly is advantageous. The method involves the production of haploid plants through various techniques such as anther culture, ovary culture and microspore culture, followed by chromosome doubling to obtain homozygous lines. The applications of doubled haploids in rice breeding are diverse and impactful. DH technology allows for the rapid fixation of desirable traits, shortening the breeding cycle and enabling the development of new cultivars with improved yield, resistance to diseases, and tolerance to various environmental stresses. This technology also facilitates the study of genetic mechanisms and accelerates the identification of important genes associated with specific traits. Despite its potential, the application of DH technology in rice faces challenges, especially in *indica* rice lines due to their recalcitrant nature. Optimizations of anther culture methods and media manipulation are necessary to enhance callusing potential and somatic embryogenesis, particularly in *indica* rice. Additionally, identifying genomic regions contributing to yield in rice hybrids is essential for effective DH utilization.

Future perspectives in DH technology for rice involve developing novel media compositions to enhance callusing potential and focusing on cost-effective pathways like direct somatic embryogenesis. Understanding the mechanism of spontaneous chromosome doubling in androgenesis is critical for efficient DH production. Furthermore, comprehensively studying the genetic basis of the superior yield of DHs through the generation of a large number of DHs is essential for maximizing their potential in rice breeding programs. With these advancements and continued research efforts, DH technology holds promising prospects for accelerating varietal development and addressing the challenges of food security and agricultural sustainability.

Author Contributions M. Tamilzharasi conceptualized and designed the study and drafted the manuscript. Umabalan J. conducted the literature review. M. Tamilzharasi, Umamaheswari D., Anandhan T., and Vengadessan V. reviewed and approved the final manuscript for submission.

References

AfricaRice (2022) Stress-tolerant rice varieties. https://www.africarice.org/stress-tolerant-rice. Accessed 17 Mar 2022

Baldursson S, Norgaard JV, Krogstrup P et al (1993) Factors influencing haploid callus initiation and proliferation in megagametophyte cultures of Sitka spruce (*Picea sitchensis*). Silvae Genet 42(2–3):79–86

Barany I, Testillano PS, Mityko J, Risueno MC et al (2001) The switch of the microspore developmental programme in capsicum involves HSP70 expression and leads to the production of haploid plants. Int J Dev Biol 45:S39–S40

Barnabás B (2003) Protocol for producing doubled haploid plants from anther culture of wheat (Triticum aestivum L.). In: Maluszynski M, Kasha KJ, Forster BP, Szarejko I et al (eds) Doubled haploid production in crop plants—a manual. Kluwer, Dordrecht, Boston, London, pp 65–70

Barnabás B, Obert B, Kovacs G (1999) Colchicine, an efficient genome-doubling agent for maize (*Zea mays* L.) microspores cultured in anthero. Plant Cell Rep 18:858–962

Bhat JG, Murthy HN (2007) Factors affecting in-vitro gynogenic haploid production in Niger (*Guizotia abyssinica* (L. f.) Cass.). Plant Growth Regul 52(3):241–248. https://doi.org/10.1007/s10725-007-9196-9

Bosland PW, Barchenger DW (2024) Resistance: classical breeding methods. In: Bosland PW, Barchenger DW (eds) . Academic Press, Breeding disease-resistant horticultural crops, pp 195–204. ISBN: 9780443152788. https://doi.org/10.1016/B978-0-443-15278-8.00008-5

Calayugan MIC, Formantes AK, Amparado A, Descalsota-Empleo GI, Nha CT, Inabangan-Asilo MA, Swe ZM, Hernandez JE, Borromeo TH, Lalusin AG, Mendioro MS et al (2020) Genetic analysis of agronomic traits and grain iron and zinc concentrations in a doubled haploid population of rice (*Oryza sativa* L.). Sci Rep 10(1):2283

Central Rice Research Institute (CRRI) (2007–2008) Annu Rep, 34. www.crri.nic.in/

Central Rice Research Institute (CRRI) (2009–2010) Annu Rep: 35. www.crri.nic.in/

Chaikam V, Nair SK, Babu R, Martinez L, Tejomurtula J, Boddupalli PM et al (2015) Analysis of effectiveness of R1-nj anthocyanin marker for in vivo haploid identification in maize and molecular markers for predicting the inhibition of R1-nj expression. Theor Appl Genet 128(1):159–171. https://doi.org/10.1007/s00122-014-2419-3

Chaikam V, Gowda M, Martinez L, Ochieng J, Omar HA, Prasanna BM (2020) Improving the efficiency of colchicine-based chromosomal doubling of maize haploids. Plan Theory 9(4):459. https://doi.org/10.3390/plants9040459
Chang MT, Coe EH Jr (2009) Molecular genetic approaches to maize improvement. Biotechnol Agric For 63:127–138
Chaudhary HK, Badiyala A, Jamwal NS (2015) New frontiers in doubled haploidy breeding in wheat. Agric Res J 52(4):1–12. https://doi.org/10.5958/2395-146X.2015.00053.8
Chen CC (1976) Studies on the anther culture of rice: pollen stage and low temperature treatment. Natl Sci Council Monthly (Taipei) 4(2):2187–2190
Chen CC, Lin CM (1981) Genotypic differences in plant production in anther culture of rice. In: Chang WC (ed) Proceedings of the symposium on plant cell and tissue culture. Academia Sinica, Taipei, pp 199–203
Chen QF, Wang CL, Lu YM, Shen M, Afza R, Duren MV et al (2001) Anther culture in connection with induced mutations for rice improvement. Euphytica 120(3):401–408
Chen JF, Cui L, Malik AA, Mbira KG et al (2011) In vitro haploid and dihaploid production via unfertilized ovule culture. Plant Cell Tissue Organ Cult 104(3):311–319. https://doi.org/10.1007/s11240-010-9874-6
Comai L, Tan EH (2019) Haploid induction and genome instability. Trends Genet 35(11):791–803. https://doi.org/10.1016/j.tig.2019.07.005
Conner JA, Podio M, Ozias-Akins P (2017) Haploid embryo production in rice and maize induced by PsASGR-BBML transgenes. Plant Reprod 30(1):41–52. https://doi.org/10.1007/s00497-017-0298-x
Dash AK, Rao GJN, Rao RN (2014) Effect of genotype on anther culture response in indica rice hybrids of maintainer lines. Oryza 51(2):165–167
Davies PA (2003) Barley isolated microspore culture (IMC) method. In: Maluszynski M, Kasha KJ, Forster BP, Szarejko I et al (eds) Doubled haploid production in crop plants: a manual. Kluwer Academic, Dordrecht, pp 49–92
Dias JCD (2001) Effect of incubation temperature regimes and culture medium on broccoli microspore culture embryogenesis. Euphytica 119(3):389–396
Dong X, Xu X, Miao J, Li L, Zhang D, Mi X et al (2013) Fine mapping of qhir1 influencing in vivo haploid induction in maize. Theor Appl Genet 126(7):1713–1720. https://doi.org/10.1007/s00122-013-2086-9
Dong L, Li L, Liu C, Geng S, Li X, Xie C et al (2018) Genome editing and double-fluorescence proteins enable robust maternal haploid induction and identification in maize. Mol Plant 11(9):1214–1217. https://doi.org/10.1016/j.molp.2018.06.011
Dwivedi SL, Britt AB, Tripathi L, Sharma S, Upadhyaya HD, Ortiz R et al (2015) Haploids: constraints and opportunities in plant breeding. Biotechnol Adv 33(6):812–829. https://doi.org/10.1016/j.biotechadv.2015.07.001
Enriquez EC, Brar DS, Rosario TL, Jones M, Khush GS et al (2000) Production and characterization of doubled haploids from anther culture of the F1s of *Oryza sativa* L. × *O. Glaberrima* Steud. Rice Genet Newslett 17:67–69
FAO, IFAD, UNICEF, WFP, WHO (2021) The state of food security and nutrition in the world 2021. FAO, IFAD, UNICEF, WFP, and WHO. https://doi.org/10.4060/cb4474en
Ferrie AMR, Caswell KL (2011) Isolated microspore culture techniques and recent progress for haploid and doubled haploid plant production. Plant Cell Tissue Organ Cult 104:301–309. https://doi.org/10.1007/s11240-010-9800-y
Ferrie AMR, Epp DJ, Keller WA (1995) Evaluation of *Brassica rapa* L genotypes for microspore culture response and identification of a highly embryogenic line. Plant Cell Rep 14(9):580–584
Geethanjali S, Kadirvel P, Gunathilagaraj K, Maheswaran M (2009) Detection of quantitative trait loci (QTL) associated with resistance to white-backed planthopper (*Sogatella furcifera*) in rice (Oryza sativa). Plant Breed 128:130–136
Geiger HH (2009) Doubled haploids. In: Jeffrey LB, Sarah H (eds) Handbook of maize. Springer, New York, pp 641–657. https://doi.org/10.1007/978-0-387-77863-1_32
Germana MA (2011) Anther culture for haploid and doubled haploid production. Plant Cell Tissue Organ Cult 104:283–300

Gosal SS, Sindhu AS, Sandhu JS, Sandhu-Gill R, Singh B, Khehra GS et al (1997) Haploidy in rice. In: Jain SM, Sopory SK, Veilleux RE (eds) Vitro haploid production in higher plants. Springer, Dordrecht, pp 1–35. https://doi.org/10.1007/978-94-017-1862-2_1

Gosal SS, Pathak D, Wani SH, Vij S, Pathak M et al (2020) Accelerated breeding of plants: methods and applications. In: Accelerated plant breeding, cereal crops, vol 1, pp 1–29

Gueye T, Ndir KN (2010) In vitro production of double haploid plants from two rice species (*Oryza sativa* L. and *Oryza glaberrima* Steudt.) for the rapid development of new breeding material. Sci Res Essays 5(7):709–713. https://doi.org/10.5897/SRE.9000581

Guiderdoni E, Galinato E, Luistro J et al (1992) Anther culture of tropical japonica × indica hybrids of rice (*Oryza sativa* L.). Euphytica 62:219–224

Heberle-Bors E (1984) Pollen embryogenesis: a model system for the life cycle of higher plants. J Embryol Exp Morphol 82:17

Herath HMI, Bandara DC, Samarajeewa PK et al (2009) Effect of low temperature pretreatment on anther culture in selected Indica, japonica rice varieties and their inter subspecific hybrids. Ceylon J Sci 38:11–16

Hittalmani S, Parco A, Mew TV, Zeigler RS, Huang N (2000) Fine-mapping and DNA marker-assisted pyramiding of the three major genes for blast resistance in rice. Theor Appl Genet 100(7):1121–1128

Hofinger BJ, Huynh OA, Jankowicz-Cieslak J, Müller A, Otto I, Kumlehn J, Till BJ et al (2013) Validation of doubled haploid plants by enzymatic mismatch cleavage. Plant Methods 9(43):1–10

Hooghvorst I, Ribas P, Nogués S (2020) Chromosome doubling of androgenic haploid plantlets of rice (Oryza sativa) using antimitotic compounds. Plant Breed 139(4):754–761. https://doi.org/10.1111/pbr.12824

Hu H, Schrag TA, Peis R, Unterseer S, Schipprack W, Chen S, Melchinger AE et al (2016) The genetic basis of haploid induction in maize identified with a novel genome-wide association method. Genetics 202(4):1267–1276. https://doi.org/10.1534/genetics.115.184234

Hua K, Zhang J, Botella JR, Ma C, Kong F, Liu B et al (2019) Perspectives on the application of genome-editing technologies in crop breeding. Mol Plant 12(8):1047–1059. https://doi.org/10.1016/j.molp.2019.06.009

Hunter CP (1988) Plant regeneration from microspores of barley, *Hordeum vulgare* L. [Ph.D. thesis]. Wye College, University of London, London

Hussain T, Franks C (2019) Discovery of sorghum haploid induction system. Methods Mol Biol (Clifton, NJ) 1931:49–59. https://doi.org/10.1007/978-1-4939-9039-9_4

Immonen S (1999) Androgenic green plants from winter rye, *Secale cereale* L., of diverse origin. Plant Breed 118(4):319–322

Ishii T, Karimi-Ashtiyani R, Houben A (2016) Haploidization via chromosome elimination: means and mechanisms. Annu Rev Plant Biol 67:421–438. https://doi.org/10.1007/s00122-018-3261-9

Jackson D (2017) No sex please we're (in) breeding. EMBO J 36(6):703–704. https://doi.org/10.15252/embj.201796735

Jacquier NM, Gilles LM, Pyott DE, Martinant JP, Rogowsky PM, Widiez T et al (2020) Puzzling out plant reproduction by haploid induction for innovations in plant breeding. Nat Plants 6(6):610–619. https://doi.org/10.1038/s41477-020-0664-9

Jeong OY, Bombay M, Jo S, Ancheta MB, Lee JH et al (2020) QTL for the iron and zinc contents of the milled grains of a doubled-haploid rice (*Oryza sativa* L.) population grown over two seasons. J Crop Sci Biotechnol 23:291–299

Kalinowska K, Chamas S, Unkel K, Demidov D, Lermontova I, Dresselhaus T et al (2019) State-of-the-art and novel developments of in vivo haploid technologies. Theor Appl Genet 132(3):593–605. https://doi.org/10.1007/s00122-018-3261-9

Kasha KJ, Kao KN (1970) High frequency haploid production in barley (*Hordeum vulgare* L.). Nature 225(5235):874–876. https://doi.org/10.1038/225874a0

Kasha KJ, Simion E, Oro R, Yao QA, Hu TC, Carlson AR (2001) An improved in vitro technique for isolated microspore culture of barley. Euphytica 120:379–385

Kelliher T, Starr D, Richbourg L, Chintamanani S, Delzer B, Nuccio ML et al (2017) Matrilineal, a sperm-specific phospholipase, triggers maize haploid induction. Nature 542(7639):105–109. https://doi.org/10.1038/nature20827

Kelliher T, Starr D, Su X, Tang G, Chen Z, Carter J et al (2019) One-step genome editing of elite crop germplasm during haploid induction. Nat Biotechnol 37(3):287–292. https://doi.org/10.1038/s41587-019-0038-x

Khanday I, Skinner D, Yang B, Mercier R, Sundaresan V et al (2019) A male-expressed rice embryogenic trigger redirected for asexual propagation through seeds. Nature 565(7737):91–95. https://doi.org/10.1038/s41586-018-0785-8

Khush GS, Virmani SS (1996) Haploids in plant breeding. In: Jain SM, Sopory SK, Veilleux RE (eds) In vitro haploid production in higher plants. Current plant science and biotechnology in agriculture, vol 23. Springer, Dordrecht, pp 11–33. https://doi.org/10.1007/978-94-017-1860-8_2

Kleiber D, Prigge V, Melchinger AE, Burkard F, San Vicente F, Palomino G, Gordillo GA et al (2012) Haploid fertility in temperate and tropical maize germplasm. Crop Sci 52:623–630

Kumari S (2019) Production of doubled haploids in rice (*Oryza sativa* L.) through anther culture and wide crosses and their molecular characterization., Doctoral dissertation,. Punjab Agricultural University, Ludhiana

Kunihiro Y, Qian Q, Sato H, Teng S, Zeng DL, Fujimoto K, Zhu LH et al (2002) QTL analysis of sheath blight resistance in rice (*Oryza sativa* L.). Chin J Genet 29:50–55

Laurie DA, Bennett MD (1988) The production of haploid wheat plants from wheat x maize crosses. Theor Appl Genet 76(3):393–397. https://doi.org/10.1007/BF00265339

Lee SY, Lee JH, Kwon TO et al (2003) Selection of salt-tolerant doubled haploids in rice anther culture. Plant Cell Tissue Organ Cult 74(2):143–149

Lee SM, Kang JW, Lee JY, Seo JH, Shin DJ, Cho JH, Jo SM, Song YC, Park DS, Ko JM, Koh HJ et al (2020) QTL analysis for Fe and Zn concentrations in rice grains using a doubled haploid population derived from a cross between rice (Oryza sativa) cultivar 93-11 and Milyang 352

Levan A (1945) A haploid sugar beet after colchicine treatment. Hereditas 31:399–410

Liu W, Zheng MY, Polle EA, Konzak CF et al (2002) Highly efficient doubled-haploid production in wheat (*Triticum aestivum* L.) via induced microspore embryogenesis. Crop Sci 42:686–692

Mandal N, Gupta S (1997) Anther culture of an interspecific rice hybrid and selection of fine grain type with submergence tolerance. Plant Cell Tissue Organ Cult 51:79–82

Meng D, Liu C, Chen S, Jin W (2021) Haploid induction and its application in maize breeding. Mol Breed 41(3):1–9. https://doi.org/10.1007/s11032-021-01204-5

Mercy ST, Zapata FJ (1987) Position of anthers at plating and its influence on anther callusing in rice. Plant Cell Rep 6:318–319

Miah MAA, Earle ED, Khush GS (1985) Inheritance of callus formation ability in anther cultures of rice, *Oryza sativa* L. Theor Appl Genet 70(2):113–116

Mieulet D, Jolivet S, Rivard M, Cromer L, Vernet A, Mayonove P et al (2016) Turning rice meiosis into mitosis. Cell Res 26(11):1242–1254. https://doi.org/10.1038/cr.2016.117

Min ZY, Li H, Zou T, Tong L, Cheng J, Sun XW et al (2016) Studies of in vitro culture and plant regeneration of unfertilized ovary of pumpkin. Chinese Bullet Bot 51(1):74–80

Mishra R, Rao GJN (2016) In-vitro androgenesis in rice: advantages, constraints and future prospects. Rice Sci 23(2):57–68. https://doi.org/10.1016/j.rsci.2016.02.001

Mishra R, Rao GJN, Rao RN, Kaushal P et al (2015) Development and characterization of elite doubled haploid lines from two indica rice hybrids. Rice Sci 22(6):290–299

Mohd Kyum K, Kaur H, Kamboj A, Goyal L, Bhatia D et al (2023) Strategies and prospects of haploid induction in rice (*Oryza sativa*). Plant Breed 141:1. https://doi.org/10.1111/pbr.12971

Naik N, Rout P, Umakanta N, Verma RL, Katara JL, Sahoo KK et al (2017) Development of doubled haploids from an elite indica rice hybrid (BS6444G) using anther culture. Plant Cell Tissue Organ Cult 128(3):679–689

Nitsch JP, Nitsch C (1969) Haploid plants from pollen grains. Science 163:85–87

Ochatt S, Pech C, Grewal R, Conreux C, Lulsdorf M, Jacas L (2009) Abiotic stress enhances androgenesis from isolated microspores of some legume species (Fabaceae). J Plant Physiol 166(15):1314–1328

Osman KA, Kang KH, El-Siddg AA, Ahmed YM, Abdalla SM et al (2020) Assessment of genetic variability for yield and attributed traits among rice doubled haploid (DH) lines in semi-arid zone Sudan. Afr J Agric Res 16(7):939–946. https://doi.org/10.5897/AJAR2019.14496

Palanisamy D, Marappan S, Ponnuswamy RD, Mahalingam PS, Bohar R, Vaidyanathan S et al (2019) Accelerating hybrid rice breeding through the adoption of doubled haploid technology for R-line development. Biologia 74(10):1259–1269. https://doi.org/10.2478/s11756-019-00300-4

Palmer CE, Keller WA, Arnison PG (1996) Utilization of brassica haploids. In: Jain SM, Sopory SK, Veilleux RE (eds) In vitro haploid production in higher plants. Kluwer, Academic Publishers, Dordrecht, pp 173–192

Patil VD, Nerkar YS, Misal MB, Harkal SR (1997) Parag 401 a semi dwarf rice variety developed through anther culture. Int Rice Res Notes 22(2):19

Patra S, Mohanty P, Tripathy SK (2019) Scope of androgenesis for genetic improvement in rice. Emerg Life Sci Res 5:23–28. https://doi.org/10.31783/elsr.2019.512328

Pauk J, Jancsó M, Simon-Kiss I et al (2009) Rice doubled haploids and breeding. In: Touraev A, Forster BP, Jain SM (eds) Advances in haploid production in higher plants. Springer, Dordrecht, pp 189–197. https://doi.org/10.1007/978-1-4020-8854-4_16

Raina SK, Irfan ST (1998) High-frequency embryogenesis and plantlet regeneration from isolated microspores of indica rice. Plant Cell Rep 17(12):957–962

Ravi M, Chan SW (2010) Haploid plants produced by centromere-mediated genome elimination. Nature 464(7288):615–618. https://doi.org/10.1038/nature08842

Reddy VS, Leelavathi S, Sen SK et al (1985) Influence of genotype and culture medium on microspore callus induction and green plant regeneration in anthers of *Oryza sativa*. Physiol Plant 63:309–314

Rout P, Naik N, Ngangkham U, Verma RL, Katara JL, Singh ON et al (2016) Doubled haploids generated through anther culture from an elite long duration rice hybrid, CRHR32: method optimization and molecular characterization. Plant Biotechnol 33(3):177–186

Rukmini M, Rao GJN, Rao RN (2013) Effect of cold pretreatment and phytohormones on anther culture efficiency of two indica rice (*Oryza sativa* L.) hybrids—Ajay and Rajalaxmi. J Exp Biol Agric Sci 2:69–76

Samantaray S, Dash B, Bhuyan SS, Devanna BN, Parameswaran C, Katara JL, Verma RL, Rout P et al (2021) Doubled haploid technologies for rice improvement: technical Progress and prospects. In: Advances in rice breeding: stress tolerance, climate resilience, quality & high yield. ICAR-National Rice Research Institute, Cuttack, Odisha, India, pp 328–344

Sankof D, Zheng C, Lenert A (2020) Poverty and shared prosperity 2020: reversals of fortune. World Bank. https://doi.org/10.1596/978-1-4648-1602-4

Sarao NK, Gosal SS (2018) In vitro Androgenesis for accelerated breeding in Rice. In: Gosal S, Wani S (eds) Biotechnologies of crop improvement, vol 1. Springer, Cham, pp 407–435. https://doi.org/10.1007/978-3-319-78283-6_12

Sasmita P (2011) Evaluation of uniformity, variability, and stability of agronomic traits of doubled haploid rice lines resulting from anther culture. Nusantara Biosci 2(2):67–72

Sathish P, Gamborg OL, Nabors MW et al (1995) Rice anther culture: callus initiation and androclonal variation in progenies of regenerated plants. Plant Cell Rep 14:432–436

Seguí-Simarro JM (ed) (2021) Doubled haploid technology: volume 1: general topics, alliaceae, cereals, Methods in molecular biology, vol 2287. New York, NY, Springer US. https://doi.org/10.1007/978-1-0716-1315-3_3

Segui-Simarro JM, Nuez F (2008) Pathways to doubled haploidy: chromosome doubling during androgenesis. Cytogenet Genome Res 120:358–369

Shariatpanahi ME, Bal U, Heberle-Bors E, Touraev A et al (2006) Stresses applied for the reprogramming of plant microspores towards in vitro embryogenesis. Physiol Plant 127:519–534

Shen J, Li M, Chen Y et al (1983) Improving rice by anther culture. In: Cell and tissue culture techniques for cereal improvement. Sci Press, Beijing, pp 183–205

Sood S, Dwivedi S (2015) Doubled haploid platform: an accelerated breeding approach for crop improvement. In: Bahadur B, Venkat Rajam M, Sahijram L, Krishnamurthy K et al (eds) Plant biology and biotechnology. Springer, New Delhi, pp 89–111. https://doi.org/10.1007/978-81-322-2283-5_5

Soundararajan RP, Kadirvel P, Gunathilagaraj K, Maheswaran M (2004) Mapping of quantitative trait loci associated with resistance to brown plant hopper in rice by means of a doubled haploid population. Crop Sci 44:2214–2220

Statista (2024). https://statista.com/statistics/263977/world-grain-production-by-type/. Accessed 6 Feb 2024

Swamy BM, Descalsota GIL, Nha CT, Amparado A, Inabangan-Asilo MA, Manito C, Tesoro F, Reinke R et al (2018) Identification of genomic regions associated with agronomic and biofortification traits in DH populations of rice. PLoS One 13(8):e0201756

Touraev A, Forster BP, Jain SM (2009) Advances in haploid production in higher plants. Springer, Berlin. https://doi.org/10.1007/978-1-4020-8854-4. ISBN: 978-1-4020-8854-4

Tuvesson S, Von Post R, Ljungberg A (2003) Triticale anther culture. In: Maluszynski M, Kasha KJ, Forster BP, Szarejko I et al (eds) Doubled haploid production in crop plants—a manual. Kluwer Academic, Dordrecht, pp 117–121

Wang K (2020) Fixation of hybrid vigor in rice: synthetic apomixis generated by genome editing. Abiotech 1(1):15–20. https://doi.org/10.1007/s42994-019-00001-1

Wang X, Jia MH, Ghai P, Lee FN, Jia Y (2015) Genome-wide association of rice blast disease resistance and yield-related components of rice. Mol Plant-Microbe Interact 28(12):1383–1392

Wang C, Liu Q, Shen Y, Hua Y, Wang J, Lin J et al (2019a) Clonal seeds from hybrid rice by simultaneous genome engineering of meiosis and fertilization genes. Nat Biotechnol 37(3):283–286. https://doi.org/10.1038/s41587-018-0003-0

Wang B, Zhu L, Zhao B, Zhao Y, Xie Y, Zheng Z et al (2019b) Development of a haploid-inducer mediated genome editing system for accelerating maize breeding. Mol Plant 12(4):597–602. https://doi.org/10.1016/j.molp.2019.03.006

Watts A, Kumar V, Raipuria RK, Bhattacharya RC (2018) In vivo haploid production in crop plants: methods and challenges. Plant Mol Biol Report 36(5):685–694. https://doi.org/10.1007/s11105-018-1132-9

Wędzony M et al (2009) Progress in doubled haploid technology in higher plants. In: Touraev A, Forster BP, Jain SM (eds) Advances in haploid production in higher plants. Springer, Dordrecht. https://doi.org/10.1007/978-1-4020-8854-4_1

Xu Q, Yuan XP, Yu HY, Wang YP, Tang SX, Wei XH et al (2011) Mapping quantitative trait loci for sheath blight resistance in rice using double haploid population. Plant Breed 130(3):404–406

Yan G, Liu H, Wang H, Lu Z, Wang Y, Mullan D et al (2017) Accelerated generation of selfed pure line plants for gene identification and crop breeding. Front Plant Sci 8:1786. https://doi.org/10.3389/fpls.2017.01786

Yao L, Zhang Y, Liu C, Liu Y, Wang Y, Liang D et al (2018) OsMATL mutation induces haploid seed formation in indica rice. Nat Plants 4(8):530–533. https://doi.org/10.1038/s41477-018-0193-y

Zapata FJ, Torrizo LB, Romero RO et al (1982) Androgenesis in *Oryza sativa*. In: Fujiwara A (ed) Plant tissue culture. Proceedings of 5th international congress on plant tissue and cell culture, Tokyo, pp 531–532

Zarsky V, Garrido D, Eller N, Tupy J, Vicente O, Schoffl F, Heberlebors E et al (1995) The expression of a small heat-shock gene is activated during induction of tobacco pollen embryogenesis by starvation. Plant Cell Environ 18(2):139–147. https://doi.org/10.1111/j.1365-3040.1995.tb00347.x

Zeng YX, Xia LZ, Wen ZH, Ji ZJ, Zeng DL, Qian Q, Yang CD (2015) Mapping-resistant QTLs for rice sheath blight disease with a doubled haploid population. J Integr Agric 14(5):801–810

Zheng X, Chen X, Zhang X, Lin Z, Shang J, Xu J, Zhai W et al (2004) Isolation and identification of a gene in response to rice blast disease in rice. Plant Mol Biol 54(1):99–109

Zhong Y, Liu C, Qi X et al (2019) Mutation of ZmDMP enhances haploid induction in maize. Nat Plants 5(6):575–580

Zhou C, Yang H (1980) In vitro induction of haploid plantlets from unpollinated young ovaries of *Oryza sativa*. Acta Genet Sin 7(3):287–288

Marker-Assisted Breeding for Rice Improvement

5

Sadras Bhavana, Sheshanka Dugyala, Rachana Bagudam, C. Subha Lakshmi, and K. Krishna Mohan

Abstract

The changing climatic scenario set forth numerous crop production constraints, including abiotic and biotic stresses, which are causing significant losses to farmers. Rice researchers throughout the world are making a continuous effort to furnish significant solutions to counter climate-imposed challenges. In association with several breeding strategies, advances in molecular biology and biotechnology have led to the successful development and release of efficient rice cultivars tolerant to abiotic and biotic stresses. Further, the molecular studies entailed detecting thousands of genes that play a vital role in bringing forth the tolerance mechanism to these stresses. Enhancing knowledge about the connections between agricultural traits and DNA sequences may help rice breeding programs. Alongside, various Quantitative Trait Loci (QTLs) associated with increased stress tolerance have been cloned. MAS (marker-assisted selection) refers to the use of DNA markers that are tightly linked to target loci as a substitute for or to assist phenotypic screening. MAS uses molecular markers to accelerate rice breeding by enabling precise selection of traits like disease resistance and stress tolerance, bypassing labor-intensive phenotypic evaluations and facili-

S. Bhavana (✉) · S. Dugyala
Department of Crop Physiology & Plant Pathology, College of Agriculture, Kaveri University, Gowraram, Telangana, India

R. Bagudam
Department of Genetics and Plant Breeding, Research Fellow, ICRISAT, Hyderabad, Telangana, India

C. S. Lakshmi
ICAR-Indian Institute of Millets Research, Rajendranagar, Hyderabad, Telangana, India

K. K. Mohan
Department of Botany, Government Degree College, Khairatabad, Telangana, India

A. Singh et al. (eds.), *Climate-Smart Rice Breeding*,
https://doi.org/10.1007/978-981-97-7098-4_5

tating rapid development of resilient, high-yielding varieties. The use of DNA markers in plant breeding called marker-assisted breeding (MAB), has been practiced in recent decades. The book chapter focuses on research areas of utmost importance for overcoming the limitations of rice production. It highlights the developments in physiology, molecular breeding and genetic research on rice, particularly with an emphasis on boosting yield, enhancing stress tolerance and improving the nutritional quality of rice.

Keywords

Abiotic and biotic stress · Molecular breeding · QTLs · Molecular markers · Rice

5.1 Introduction

Rice (*Oryza sativa* L.) stands as one of the most crucial food crops globally, and ensuring consistent rice production is of utmost significance to sustain the continually expanding human population amidst a changing climate (Maclean et al. 2002). Globally rice is cultivated in an area of 155 million ha with an annual production of around 600 million tons and an average productivity of 4.10 tha^{-1}. As per the FAO report, the projected global demand for rice by 2025 is anticipated to reach 800 million metric tons. However, the current population growth rate is expected to fall short, reaching <600 million metric tons. Consequently, a projected deficit of 200 million metric tons necessitates increased productivity per unit area despite the diminishing resources (FAOSTAT 2020).

The cultivation of rice is frequently confronted by various environmental stresses, broadly categorized into abiotic factors such as salinity, heat, drought, cold, submergence, radiation, and heavy metals, as well as biotic factors involving pathogens and herbivores, significantly impacted average yield losses in global agriculture (Ditta 2013).

Several breeding approaches for increasing the yield potential include conventional breeding, i.e., pure line selection, pedigree selection, bulked method, single seed decent method (SSD); ideotype breeding; hybrid breeding; transgenic rice breeding; marker-assisted breeding (MAB) etc. MAB stands out as a powerful tool that combines traditional breeding methods with molecular technologies to enhance the efficiency and precision of crop improvement.

In the context of rice, MAB plays a pivotal role in developing improved varieties by identifying and selecting plants with specific traits at the molecular level. Unlike conventional breeding, which relies on time-consuming and resource-intensive phenotypic selection, MAB utilizes molecular markers—DNA sequences associated with desired traits—to expedite the breeding process. Instead of relying solely on visible characteristics, breeders can now employ molecular markers to swiftly screen large rice plant populations. This precision not only accelerates the development of resistant varieties but also allows for the simultaneous incorporation of multiple resistance genes—a process known as gene pyramiding. With the continuous advancements in genomics and biotechnology, marker-assisted breeding has become an integral part of rice improvement programs, contributing to the development of high-yielding, resilient, and sustainable rice varieties to meet the challenges of a changing agricultural landscape.

In this book chapter, we aim to provide a comprehensive overview of marker-assisted breeding (MAB) in rice, covering its history, methodologies, integration with other techniques, applications, and future directions. Starting with the evolution of breeding techniques in rice and advancements in molecular and biotechnological studies, we delve into the intricacies of MAB, including the identification of target traits, molecular marker development, genetic mapping, and marker-trait association analysis. We also explore the integration of MAB with other techniques, highlighting the synergies between these approaches. Furthermore, we discuss the practical applications of MAB in enhancing productivity, combating biotic and abiotic stresses, and improving nutritional quality in rice. Finally, we outline future directions for MAB research and conclude with insights into the transformative potential of this approach in rice breeding programs.

5.1.1 History of Breeding Techniques in Rice

The history of rice breeding reflects a dynamic interplay between traditional knowledge and cutting-edge science. From early domestication to modern biotechnological innovations, the journey of rice breeding has been marked by continuous adaptation to the evolving needs of agriculture and the global population.

The history of breeding techniques in rice spans thousands of years and has evolved from traditional, empirical methods to sophisticated, scientifically guided approaches. Here is a detailed overview of the history of breeding techniques in rice (Fig. 5.1):

5.1.2 Advancement of Molecular and Biotechnological Studies in Rice

The advancement of molecular and biotechnological studies in rice has significantly transformed rice breeding and crop improvement over the past few decades. Here's a detailed overview of key milestones and advancements in molecular and biotechnological studies related to rice (Fig. 5.2):

The continued progress in molecular and biotechnological studies in rice underscores the importance of innovation in addressing global challenges, including food security, climate change, and sustainability. These advancements have not only expanded our understanding of the rice genome but have also empowered scientists and breeders to develop rice varieties with improved traits, contributing to the resilience and productivity of rice agriculture worldwide.

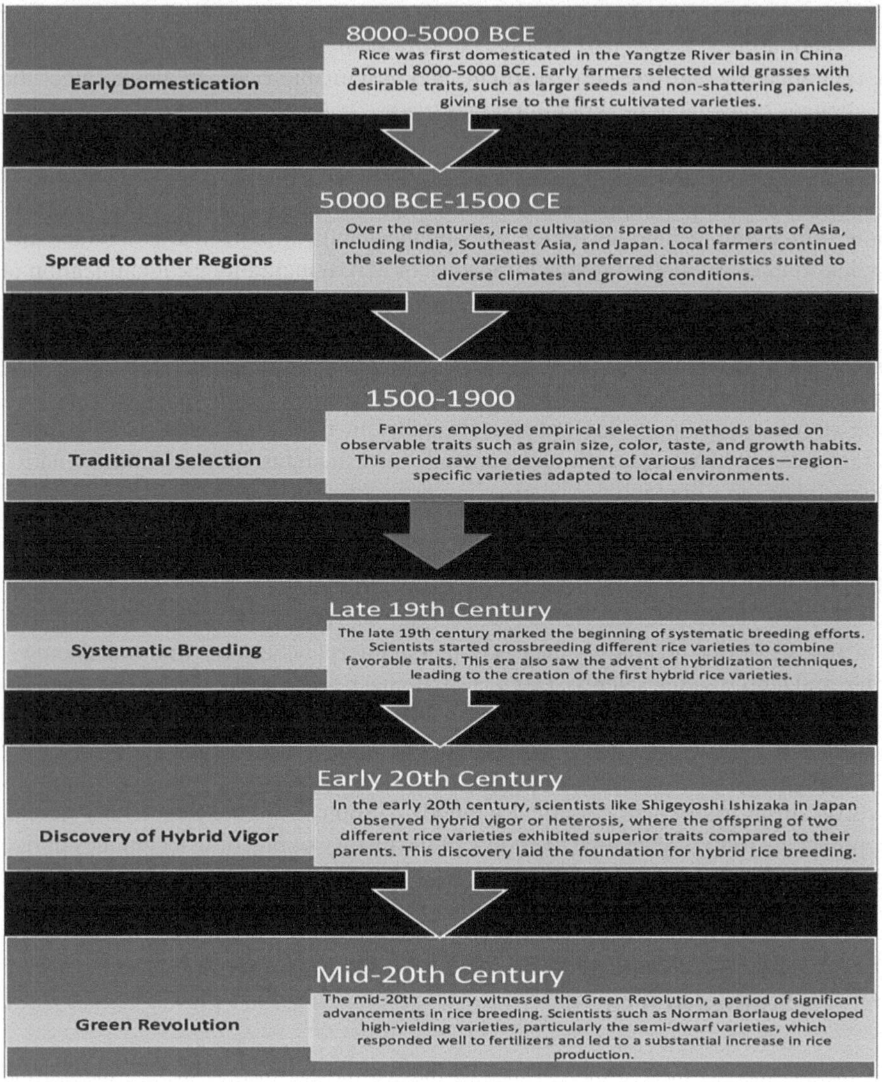

Fig. 5.1 Overview of the history of breeding techniques in rice

Fig. 5.2 Overview of key milestones and advancements in molecular and biotechnological studies

5.2 Marker-Assisted Breeding (MAB)—Its Methodology

Marker-assisted breeding (MAB) in rice involves the use of molecular markers to select plants with desired traits during the breeding process (Fig. 5.3). This technique allows for a more efficient and precise selection of traits compared to traditional breeding methods. Here's a detailed overview of the methodologies involved in marker-assisted breeding in rice:

5.2.1 Identification of Target Traits

The first step in marker-assisted breeding is identifying the target traits, such as disease resistance, tolerance to abiotic stresses, or improved yield. Researchers focus on traits that are important for rice cultivation and have a significant impact on productivity.

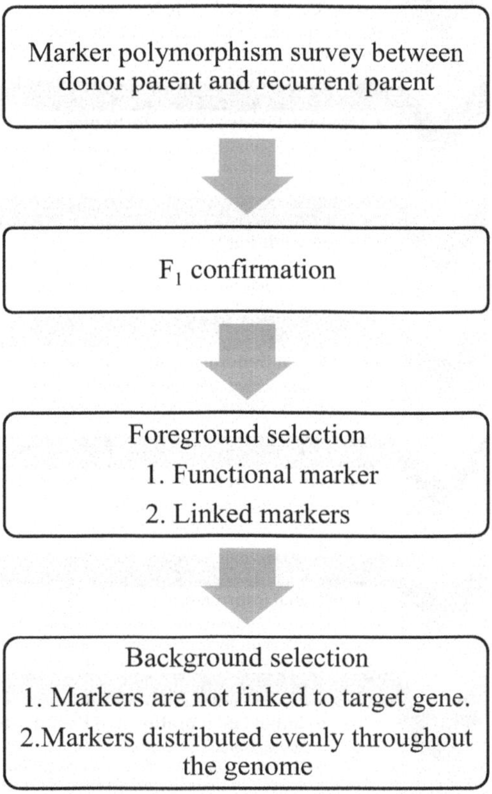

Fig. 5.3 General steps of MAS in a breeding program

5.2.2 Molecular Marker Development

Molecular markers are DNA sequences associated with specific traits or genes. Various types of molecular markers (as shown in Table 5.1) are used, including Single Nucleotide Polymorphisms (SNPs), Simple Sequence Repeats (SSRs), and Insertion/Deletion (InDel) markers. These markers are selected based on their polymorphism and linkage to the target trait.

5.2.3 Genetic Mapping

Genetic mapping involves determining the location of genes on a rice chromosome. Linkage mapping and association mapping are common techniques used to identify the genomic regions associated with the target traits. This information is crucial for developing markers linked to the desired traits. Research studies that identified different markers on specific chromosomes of rice varieties are listed in Table 5.2.

5.2.4 Marker-Trait Association Analysis

Once the molecular markers are developed, researchers conduct marker-trait association analysis. This involves genotyping a population of rice plants to identify the correlation between the presence of specific markers and the expression of the target traits.

5.2.5 Marker-Assisted Selection (MAS)

MAS is the core of marker-assisted breeding. In this step, plant breeders use the identified molecular markers to select plants with the desired traits in early

Table 5.1 Types of markers used for marker-assisted breeding in rice

Marker type	Description	Examples
SNP	Single nucleotide polymorphism, variation at a single nucleotide in a DNA sequence	Sub1, qLTG3–1, sd1
SSR	Simple sequence repeat, short, repeated DNA sequences with variations in the number of repeats	Pup1, Gn1a, Wx
RFLP	Restriction fragment length polymorphism, variation in the length of DNA fragments	Xa21, Pi-ta
AFLP	Amplified fragment length polymorphism, PCR-based method to detect polymorphisms	–
CAPS	Cleaved amplified polymorphic sequences, variation detected by restriction enzyme cleavage	–
DArT	Diversity arrays technology, microarray-based marker system	–
InDel	Insertion/deletion polymorphism, variation due to the insertion or deletion of DNA sequences	–

Table 5.2 Types of marker and chromosome number of specific genes in marker-assisted breeding of rice

Marker	Trait associated	Marker type	Chromosome location	Notable varieties/lines	Application	References
Sub1	Submergence tolerance	SNP	Chromosome 9	Swarna Sub1, samba Sub1	Improves flood tolerance	Septiningsih et al. (2009)
Pup1	Phosphorus uptake efficiency	SSR	Chromosome 12	Kasalath, Nipponbare	Enhances phosphorus uptake efficiency	Wissuwa et al. (2002)
Xa21	Bacterial blight resistance	RFLP	Chromosome 11	IRBB21, IR24	Provides resistance to bacterial blight	Song et al. (1995)
Xa4	Bacterial blight resistance	SSR	Chromosome 4	IRBB4, IR24	Conveys resistance against bacterial blight	Yoshimura et al. (1998)
Pi-ta	Blast resistance	RFLP	Chromosome 12	CO39, IR64	Confers resistance against rice blast	Chen et al. (1997)
qLTG3–1	Drought tolerance	SNP	Chromosome 3	IRAT109, Vandana	Enhances drought tolerance	Vikram et al. (2011)
sd1	Semi-dwarf stature	SNP	Chromosome 1	Dee-geo-woo-gen, IR8	Reduces plant height for lodging resistance	Monna et al. (2002)
Gn1a	Grain number	SSR	Chromosome 1	Kasalath, IR64	Increases the number of grains per panicle	Ashikari et al. (2005)
Wx	Amylose content in endosperm	SSR	Chromosome 6	Basmati, Nipponbare	Modulates starch composition in rice grains	Isshiki et al. (1998)
Alk	Alkali tolerance	SSR	Chromosome 1	Pokkali, Nonabokra	Improves tolerance to alkaline soils	Ha and Fukai (1994)
qDTY	Drought tolerance	SSR, SNP	Chromosome 1, 2, 3	IR64, Swarna Sub1, BRRI dhan 56	Enhancing drought tolerance in rice breeding programs	Kumar et al. (2014a)
SALTOL	Salinity tolerance	SSR, SNP	Chromosome 1	FL478, IR29, Pokkali	Developing salinity-tolerant rice varieties	Thomson et al. (2010)

generations. This reduces the time required for phenotypic selection and allows for the identification of desirable traits before the plants reach maturity.

5.2.6 Gene Pyramiding

Gene pyramiding involves combining multiple genes or traits in a single plant to enhance its resistance or resilience. Using molecular markers, breeders can pyramid genes linked to different traits in a single rice variety, creating plants with a broader spectrum of resistance.

5.2.7 Backcrossing

After selecting plants with the desired traits, backcrossing is often employed to transfer these traits into elite rice varieties. Backcrossing involves crossing the selected plants with the elite variety and repeatedly backcrossing the resulting progeny with the elite variety while selecting the desired traits.

5.2.8 Validation Through Field Trials

The selected plants are subjected to field trials to evaluate their performance under natural conditions. This step ensures that the identified markers and selected traits are effective in real-world environments.

5.2.9 Integration with Conventional Breeding

Marker-assisted breeding is often integrated with conventional breeding methods. The use of molecular markers complements traditional approaches, allowing for the optimization of breeding strategies.

5.2.10 Continuous Improvement

Advances in genomic technologies and the discovery of new markers lead to continuous improvement in marker-assisted breeding methodologies. Researchers stay abreast of technological developments to enhance the efficiency and accuracy of breeding programs.

MAB enhances precision by utilizing functional markers, particularly perfect markers from map-based cloning, surpassing closely linked markers. MAS employs tightly linked markers to pinpoint individuals carrying the desired gene or allele. In marker-assisted backcross breeding (MABB), functional markers aid foreground selection, identifying individuals with the target allele (Collard and Mackill 2008a).

Foreground selection is a time-efficient alternative to labor-intensive phenotypic screening. Recombinant selection, the second stage of MABB, involves markers within 5 cm flanking the target gene, reducing linkage drag and unwanted traits (Hospital 2005). Background selection, the third stage, uses unrelated markers to choose the recurrent parent genome, expediting recovery in 2–4 backcross generations and advancing MABB efficiency.

5.3 Integration of MAB and Other Techniques in Rice

Multi-omics approaches, including metabolomics, transcriptomics, and proteomics, can be integrated with Molecular MAB to enhance the precision and efficiency of rice breeding programs. These integrations provide a more comprehensive understanding of the genetic and molecular basis of traits, allowing breeders to make informed decisions during the selection process. Here's a detailed overview of the integration of MAB with other molecular techniques in rice, along with examples:

5.3.1 Transcriptomics and MAB Integration

Description: Transcriptomics involves studying the complete set of RNA transcripts produced by the genome under specific conditions. Integration with MAB allows for the identification of key genes associated with desirable traits.

Example: Identifying and validating candidate genes for disease resistance in rice. MAB can be enhanced by selecting individuals with specific transcriptomic profiles related to resistance, as in a study by Ngernmuen et al. (2020) where the research examines defense mechanisms in "Jao Hom Nin" rice, contrasting responses related to the blast-resistant genes Pish and Pik through transcriptome analysis, shedding light on genetic resistance against pathogens.

5.3.2 Proteomics and MAB Integration

Description: Proteomics focuses on studying the entire set of proteins expressed by a genome. Integrating proteomics with MAB helps understand how gene expression at the protein level influences traits.

Example: Identifying proteins associated with high grain quality in rice. MAB can then be employed to select plants with the desired protein expression patterns. A study by Zhang et al. (2020) discusses the identification of differentially expressed proteins in rice related to grain quality, particularly focusing on amylose accumulation, which is a crucial factor for rice quality in terms of cooking and eating. The findings from proteomic profiling can be used to enhance marker-assisted breeding strategies by selecting individuals with desirable protein expression patterns.

5.3.3 Metabolomics and MAB Integration

Description: Metabolomics involves the study of the complete set of small molecules (metabolites) present in a biological system. Integration with MAB helps in understanding the metabolic pathways associated with important traits.

Example: Identifying metabolites associated with drought tolerance in rice, as seen in a study by Kumar et al. (2021), where the emphasis was on the role of primary and secondary metabolites in drought tolerance and how metabolomics can be used to profile these metabolites. It also explores how such profiles can be leveraged to enhance breeding programs through marker-assisted selection.

5.3.4 Genomic Selection (GS) and MAB Integration

Description: Genomic Selection involves predicting the performance of individuals based on their entire genomic information. Integrating GS with MAB enhances the accuracy of predicting trait performance.

Example: Predicting grain yield in rice based on genome-wide markers. MAB can then be employed to select individuals with the highest predicted yield. As demonstrated in a study by Crossa et al. (2014), using genomic prediction models to enhance grain yield in cereal crops can be effectively applied to rice breeding programs. Marker-assisted breeding can leverage these predictions to select individuals with the highest predicted yield based on their genome-wide markers.

5.3.5 Marker-Assisted Backcrossing (MABC) and MAB Integration

Description: MABC involves backcrossing a plant with a desirable trait to the recurrent parent while selecting the trait of interest using molecular markers. Integrating other molecular techniques enhances the precision of selecting individuals in the backcrossing process.

Example: Incorporating disease resistance genes into a high-yielding rice variety through MABC, with the assistance of transcriptomics to identify and prioritize resistance-related genes, as observed in the study by Pradhan et al. (2022). The study was carried out using MABC to incorporate two bacterial blight resistance genes, Xa21 and xa13, into the popular rice variety Ranidhan. MABC was employed to efficiently introduce these resistance genes into the genetic background of Ranidhan while preserving its high-yielding characteristics. The use of MABC allowed for precise selection and integration of the desired traits, showcasing the effectiveness of this breeding approach in developing disease-resistant crop varieties.

5.3.6 Functional Genomics and MAB Integration

Description: Functional genomics involves studying gene function on a genome-wide scale. Integration with MAB facilitates the identification of genes directly affecting target traits.

Example: Using CRISPR/Cas9 technology to knock out or modify specific genes associated with abiotic stress tolerance in rice, followed by MAB to select lines with the desired modifications.

One notable example is the modification of the OsERF922 gene in rice using CRISPR/Cas9 technology. OsERF922 is a gene involved in ethylene signaling, which negatively affects drought tolerance. Researchers employed CRISPR/Cas9 to create knockouts of OsERF922, resulting in rice plants with enhanced drought tolerance. After the CRISPR/Cas9-induced mutations were generated, marker-assisted breeding was used to select lines (Jiang et al. 2019) that carried the knockout alleles while ensuring that the rest of the plant's genome remained optimal for agricultural performance.

5.3.7 Epigenomics and MAB Integration

Description: Epigenomics studies heritable changes in gene function that do not involve changes in the DNA sequence. Integration with MAB provides insights into epigenetic modifications influencing trait expression.

Example: Investigating DNA methylation patterns associated with yield-related genes in rice. MAB can then be used to select plants with favorable epigenetic modifications. A study by Secco et al. (2015) explores how stress-induced gene expression can lead to transient DNA methylation changes, potentially impacting yield-related genes in rice. Marker-assisted breeding (MAB) can then be employed to select plants exhibiting these favorable epigenetic modifications.

5.3.8 Marker-Assisted Pyramiding

This approach combines multiple genes or QTLs associated with different traits or the same trait from various sources into a single genotype. Marker-assisted pyramiding is used to develop crops with enhanced resistance to multiple diseases or improved tolerance to various environmental stresses by ensuring that all targeted genes are present in the new variety.

Example: One notable example of marker-assisted pyramiding is the development of rice varieties with multiple disease resistance. Researchers have successfully combined several blast resistance genes (such as Pi1, Pi2, Pi33, and Pita) into a single rice variety using marker-assisted pyramiding (Hittalmani et al. 2000). This multi-gene approach has led to the creation of rice varieties that possess durable resistance to different strains of the blast fungus, significantly reducing the risk of disease outbreaks and ensuring stable yields under various environmental conditions.

This process typically involves using molecular markers to track the presence of each resistance gene during the breeding process, ensuring that all desired genes are integrated into the final genotype. The resulting rice variety exhibits enhanced resistance to multiple pathogens, offering better protection against disease and improving overall crop resilience.

5.3.9 Early Generation Marker-Assisted Selection

This method uses molecular markers to screen and select plants with desirable traits at the early stages of the breeding cycle, such as the F2 or F3 generation. Early-generation MAS helps quickly identify and advance promising lines, thereby reducing the time and resources required for later-generation phenotypic evaluations.

Example: An example of early-generation marker-assisted selection is its use in developing wheat varieties with improved resistance to Fusarium head blight (FHB) (Liu et al. 2006). Researchers utilize molecular markers linked to FHB resistance genes, such as Fhb1, to screen F2 populations derived from crosses between resistant and susceptible parents. By identifying plants carrying the Fhb1 gene early in breeding, breeders can select and advance these resistant lines to subsequent generations without waiting for field testing to assess resistance.

This approach accelerates the breeding process, ensuring that only those plants with the desired resistance gene are progressed, thus saving time and resources while improving the efficiency of developing FHB-resistant wheat varieties.

By integrating these molecular techniques with MAB, researchers and breeders can gain a deeper understanding of the genetic and molecular mechanisms underlying important traits in rice, ultimately accelerating the breeding process and improving the efficiency of developing new and improved rice varieties.

5.4 MAB—Its Application

To improve rice varieties, there is a need to address numerous issues such as disease resistance, drought tolerance, submergence tolerance, salt tolerance, early maturity and yield improvement using marker-assisted breeding is essential.

5.4.1 MAB for Improving Productivity in Rice

Improving rice productivity is a critical imperative driven by various factors that are essential for ensuring global food security and fostering sustainable agriculture. The increasing global population amplifies the demand for food, particularly rice, making productivity enhancements crucial to meet the expanding needs of a larger populace. Limited arable land underscores the importance of boosting productivity to achieve higher yields without expanding cultivation areas. The challenges posed by climate change, including extreme weather events and temperature variations, necessitate the development of resilient rice varieties and the adoption of climate-smart agricultural practices to safeguard yields. As dietary patterns evolve, improving rice productivity becomes instrumental in diversifying food production, fostering a more balanced and resilient diet. Additionally, given rice's significant role in global trade, productivity improvements not only ensure food security within countries but also influence global trade dynamics, impacting economies and international relations.

MAB plays a pivotal role in elevating productivity in rice cultivation through its precision and efficiency in the breeding process. This method accelerates the traditional breeding timeline. The technique also contributes to improved yield and quality by targeting markers related to traits such as increased grain size, enhanced nutritional content, and superior milling quality (Table 5.3). In the context of climate change, MAB plays a key role in developing climate-adapted rice varieties (Singh et al. 2022; Majhi et al. 2022). Through targeted trait selection, MAB contributes to sustainable and resource-efficient farming, mitigating the environmental impact of agriculture. Additionally, MAB facilitates the customization of rice varieties to local conditions, resulting in region-specific cultivars.

In essence, MAB's precision breeding techniques contribute significantly to advancing rice productivity by developing varieties with enhanced yield, quality, resilience, and resource utilization traits.

It's important to note that the effectiveness of these genes may vary based on the variety of rice, environmental conditions, and specific breeding goals.

5.4.2 MAB for Biotic and Abiotic Stress in Rice

Rice faces various environmental stresses categorized into two main areas: abiotic (salinity, heat, drought, cold, submergence, radiation, and heavy metals) and biotic (pathogens and herbivores) factors (Ditta 2013). Environmental stressors like heat, cold, drought, and salinity significantly impact average yield losses in global agriculture (Wang et al. 2003). Among these, drought, characterized by low water supply, stands out as the most prevalent limiting abiotic stress. Heat stress is another environmental challenge impeding the biomass productivity and grain production of crops (Boyer 1982). Salinity, acidity, and low temperature are also noteworthy abiotic stresses restricting the growth of forage grasses and cereals.

Additionally, biotic stresses such as pests and pathogens substantially impact plants. In rice, bacterial blight, blast, and sheath blight are considered the most devastating diseases, while stem borers, leaf folders, and plant hoppers are identified as highly damaging pests (Jiang et al. 2012). The infestation of the white-backed plant hopper (WBPH), a phloem-sucking insect, further hampering rice growth and productivity. Developing disease-resistant and stress-tolerant plants is a critical goal in rice breeding programs due to the constant interference of major abiotic and biotic stresses with rice production.

Table 5.3 Several genes that have been the focus of research for enhancing productivity in rice

S. no.	Genes	Function	Productivity impact	Reference
1.	DEP1 (dense and erect panicle 1)	Associated with increased grain number and denser panicles	Enhances grain yield by promoting compact panicle architecture	Huang et al. (2009)
2.	GS3 (grain size 3)	Regulates grain size and weight	Modulating GS3 can lead to larger and heavier grains, contributing to increased yield	Zhang et al. (2021)
3.	Gn1a (grain number 1a)	Regulates cytokinin levels, influencing grain number	Overexpression of Gn1a is associated with increased grain number per panicle, contributing to higher yield	Lu et al. (2022a)
4.	IPA1 (ideal plant architecture 1)	Influences plant architecture, including tiller angle and number	Contributes to improved canopy structure and increased yield potential	Liu et al. (2019)
5.	Pup1 (phosphorus uptake 1)	Confers tolerance to phosphorus deficiency	Enhances the plant's ability to take up phosphorus, vital for overall growth and productivity	Shin et al. (2021)
6.	SUB1A (submergence 1A)	Provides tolerance to submergence stress	Allows rice plants to survive and recover after temporary submergence, preventing yield losses	Schmitz et al. (2013)
7.	qDTY (quantitative trait loci for drought tolerance)	Various qDTY genes contribute to drought tolerance	Improves the ability of rice plants to withstand water scarcity, maintaining productivity under drought conditions	Yi et al. (2023)
8.	Pita (pyramiding of major blast resistance genes)	Confers resistance to rice blast disease caused by Magnaporthe oryzae	Prevents yield losses due to blast infection, ensuring a stable and productive harvest	Zampieri et al. (2023)
9.	OsSPL14 (squamosa promoter binding protein-like 14)	Regulates panicle architecture and grain number	Overexpression of OsSPL14 is associated with increased grain yield	Lian et al. (2020)
10.	OsNAC5 (NAC domain-containing protein 5)	Associated with drought tolerance	Enhances the plant's ability to withstand water scarcity, maintaining productivity under drought conditions	Takasaki et al. (2010)
11.	OsCKX2 (cytokinin oxidase 2)	Involved in cytokinin degradation, influencing grain size	Modulating OsCKX2 can lead to larger grains and increased yield potential	Yeh et al. (2015)

5.4.2.1 MAB for Biotic Stress in Rice

Marker-assisted breeding (MAB) for biotic stress in rice is a powerful approach that integrates molecular genetics and traditional breeding to enhance disease resistance in rice cultivars. This method involves identifying and using genetic markers linked to biotic stress resistance traits, such as resistance to bacterial blight, blast disease, and various viral infections, to guide the selection process in breeding programs. By employing markers associated with specific resistance genes, MAB enables the precise incorporation of these genes into high-yielding rice varieties without the unwanted introduction of undesirable traits (listed in Table 5.5). Advanced genomic tools, such as QTL mapping and genome-wide association studies (GWAS), are instrumental in identifying these markers and understanding the genetic basis of resistance. The integration of MAB in rice breeding programs exemplifies the convergence of traditional agricultural practices and modern biotechnology to address global challenges in crop production. Table 5.4 gives comprehensive examples illustrating each stage of marker-assisted breeding for biotic stress in rice, offering a detailed exploration of the application of this technique across various phases of the breeding process.

5.4.2.2 MAB for Abiotic Stress in Rice

Marker-assisted breeding (MAB) for abiotic stress in rice involves the utilization of molecular markers to enhance the selection process for traits related to abiotic stress tolerance. This approach aims to expedite breeding rice varieties that exhibit resilience to environmental stresses such as drought, salinity, and extreme temperatures (Table 5.7). By employing molecular markers linked to key genes associated with abiotic stress tolerance, breeders can accurately identify and select plants with desirable traits at early stages of development. Comprehensive examples illustrating each stage of marker-assisted breeding for biotic stress in rice, offering a detailed exploration of the application of this technique across various phases of the breeding process, is listed in Table 5.6.

Marker-assisted breeding (MAB) for biotic and abiotic stress in rice has revolutionized developing resilient and high-yielding varieties by leveraging molecular markers linked to resistance against diseases caused by pathogens and pests and tolerance to stresses caused by environmental changes.

5.4.2.3 Significance of Marker-Assisted Breeding in Rice for Biotic and Abiotic Stress

1. **Precision and Efficiency:** MAB enables precise selection of plants possessing desired resistance traits, expediting the breeding process by eliminating the need for time-consuming and resource-intensive traditional methods.
2. **Enhanced Genetic Gain:** Identification and incorporation of specific genes or genomic regions associated with resistance allow breeders to stack multiple desirable traits efficiently, developing varieties with enhanced resistance to various biotic stresses.

Table 5.4 Examples at each stage of marker-assisted breeding for biotic stress in rice

Stage in marker-assisted breeding	Description	Examples	Software tools	References
Trait identification and marker discovery	Identification of specific biotic stress traits and discovery of associated genetic markers	Identifying blast resistance genes (e.g., Pi-ta, Pi9, Pi2, Pi40)	QTL mapping software (e.g., QTL IciMapping, MapQTL)	Varshney et al. (2020), Jena and Mackill (2008)
		Discovery of bacterial blight resistance QTLs	Genome-wide association analysis tools (e.g., TASSEL, GAPIT)	
		Identifying stem borer resistance genes and associated markers	Genomic databases (e.g., NCBI), comparative genomics tools	
Marker validation and selection	Validation of markers to confirm association with desired traits	Validating the association of Xa21, Xa4, Xa5, Xa27 with blight resistance	Statistical analysis software (e.g., R, PLINK)	Collard et al. (2005)
		Selection of reliable markers demonstrating strong association	Genomic analysis tools (e.g., TASSEL, GenomeStudio)	
		Validation of SSR markers associated with sheath blight resistance	Marker validation software (e.g., flapjack, PolyMarker)	
Breeding material development	Introducing resistance genes into elite rice varieties	Introgression of Xa21, Xa4, Xa5, Xa27 genes into high-yielding varieties	Marker-assisted breeding software platforms (e.g., flapjack, breeding view)	Collard and Mackill (2008b)
		Utilizing MAB for backcrossing to incorporate resistance genes	Breeding simulation software (e.g., BreedWheat, BreedR)	
		Incorporating genes from wild rice species for resistance	Genetic transformation tools (e.g., agrobacterium-mediated transformation)	

(continued)

Table 5.4 (continued)

Stage in marker-assisted breeding	Description	Examples	Software tools	References
Genotyping and phenotyping	Genotyping to screen for resistance genes or genomic regions	Genotyping for the presence of Pi-ta, Xa21, and other resistance genes	Genotyping platforms (e.g., KASP, Illumina, TaqMan)	Cobb et al. (2013)
		Phenotyping for resistance under pathogen exposure	Disease screening facilities, controlled environment chambers	
		High-throughput phenotyping for blast, blight, and sheath blight resistance	Phenotyping software (e.g., FieldBook, LemnaTec)	
Selection and advanced breeding	Selection based on both markers and observed resistance	Selecting lines with both markers and observed resistance	Genomic selection software (e.g., GBSA, RR-BLUP)	Ribaut and Ragot (2007)
		Advancing superior lines through breeding cycles	Marker-assisted breeding analysis tools (e.g., breeders toolbox)	
		Incorporating multi-stress tolerance (blast, blight, drought)	Multi-trait selection tools, breeding decision support systems	
Field trials and variety release	Multi-location field trials to evaluate performance and stability	Evaluating resistance in different agro-climatic zones	Field trial management software, geographic information systems (GIS)	International Rice Research Institute (IRRI) (2013))
		Releasing superior, disease-resistant rice varieties	Decision support systems for variety release. Release management tools	
		Evaluation of farmer-preferred traits in released varieties	Participatory varietal selection (PVS), farmer field days	

(continued)

Table 5.4 (continued)

Stage in marker-assisted breeding	Description	Examples	Software tools	References
Monitoring and improvement	Continuous monitoring and iterative improvement based on field performance	Monitoring resistance and productivity in farmers' fields	Field data collection and analysis tools, farm management software	Witcombe et al. (2005)
		Incorporating additional resistance traits based on feedback	Participatory plant breeding tools, decision support systems	
		Iterative improvements based on farmer feedback	Farmer surveys, participatory rural appraisal (PRA)	

3. **Diverse Marker Toolkit:** The wide array of molecular markers, including functional genes, QTLs, SNPs, SSRs, and CRISPR-edited markers, provide a diverse toolkit for breeders to target and select for specific resistances.
4. **Minimization of Resource Expenditure:** MAB minimizes costs and resources by enabling the identification of superior plants at the early stages, reducing the need for extensive field evaluations.
5. **Broad Applicability:** The markers used in MAB are applicable across different regions, allowing for the development of tailored rice varieties adapted to various agro-climatic conditions.
6. **Stacking Multiple Traits:** MAB facilitates stacking multiple resistance genes, providing comprehensive protection against multiple diseases and pests.

5.4.3 MAB for Nutritional and Quality Enhancement in Rice (Bio-Fortification)

For almost 3.5 billion people worldwide, rice serves as their main source of calories. About 92% of all land is used for rice cultivation, which provides more than 23% of global calorie demands (Sharma et al. 2012). It is also a plant-based source of dietary fiber, vitamins, minerals, and other disease-fighting antioxidants. A greater awareness of molecular breeding techniques makes rice genetic improvement easier to achieve. Biofortification is the process of generating nutrient-dense staple crops using both traditional breeding techniques and recent advances in biotechnology. Instead of adding nutrients to plant foods during processing, it enhances food nutrition while crops are advancing.

Table 5.5 Examples of previous work in marker-assisted breeding for biotic stress in rice

Type of work	Example	Notable achievements	References
Blast disease resistance	Identification and utilization of the *Pi-ta* gene	– Incorporation of *Pi-ta* gene into IRRI breeding lines for blast-resistant rice varieties	Huang et al. (2008)
	Utilization of *Pi9*, *Pi2*, and *Pi40* genes	– Incorporation of Pi9, Pi2, and Pi40 genes into rice varieties for diverse blast resistance	Ning et al. (2020)
	Development of QTLs (quantitative trait loci) for blast resistance	– Identification and deployment of multiple QTLs associated with blast resistance in various rice lines	–
	Utilization of *Piz-t* gene	– Development of *Piz-t* gene-based varieties with broad-spectrum resistance to blast disease	Ning et al. (2020)
Bacterial blight resistance	Utilization of *Xa21* gene	– Incorporation of *Xa21* gene into *Sub1* rice varieties for resistance against bacterial blight	Mohapatra et al. (2023)
	Utilization of *Xa4*, *Xa5*, and *Xa27* genes	– Stacking multiple *Xa* genes for enhanced resistance to different strains of bacterial blight	Singh et al. (2001)
	Utilization of *Xa23* and Xa33 genes	– Incorporation of Xa23 and Xa33 genes into rice varieties, enhancing resistance against bacterial blight	Yang et al. (2022)
	Development of Xoo-resistant QTLs	– Identification and utilization of QTLs linked to resistance against Xanthomonas oryzae pv. Oryzae (Xoo)	Lu et al. (2022b)
Sheath blight resistance	Integration of genes conferring resistance against sheath blight	– Development of rice varieties with improved resistance to sheath blight, incorporating various resistance genes	Chen et al. (2023)
	Utilization of major QTLs (quantitative trait loci) for sheath blight resistance	– Identification and utilization of QTLs associated with resistance against the sheath blight pathogen	Yuan et al. (2019)

(continued)

Table 5.5 (continued)

Type of work	Example	Notable achievements	References
Brown planthopper (BPH) resistance	Introgression of Bph genes for brown planthopper resistance	– Successful transfer of Bph genes into rice varieties via MAB, providing resistance against brown planthopper	Yuexiong et al. (2020)
	Incorporation of BPH-resistant QTLs	– Utilization of QTLs associated with brown planthopper resistance to develop BPH-resistant rice varieties	–
Stem borer resistance	Utilization of genes (e.g., CYP genes) for resistance against stem borers	– Incorporation of specific genes conferring resistance against stem borers into elite rice cultivars	Wang et al. (2014)
	Integration of QTLs for stem borer resistance	– Successful deployment of QTLs linked to stem borer resistance, enhancing pest resistance in rice varieties	Makkar and Bentur (2017)
Development of multi-stress-tolerant varieties	Creation of rice varieties resilient to blast, bacterial blight, and drought	– Development of multi-stress-tolerant rice varieties, like BRRI Dhan varieties in Bangladesh	–
	Introduction of genes for resistance against blast, bacterial blight, and sheath blight	– Creation of multi-stress-tolerant varieties through the integration of various disease resistance genes	–
CRISPR-Cas9 for pest resistance	Editing of target genes (e.g., Bph14) for pest resistance using CRISPR-Cas9 technology	– CRISPR-Cas9-mediated modification of genes associated with pest resistance, improving resistance in rice	–
	Engineering OsSERK2 for enhancing resistance against multiple pests	– CRISPR-Cas9 editing of OsSERK2 gene, resulting in improved pest resistance across different rice pests	–
Marker-assisted backcrossing (MAB) success	Integration of molecular markers in backcross breeding	– Implementation of MAB in Chinese rice varieties, incorporating disease-resistant genes efficiently	–
	Introgression of *Pi54* gene for blast resistance	– Successful transfer of *Pi54* gene into elite rice cultivars using MAB for enhanced blast resistance	–
	Introgression of Bph genes for brown planthopper resistance	– Successful transfer of Bph genes into rice varieties via MAB, providing resistance against brown planthopper	–
	Introgression of Bph genes for brown planthopper resistance	– Successful transfer of Bph genes into rice varieties via MAB, providing resistance against brown planthopper	–

(continued)

Table 5.5 (continued)

Type of work	Example	Notable achievements	References
Utilization of CRISPR-Cas9 for precision breeding	Targeted gene editing for disease resistance	– CRISPR-Cas9 editing of rice genes, such as editing *EFR* gene for enhanced bacterial pathogen resistance	–
	Editing of SWEET genes for resistance to bacterial blight	– CRISPR-Cas9-based modification of *SWEET* genes, reducing susceptibility to bacterial blight	–
	Editing of OsSWEET11 for resistance to rice blast	– CRISPR-Cas9-mediated editing of OsSWEET11 gene, reducing susceptibility to rice blast disease	–
	Engineering of NPR1 for enhancing broad-spectrum disease resistance	– Modification of NPR1 gene using CRISPR-Cas9, resulting in improved resistance against various pathogens	–

5.4.3.1 Improvement of the Grain Nutritive Value

Rice grain consists of 80% starch and its quality is dependent on a combination of several traits. The rice genome has been fully sequenced since the advent of DNA sequencing technology, which facilitates gene annotation (International Rice Genome Sequencing Project (IRGSP) 2005). Therefore, these recent advances could assist rice growers in improving the nutritional value and grain quality of rice (Fig. 5.4).

5.4.3.2 Candidate Genes for Nutritional Traits

Several transporter proteins are necessary for most of the nutrients stored in grains, and these proteins have recently been characterized, cloned, and expressed (Table 5.8). Multiple studies have identified related genes that improve the levels of Fe and Zn for rice grain biofortification. These genes can be obtained by modifying NAS genes or blocking other pathways related to nutrient translocation genes (Mahender et al. 2016).

Basically, the key characteristics that must be considered to produce an enhanced rice variety are the color, chalkiness, and size and shape of the grains without sacrificing their yield and nutritional value. Since multiple genes control very characteristic needed to create a high-yield rice variety, these traits are referred to as quantitative traits.

The micronutrient levels in rice grains are initially low and decrease further during polishing. Individuals who rely solely on rice for their vitamin and mineral intake are more susceptible to malnutrition. Lu et al. (2008) identified ten QTLs for the content of iron (Fe), calcium (Ca), zinc (Zn), manganese (Mn), and copper (Cu) in rice grains located on chromosomes 1, 2, 4, 5, 7, 9, and 11. Additionally, Garcia-Oliveira et al. (2009) mapped 26 QTLs associated with iron, zinc, manganese, copper, calcium, magnesium, phosphorus, and potassium in rice grains. Given that

Table 5.6 Examples at each stage of marker-assisted breeding for abiotic stress in rice

Stage in marker-assisted breeding	Description	Examples	Software tools	Reference
Trait identification and marker discovery	Identification of specific abiotic stress traits and discovery of associated genetic markers	Identifying submergence tolerance genes (e.g., SUB1B, SNORKEL)	QTL mapping software (e.g., QTL Ici mapping, map QTL)	Varshney et al. (2020), Jena and Mackill (2008)
		Discovery of cold tolerance QTLs	Genome-wide association analysis tools (e.g., TASSEL, GAPIT)	
		Identifying genes related to oxidative stress tolerance	Comparative genomics tools, sequence alignment software	
Marker validation and selection	Validation of markers to confirm association with desired traits	Validating the association of SUB1B, SNORKEL with submergence tolerance	Statistical analysis software (e.g., R, PLINK)	Collard et al. (2005)
		Selection of reliable markers demonstrating strong association	Genomic analysis tools (e.g., TASSEL, GenomeStudio)	
		Validation of SSR markers associated with cold tolerance	Marker validation software (e.g., flapjack, PolyMarker)	
Breeding material development	Introducing stress tolerance genes into elite rice varieties	Introgression of SUB1B, SNORKEL genes into high-yielding varieties	Marker-assisted breeding software platforms (e.g., flapjack, breeding view)	Collard and Mackill (2008b)
		Utilizing MAB for backcrossing to incorporate stress tolerance genes	Breeding simulation software (e.g., BreedWheat, BreedR)	
		Incorporating genes from wild rice species for stress tolerance	Genetic transformation tools (e.g., agrobacterium-mediated transformation)	

(continued)

Table 5.6 (continued)

Stage in marker-assisted breeding	Description	Examples	Software tools	Reference
Genotyping and phenotyping	Genotyping to screen for stress tolerance genes or genomic regions	Genotyping for presence of SUB1B, SNORKEL and other stress tolerance genes	Genotyping platforms (e.g., KASP, Illumina, TaqMan)	Cobb et al. (2013)
		Phenotyping for stress tolerance under controlled conditions	Controlled environment chambers, growth chambers	
		High-throughput phenotyping for cold, oxidative, and submergence tolerance	Phenotyping software (e.g., FieldBook, LemnaTec)	
Selection and advanced breeding	Selection based on both markers and observed stress tolerance	Selecting lines with both markers and observed tolerance	Genomic selection software (e.g., GBSA, RR-BLUP)	Ribaut and Ragot (2007)
		Advancing superior lines through breeding cycles	Marker-assisted breeding analysis tools (e.g., breeders toolbox)	
		Incorporating multi-stress tolerance (submergence, cold, oxidative)	Multi-trait selection tools, breeding decision support systems	
Field trials and variety release	Multi-location field trials to evaluate stress tolerance performance and stability	Evaluating stress tolerance in different agro-climatic zones	Field trial management software, geographic information systems (GIS)	International Rice Research Institute (IRRI) (2013))
		Releasing superior, stress-tolerant rice varieties	Decision support systems for variety release, release management tools	
		Evaluation of farmer-preferred traits in released varieties	Participatory varietal selection (PVS), farmer field days	

(continued)

Table 5.6 (continued)

Stage in marker-assisted breeding	Description	Examples	Software tools	Reference
Monitoring and improvement	Continuous monitoring and iterative improvement based on field performance	Monitoring stress tolerance and productivity in farmers' fields	Field data collection and analysis tools, farm management software	Witcombe et al. (2005)
		Incorporating additional stress tolerance traits based on feedback	Participatory plant breeding tools, decision support systems	
		Iterative improvements based on farmer feedback	Farmer surveys, participatory rural appraisal (PRA)	

Table 5.7 Examples of previous work in marker-assisted breeding for abiotic stress in rice

Abiotic stress	Markers/genes	Trait/resistance conferred	Description/examples	References
Drought tolerance	SUB1A	Tolerance to drought stress	SUB1A gene confers tolerance to both submergence and drought. Varieties with SUB1A show improved drought resistance	Fukao and Yeung (2012)
	PSTOL1	Tolerance to phosphorus deficiency	PSTOL1 gene enhances phosphorus uptake, aiding in drought tolerance due to improved nutrient utilization	Chithrameenal et al. (2018)
Salinity tolerance	Saltol QTLs	Tolerance to high salinity	Saltol QTLs are associated with increased tolerance to salinity stress, enabling rice to grow in salt-affected soils	Lokeshkumar et al. (2023)
	HKT1;5	Sodium exclusion in roots	HKT1; 5 gene reduces sodium accumulation in rice roots, improving tolerance to saline soils	Shohan et al. (2019)

(continued)

Table 5.7 (continued)

Abiotic stress	Markers/genes	Trait/resistance conferred	Description/examples	References
Heat stress tolerance	Heat shock proteins (HSPs)	Tolerance to high temperatures	Breeding involving HSP genes has led to varieties capable of withstanding high-temperature stress	Chakraborty et al. (2024)
	Thermotolerance genes	Enhanced tolerance to high temperatures	Varieties with thermo-tolerance genes exhibit better resilience to extreme heat conditions	–
Cold tolerance	CBF genes	Tolerance to low temperatures	CBF genes play a role in conferring cold tolerance, allowing rice to thrive in colder regions	Sanghera et al. (2011)
	ICE1, CBF/DREB genes	Enhanced tolerance to cold stress	ICE1 and CBF/DREB genes contribute to improved cold tolerance, allowing rice to thrive in colder climates	Deng et al. (2017)
Submergence tolerance	SUB1B, SNORKEL	Tolerance to prolonged submergence	Genes like SUB1B and SNORKEL aid in developing varieties resilient to extended submergence	Kaspary and Merotto (2020)
	SUB1C, SNORKEL2	Enhanced tolerance to submergence	SUB1C and SNORKEL2 genes offer increased resilience to extended periods of flooding, sustaining yield	Kaspary and Merotto (2020)
Oxidative stress	Antioxidant genes	Tolerance to oxidative stress	Genes associated with ROS scavenging and antioxidants provide protection against oxidative stress	Tsukamoto et al. (2004)
	CAT, SOD, APX genes	Enhanced antioxidant enzyme activity	CAT, SOD, and APX genes elevate antioxidant levels, reducing oxidative stress effects in rice	Rossatto et al. (2017)

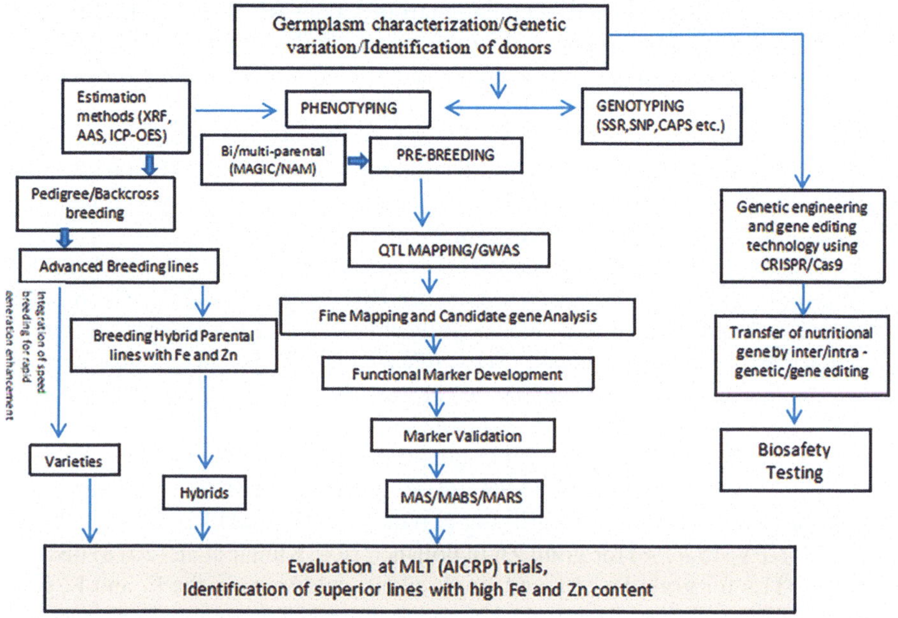

Fig. 5.4 Integrated breeding methodologies for enhancing the nutritional content of rice

Table 5.8 Genes for breeding nutritional traits in rice

S. no.	Genes for nutritional factors	Functions	References
1	OsYSL2, OsNAAT1 and OsNAC	High grain Zn content	Chandel et al. (2011)
2	OsZIP8a, OsZIP8c and OsZIP4b	Zn content	Gande et al. (2021)
3	RLRH1, RLRH2	Lysine biofortification	Wong et al. (2015)
4	OsNRAMP1	Iron transport	Curie et al. (2000)
5	Aspartate kinase (AK) and dihydrodipicolinate synthase (DHPS)	Lysine biosynthesis	Galili et al. (2002)
6	Granule-bound starch synthase I (GBSS I)	Determine amylose content	Tripathy et al. (2017)

these QTLs are scattered throughout the genome, it is impractical to utilize MAS to introduce a limited number of elements with numerous minor QTLs.

The application of molecular marker techniques has facilitated the mapping of numerous QTLs related to zinc, iron, and other nutritional quality traits in rice. In a population descended from the high-yielding PAU 201 and iron-rich Palman 579 varieties, transgressive segregation for grain Fe and Zn was observed, resulting in higher percentages of Fe (475 mg/g) and zinc (157 mg/g). Kumar et al. (2014b) identified three QTLs for zinc and 11 potential QTLs for iron in rice grains on chromosomes 2, 3, 7, 10, and 12 within this population.

Dixit et al. (2019) generated a population from parents RP-Bio 226 and Sampada, revealing shared genomic regions on chromosome 1 linked to grain Fe and Zn contents. Through the creation of BC_2F_3 mapping populations by crossing Swarna with *O. nivara* accessions, consistent detection of QTLs (qFe2.1, qFe3.1, qFe8.2, and qZn12.1) accounting for over 15% of the variation in phenotype was observed in both populations. Islam et al. (2020) identified two Zn-related QTLs on chromosomes 4 and 5 (QTL.Zn.4 and QTL.Zn.5) and a Fe-related QTL (QTL.Fe.9) on chromosome 9 in 113 aromatic rice germplasm samples.

Mohiuddin et al. (2020) mapped QTLs associated with higher Zn levels in rice grain in an F_2 population from a cross between BRRI Dhan 28 and the Zn-enriched cultivar Kalobokri. They discovered a significant QTL, qGZn3, on chromosome 3. Lee et al. (2020) found associated QTLs, qFe3-1 and qZn3-1, in a double-haploid population derived from rice cultivars 93-11 and Milyang 352, located on chromosome 3 between markers RM5419 and RM1164.

Suman et al. (2021) utilized genotyping by sequencing (GBS) data from 44 RILs to map 21 QTLs and 102 SSRs from 190 RIL populations derived from PR116 and Ranbir Basmati, identifying two major QTLs (qZPR.1.1-PV, 37.84%; qZPR.11.1-PV, 15.47%) for grain Zn in polished rice. Kumar et al. (2018) discovered 16 QTLs for grain Fe, Zn, and single-plant yield traits in the F_5 and F_6 generations of RILs derived from PAU 201 and Palman 579, noting their high Fe and Zn contents.

Planned crosses involving high × high (for Zn/Fe) may yield progenies with elevated micronutrient content compared to the parent with the lowest amount (assuming additive functions). Nevertheless, it is advisable to select progenies surpassing mid-parent values.

Improving marker-based selection of micronutrient traits in rice involves identifying potential KASP markers (SNPs), such as NAS3, using robust phenotyping techniques. The development of a desirable genotype with superior grain quality traits relies significantly on compatible QTLs and genes.

Techniques (as shown in Fig. 5.4) include AAS (atomic absorption spectroscopy), XRF/ED-XRF (energy-dispersive X-ray fluorescence), and ICP-OES (inductively coupled plasma–optical emission spectroscopy) for precise elemental analysis. Genetic studies involve QTL (quantitative trait loci) mapping and GWAS (genome-wide association study). Field trials are conducted through AICRIP (All India Coordinated Rice Improvement Programme) and MLT (Multi Locational Trial). Cutting-edge genetic modification is achieved with CRISPR/Cas9 technology. Molecular markers such as SSRs (simple sequence repeats), SNPs (single-nucleotide polymorphisms), and CAPS (cleaved amplified polymorphic sequences) are employed for MAS (marker-assisted selection), MABC (marker-assisted backcrossing), and MARS (marker-assisted recurrent selection).

The application of molecular markers in the development of rice varieties with improved traits such as grain size, shape, cooking quality, and nutritional composition is given in Table 5.9., facilitating efficient selection for enhanced nutritional value and grain characteristics, thereby contributing to improved food security and dietary health.

Table 5.9 Examples of marker-assisted breeding approaches to improve rice grain and nutritional qualities

S. no.	Traits	Breeding approach	Genes/QTL involved	Reference
1.	Amylose content	– MAS – CRISPR/CAS9-mediated targeted mutagenesis	– Wx gene – SBEI and SBEIIb	Yu et al. (2009), Sun et al. (2017)
2.	Gel consistency		QTL for amylose content and gel consistency	Yu et al. (2009)
3.	Gelatinization temperature	MAS	SSIIa gene	Waters et al. (2006)
4.	Fragrance	MAS	BADH2 gene	Luo et al. (2016)
5.	Iron and zinc	Genetic engineering	– SoyferH1 – Rice ferritin gene – OsNAS3 – Rice iron-regulated transporter 1 (OsIRT1) gene	Goto et al. (1999), Paul et al. (2012), Lee et al. (2009), Lee and An (2009)

5.5 Future Directions

1. **Genomic Advances:** Continued advancements in genomic technologies will further enhance the precision and efficiency of marker-assisted breeding, enabling more accurate selection of desired traits.
2. **Integration of Omics Data:** Incorporating omics data (genomics, transcriptomics, and proteomics) will offer a holistic understanding of gene expression and interaction, aiding in more informed breeding decisions.
3. **Accelerated Breeding Programs:** Implementing marker-assisted approaches in breeding programs worldwide will expedite developing and deploying new varieties, addressing emerging biotic stresses.
4. **Focus on Multiple Stress Tolerance:** There's a growing emphasis on developing rice varieties resilient to multiple stresses (biotic and abiotic), ensuring food security in changing environmental conditions.
5. **Climate Change Challenges**: With climate change posing new challenges to rice production, including increased pests, diseases, and abiotic stresses, MAB offers a promising solution for developing climate-resilient rice varieties with improved tolerance to these stressors.
6. **Growing Demand for High-Quality Rice:** The increasing demand for high-quality rice with enhanced nutritional content, cooking quality, and other desirable traits will drive the adoption of MAS to accelerate the development of superior rice varieties, meeting consumer preferences.

7. **Global Food Security Concerns:** As global population growth continues to strain food resources, MAS presents a viable strategy for enhancing rice productivity and ensuring food security by expediting the breeding of high-yielding, stress-tolerant rice varieties.
8. **Public and Private Sector Collaboration**: Collaborative efforts between public research institutions, private companies, and international organizations will facilitate the exchange of knowledge, resources, and germplasm, promoting the widespread adoption of MAS in rice breeding programs.
9. **Policy Support and Investment:** Supportive policies and increased investment in agricultural research and development, particularly in the field of MAS, will create an enabling environment for its adoption, fostering innovation and technological advancements in rice breeding.

In conclusion, marker-assisted breeding in rice for biotic stress stands as a pivotal technique, contributing significantly to the creation of disease-resistant, high-yielding, and resilient rice varieties. As technology progresses, MAB will continue to play a crucial role in ensuring global food security by combating biotic stresses affecting rice production worldwide.

5.6 Conclusion

Nevertheless, the substantial impact of adverse climate changes, encompassing severe drought, heat, salinity, cold, alterations in precipitation patterns, and heightened occurrences of diseases and insect pests, is markedly affecting both the yield and nutritional quality of rice. Projections indicate an anticipated 8 °C temperature rise, a surge in the average drought index to 129 from the current 52.45, and a potential sea-level rise by 2100, thereby posing the risk of impending food crises. Additionally, for each degree increase in temperature, there is an expected 3% and 7% elevation in average precipitation and humidity, respectively, fostering the prevalence of diseases and insect attacks on crops. Consequently, the world may encounter challenges in its food production system, leading to substantial yield losses and ultimately contributing to global food insecurity.

References

Ashikari M, Sakakibara H, Lin S, Yamamoto T, Takashi T, Nishimura A, Angeles ER, Qian Q, Kitano H, Matsuoka M (2005) Cytokinin oxidase regulates rice grain production. Science 309(5735):741–745

Boyer JS (1982) Plant productivity and environment. Science 218:443–448

Chakraborty A, Choudhury S, Kar SR, Deb P, Wylie SJ (2024) Gene editing for tolerance to temperature stress in plants: a review. Plant Gene 37:100439

Chandel GP, Dubey SM, Meena R (2011) In silico expression analysis of QTL specific candidate genes for grain micronutrient (Fe/Zn) content using ESTs and MPSS signature analysis in rice (*Oryza sativa* L.). J Plant Genet Transgenics 2(1):11–22

Chen X, Temnykh S, Xu Y, Cho YG, McCouch SR (1997) Development of a microsatellite framework map providing genome-wide coverage in rice (Oryza sativa L.). Theor Appl Genet 95(4):553–567

Chen J, Xuan Y, Yi J, Xiao G, Yuan DP, Li D (2023) Progress in rice sheath blight resistance research. Front Plant Sci 14:1141697

Chithrameenal K, Alagarasan G, Raveendran M, Robin S, Meena S, Ramanathan A, Ramalingam J (2018) Genetic enhancement of phosphorus starvation tolerance through marker assisted introgression of OsPSTOL1 gene in rice genotypes harbouring bacterial blight and blast resistance. PLoS One 13(9):e0204144

Cobb JN, DeClerck G, Greenberg A, Clark R, McCouch S (2013) Next-generation phenotyping: requirements and strategies for enhancing our understanding of genotype-phenotype relationships and its relevance to crop improvement. Theor Appl Genet 126(4):867–887

Collard BC, Mackill DJ (2008a) Marker-assisted selection: an approach for precision plant breeding in the twenty-first century. Philos Trans R Soc Lond Ser B Biol Sci 363(1491):557–572

Collard BC, Mackill DJ (2008b) Marker-assisted selection: an approach for precision plant breeding in the twenty-first century. Philos Trans Royal Soc B: Biol Sci 363(1491):557–572

Collard BC, Jahufer MZ, Brouwer JB, Pang EC (2005) An introduction to markers, quantitative trait loci (QTL) mapping and marker-assisted selection for crop improvement: the basic concepts. Euphytica 142(1–2):169–196

Crossa J, Pérez P, Hickey J, Burgueño J, Ornella L, Cerón-Rojas J, Zhang X, Dreisigacker S, Babu R, Li Y, Bonnett D, Mathews K (2014) Genomic prediction in CIMMYT maize and wheat breeding programs. Heredity (Edinb) 112(1):48–60

Curie C, Alonso JM, Le Jean M, Ecker JR, Briat JF (2000) Involvement of NRAMP1 from Arabidopsis thaliana in iron transport. Biochem J 347:749–755

Deng C, Ye H, Fan M, Pu T, Yan J (2017) The rice transcription factors OsICE confer enhanced cold tolerance in transgenic arabidopsis. Plant Signal Behav 2(5):e1316442

Ditta A (2013) Salt tolerance in cereals-molecular mechanisms and applications. In: Rout GR, Das AB (eds) Molecular stress physiology of plants. Springer, pp 133–154

Dixit S, Singh UM, Abbai TR, Singh VK, Paul A, Virk PS, Kumar A (2019) Identification of genomic region(s) responsible for high iron and zinc content in rice. Sci Rep 9:8136

FAOSTAT (2020) Food supply—crops primary equivalent

Fukao T, Yeung E (2012) The submergence tolerance gene SUB1A delays leaf senescence under prolonged darkness through hormonal regulation in rice. Plant Physiol 160(4):1795–1807

Galili G, Galili S, Lewinsohn E, Tadmor Y (2002) Genetic, molecular, and genomic approaches to improve the value of plant foods and feeds. Crit Rev Plant Sci 21:167–204

Gande NK, Pandey M, Brown C, Reddy R, Kumar V (2021) Identification and validation of zinc transporter genes influencing zinc content in rice grains. J Plant Nutr Soil Sci 184(4):533–545

Garcia-Oliveira AL, Tan L, Fu Y, Sun C (2009) Genetic identification of quantitative trait loci for contents of mineral nutrients in rice grain. J Integr Plant Biol 51(1):84–92

Goto F, Yoshihara T, Shigemoto N, Toki S, Takaiwa F (1999) Iron fortification of rice seed by the soybean ferritin gene. Nat Biotechnol 17(3):282–286

Ha BK, Fukai S (1994) Growth and yield of rice cultivars under high pH stress in the tropics. I. Field experiments. Plant Soil 167(2):287–296

Hittalmani S, Parco A, Mew TV, Zeigler RS, Huang N (2000) Fine mapping and DNA marker-assisted pyramiding of the three major genes for blast resistance in rice. Theor Appl Genet 100:1121–1128

Hospital F (2005) Selection in backcross programmes. Philos Trans R Soc B 360:1503–1511

Huang CL, Hwang SY, Chiang YC, Lin TP (2008) Molecular evolution of the Pi-ta gene resistant to rice blast in wild rice (Oryza rufipogon). Genetics 179(3):1527–1538

Huang X, Qian Q, Liu Z, Sun H, He S, Luo D, Xia G, Chu C, Li J, Fu X (2009) Natural variation at the DEP1 locus enhances grain yield in rice. Nat Genet 41:494–497

International Rice Genome Sequencing Project (IRGSP) (2005) The map-based sequence of the rice genome. Nature 436:793–800

International Rice Research Institute (IRRI) (2013) Stress-tolerant rice varieties for adaptation and risk management: progress and prospects in varietal improvement. IRRI

Islam MZ, Arifuzzaman M, Banik S, Hossain MA, Ferdous J, Khalequzzaman M, Pittendrigh BR, Tomita M, Ali MP (2020) Mapping QTLs underpin nutrition components in aromatic rice germplasm. PLoS One 15(6)

Isshiki M, Morino K, Nakajima M, Okagaki RJ, Wessler SR, Izawa T (1998) A naturally occurring functional allele of the rice waxy locus has a GT to TT mutation at the 5′ splice site of the first intron. Plant J 15(1):133–138

Jena KK, Mackill DJ (2008) Molecular markers and their use in marker-assisted selection in rice. Crop Sci 48(4):1266–1276

Jiang Y, Cai Z, Xie W, Long T, Yu H, Zhang Q (2012) Rice functional genomics research: progress and implications for crop genetic improvement. Biotechnol Adv 30:1059–1070

Jiang Y, Cheng Y, Huang J, Yuan Y, Tan L, Wu J (2019) CRISPR/Cas9-mediated targeted mutagenesis of OsERF922 gene confers enhanced drought tolerance in rice. Plant Biotechnol J 17(11):1998–2000

Kaspary TE, Merotto A (2020) Snorkeling strategy: tolerance to flooding in rice and potential application for weed management. Genes 11(9):975

Kumar A, Dixit S, Ram T, Yadaw RB, Mishra KK, Mandal NP (2014a) Breeding high-yielding drought-tolerant rice: genetic variations and conventional and molecular approaches. J Exp Bot 65(21):6265–6278

Kumar J, Jain S, Jain RK (2014b) Linkage mapping for grain iron and zinc content in F2population derived from the cross between PAU201 and Palman 579 in rice (*Oryza sativa* L.). Cereal Res Commun 42:389–400

Kumar S, Hash CT, Nepolean T, Mahendrakar MD, Satyavathi CT, Singh G, Rathore A, Yadav RS, Gupta R, Srivastava RK (2018) Mapping grain iron and zinc content quantitative trait loci in an iniadi-derived immortal population of pearl millet. Genes (Basel) 9(5):248

Kumar M, Patel MK, Kumar N, Bajpai AB, Siddique KH (2021) Metabolomics and molecular approaches reveal drought stress tolerance in plants. Int J Mol Sci 22(17)

Lee S, An G (2009) Over-expression of OsIRT1 leads to increased iron and zinc accumulations in rice. Plant Cell Environ 32(4):408–416

Lee S, Jeon US, Lee SJ, Kim K, Persson DP, Husted S, Schjorring JK, Kakei Y, Masuda H, Nishizawa NK, An G (2009) Iron fortification of rice seeds through activation of the nicotianamine synthase gene. Proc Natl Acad Sci U S A 106(51):22014–22019

Lee S, Kang J, Lee J, Seo J, Shin D, Cho J, Jo S, Song Y, Park D, Ko J, Koh H, Lee J (2020) QTL analysis for Fe and Zn concentrations in rice grains using a doubled haploid population derived from a cross between rice cultivar 93-11 and Milyang 352. Plant Breed Biotech 8:69–76

Lian L, Xu H, Zhang H, He W, Cai Q, Lin Y, Wei L, Pan L, Xie X, Zheng Y, Wei Y, Zhu Y, Xie H, Zhang J (2020) Overexpression of *OsSPL14* results in transcriptome and physiology changes in *indica* rice 'MH86'. Plant Growth Regul 90:265–278

Liu S, Zhang X, Pumphrey MO, Stack RW, Gill BS, Anderson JA (2006) Complex microcolinearity among wheat, rice, and barley revealed by the use of new genomics tools. Proc Natl Acad Sci 103(23):9073–9078

Liu M, Shi Z, Zhang X, Wang M, Zhang L, Zheng K, Liu J, Hu X, Di C, Qian Q, He Z, Yang DL (2019) Inducible over expression of ideal plant architecture—improves both yield and disease resistance in rice. Nat Plants 5(4):389–400

Lokeshkumar B, Krishnamurthy S, Rathor S, Warriach AS, Vinaykumar N, Dushyanthakumar B, Sharma PC (2023) Morphophysiological diversity and haplotype analysis of saltol QTL region in diverse rice landraces for salinity tolerance. Rice Sci 30(4):306–320

Lu K, Li L, Zheng X, Zhang Z, Mou T, Zhongli H (2008) Quantitative trait loci controlling cu, ca, Zn, Mn and Fe content in rice grains. J Genet 87:305–310

Lu Y, Chuan M, Wang H, Chen R, Tao T, Zhou Y, Xu Y, Li P, Yao Y, Xu C, Yang Z (2022a) Genetic and molecular factors in determining grain number per panicle of rice. Front Plant Sci 13:964246

Lu J, Li Q, Wang C, Wang M, Zeng D, Zhang F, Zhai W, Zhou Y (2022b) Identification of quantitative trait loci associated with resistance to Xanthomonas oryzae pv. Oryzae pathotypes prevalent in South China. Crop J 10(2):498–507

Luo YC, Ma TC, Zhang AF, Ong KH, Li ZF, Yang JB, Yin ZC (2016) Marker-assisted breeding of the rice restorer line Wanhui 6725 for disease resistance, submergence tolerance and aromatic fragrance. Rice 9(1):66

Maclean JL, Dawe DC, Hardy B, Hettel GP (2002) Rice almanac: source book for the most important economic activity on earth. CABI Publishing, Wallingford, UK

Mahender A, Anandan A, Pradhan SK, Pandit E (2016) Rice grain nutritional traits and their enhancement using relevant genes and QTLs through advanced approaches. Springerplus 5:1–18

Majhi PK, Singh SK, Anandan A, Khaire AR, Korada M, Habde SV, Singh DK, Bhagvan AP, Jayasudha S, Singh A (2022) Parental polymorphism survey for evaluation and selection of contrasting parents for drought tolerance in rice (Oryza sativa L.) by using SSR markers. Int J Environ Clim 12(11):2507–2519

Makkar GS, Bentur JS (2017) Breeding for stem borer and gall midge resistance in rice. In: Arora R, Sandhu S (eds) Breeding insect resistant crops for sustainable agriculture. Springer Nature Singapore Pte Ltd., pp 323–352

Mohapatra S, Barik SR, Dash PK, Lenka D, Pradhan KC, Raj KRR, Mohanty SP, Mohanty MR, Sahoo A, Jena BK, Panda AK, Panigrahi D, Dash SK, Meher J, Sahoo CR, Mukherjee AK, Das L, Behera L, Pradhan SK (2023) Molecular breeding for incorporation of submergence tolerance and durable bacterial blight resistance into the popular rice variety 'Ranidhan'. Biomol Ther 13(2)

Mohiuddin SJ, Haque MA, Haque MM, Islam MT, Biswas PS (2020) Genetic analysis reveals a major effect QTL associated with high grain zinc content in rice. Plant Breed Biotech 8:327–340

Monna L, Kitazawa N, Yoshino R, Suzuki J, Masuda H, Maehara Y, Tanji M, Sato M, Nasu S, Minobe Y (2002) Positional cloning of rice semi dwarfing gene, sd-1: rice "green revolution gene" encodes a mutant enzyme involved in gibberellin synthesis. DNA Res 9(1):11–17

Ngernmuen A, Suktrakul W, Kate-Ngam S, Jantasuriyarat C (2020) Transcriptome comparison of defense responses in the rice variety 'Jao Hom Nin' regarding two blast resistant genes, Pish and Pik. Plants 9(6)

Ning XI, Yunyu W, Aihong L (2020) Strategy for use of rice blast resistance genes in rice molecular breeding. Rice Sci 27(4):263–277

Paul S, Ali N, Gayen D, Datta SK, Datta K (2012) Molecular breeding of Osfer 2 gene to increase iron nutrition in rice grain. GM Crops Food 3(4):310–316

Pradhan KC, Barik SR, Mohapatra S, Nayak DK, Pandit E, Jena BK, Sangeeta S, Pradhan A, Samal A, Meher J, Behera L, Panigrahi D, Mukherjee AK, Pradhan SK (2022) Incorporation of two bacterial blight resistance genes into the popular rice variety, Ranidhan through marker-assisted breeding. Agriculture 12(9):1287

Ribaut JM, Ragot M (2007) Marker-assisted selection to improve drought adaptation in maize: the backcross approach, perspectives, limitations, and alternatives. J Exp Bot 58(2):351–360

Rossatto T, Amaral MN, Benitez LC, Vighi IL, Braga EJB, de Magalhaes Junior AM, Maia MAC, da Silva PL (2017) Gene expression and activity of antioxidant enzymes in rice plants, *cv.* BRS AG, under saline stress. Physiol Mol Biol Plants 23(4):865–875

Sanghera GS, Wani SH, Hussain W, Singh NB (2011) Engineering cold stress tolerance in crop plants. Curr Genomics 12(1):30–43

Schmitz AJ, Folsom JJ, Jikamaru Y, Ronald P, Walia H (2013) SUB1A-mediated submergence tolerance response in rice involves differential regulation of the brassinosteroid pathway. New Phytol 198(4):1060–1070

Secco D, Wang C, Shou H, Whelan J (2015) Stress induced gene expression drives transient DNA methylation changes at adjacent repetitive elements. eLife 4:e09343. https://doi.org/10.7554/eLife.09343

Septiningsih EM, Pamplona AM, Sanchez DL, Neeraja CN, Vergara GV, Heuer S, Ismail AM, Mackill DJ (2009) Development of submergence-tolerant rice cultivars: the Sub1 locus and beyond. Ann Bot 103(2):151–160

Sharma TR, Rai AK, Gupta SK, Vijayan J, Devanna BN, Ray S (2012) Rice blast management through host-plant resistance: retrospect and prospects. Agric Res 1:37–52

Shin NH, Lee ON, Han JH, Song K, Koh HJ, Yoo SC, Chin JH (2021) The effect of water level in rice cropping system on phosphorus uptake activity of *Pup1* in a *Pup1* + *Sub1* breeding line. Plants (Basel) 10(8):1523

Shohan MUS, Sinha S, Nabila FH, Dastidar SG, Seraj ZI (2019) HKT1;5 transporter gene expression and association of amino acid substitutions with salt tolerance across rice genotypes. Front Plant Sci 10:1420

Singh S, Sidhu JS, Huang N, Vikal Y, Li Z, Brar DS, Dhaliwal HS, Khush GS (2001) Pyramiding three bacterial blight resistance genes (xa5, xa13 and Xa21) using marker-assisted selection into indica rice cultivar PR106. Theor Appl Genet 102:1011–1015

Singh SK, Majhi PK, Anandan A, Korada M, Habde SV, Khaire AR, Singh DK, Jayasudha S, Singh A, Bhagvan AP (2022) Microsatellites based parental polymorphism survey for moisture stress in Rice (Oryza sativa L.) between the parental genotypes HUR-1309 and CR Dhan 801. Emergent Life Sci Res 8:248–258

Song WY, Wang GL, Chen LL, Kim HS, Pi LY, Holsten T, Ronald P (1995) A receptor kinase-like protein encoded by the rice disease resistance gene, Xa21. Science 270(5243):1804–1806

Suman K, Neeraja CN, Madhubabu P, Rathod S, Bej S, Jadhav KP, Kumar JA, Chaitanya U, Pawar SC, Rani SH, Subbarao LV, Voleti SR (2021) Identification of promising RILs for high grain zinc through genotype × environment analysis and stable grain zinc QTL using SSRs and SNPs in rice (*Oryza sativa* L.). Front Plant Sci 12:587482

Sun Y, Jiao G, Liu Z, Zhang X, Li J, Guo X, Du W, Du J, Francis F, Zhao Y, Xia L (2017) Generation of high-amylose rice through CRISPR/Cas9-mediated targeted mutagenesis of starch branching enzymes. Front Plant Sci 8:298

Takasaki H, Maruyama K, Kidokoro S, Ito Y, Fujita Y, Shinozaki K, Yamaguchi KS, Kazuo N (2010) The abiotic stress-responsive NAC-type transcription factor *OsNAC5* regulates stress-inducible genes and stress tolerance in rice. Mol Gen Genomics 284:173–183

Thomson MJ, De Ocampo M, Egdane J, Rahman MA, Sajise AG, Adorada DL, Blumwald E, Seraj ZI, Singh RK, Gregorio GB, Ismail AM (2010) Characterizing the saltol quantitative trait locus for salinity tolerance in rice. Rice 3(2):148–160

Tripathy SK, Dash M, Behera SK, Dinesh MI, Maharana M (2017) Nutrient rich quality rice—a journey to healthy life. Adv Plants Agric Res 7(5):364–367

Tsukamoto S, Morita S, Hirano E, Yokoi H, Masumura T, Tanaka K (2004) A novel cis-element that is responsive to oxidative stress regulates three antioxidant defense genes in rice. Plant Physiol 137(1):317–327

Varshney RK, Sinha P, Singh VK, Kumar A, Zhang Q, Bennetzen JL (2020) Role of genomics in breeding climate-resilient crops: an array of opportunities. Seminar

Vikram P, Swamy BM, Dixit S, Ahmed HU, Cruz MTS, Singh AK, Kumar A (2011) qDTY1. 1, a major QTL for rice grain yield under reproductive-stage drought stress with a consistent effect in multiple elite genetic backgrounds. BMC Genet 12(1):89

Wang W, Vinocur B, Altman A (2003) Plant responses to drought, salinity and extreme temperatures: towards genetic engineering for stress tolerance. Planta 218:1–14

Wang B, Shahzad MF, Zhang Z, Sun H, Han P, Li F, Han Z (2014) Genome-wide analysis reveals the expansion of cytochrome P450 genes associated with xenobiotic metabolism in rice striped stem borer, *Chilo suppressalis*. Biochem Biophys Res Commun 443(2):756–760

Waters DL, Henry RJ, Reinke RF, Fitzgerald MA (2006) Gelatinization temperature of rice explained by polymorphisms in starch synthase. Plant Biotechnol J 4:115–122

Wissuwa M, Wegner J, Ae N, Yano M (2002) Substitution mapping of Pup1: a major QTL increasing phosphorus uptake of rice from a phosphorus-deficient soil. Theor Appl Genet 105(6–7):890–897

Witcombe JR, Gyawali S, Subedi M, Virk DS, Joshi KD, Musa AM, Harris D (2005) Participatory plant breeding is better described as highly client-oriented plant breeding. I. Four indicatives. Exp Agric 41(3):299–319

Wong HW, Liu Q, Sun SSM (2015) Biofortification of rice with lysine using endogenous histones. Plant Mol Biol 87:235–248

Yang Y, Zhou Y, Sun J, Liang W, Chen X, Wang X, Zhou J, Yu C, Wang J, Wu S, Yao X, Zhou Y, Zhu J, Yan C, Zheng B, Chen J (2022) Research progress on cloning and function of Xa genes against rice bacterial blight. Front Plant Sci 13:847199

Yeh Y, Chen W, Ng Y, Lin Y, Tseng H, Li H, Ku BMS (2015) Down-regulation of cytokinin oxidase 2 expression increases tiller number and improves rice yield. Rice 8:36

Yi Y, Hassan MA, Cheng X, Li Y, Liu H, Fang W, Zhu Q, Wang S (2023) QTL mapping and analysis for drought tolerance in rice by genome-wide association study. Front Plant Sci 14:1223782

Yoshimura S, Yamanouchi U, Katayose Y, Toki S, Wang ZX, Kono I, Kurata N, Yano M, Iwata N, Sasaki T (1998) Expression of Xa1, a bacterial blight-resistance gene in rice, is induced by bacterial inoculation. Proc Natl Acad Sci 95(4):1663–1668

Yu YH, Li G, Fan YY, Zhang KQ, Min J, Zhu ZW, Zhuang JY (2009) Genetic relationship between grain yield and the contents of protein and fat in a recombinant inbred population of rice. J Cereal Sci 50(1):121–125

Yuan C, Yuxiang Z, Zhijuan J, Yan L, Zhihua W, Changdeng Y (2019) Identification of stable quantitative trait loci for sheath blight resistance using recombinant inbred line. Rice Sci 26(5):331–338

Yuexiong Z, Gang Q, Qianqian M, Minyi W, Xinghai Y, Zengfeng M, Haifu L, Chi L, Zhenjing L, Fang L, Dahui H, Rongbai L (2020) Identification of major locus Bph35 resistance to brown planthopper in rice. Rice Sci 27(3):237–245

Zampieri E, Volante A, Marè C, Orasen G, Desiderio F, Biselli C, Canella M, Carmagnola L, Milazzo J, Adreit H, Tharreau D, Nicolas P, Vaccino P, Vale G (2023) Marker-assisted pyramiding of blast-resistance genes in a *japonica* elite rice cultivar through forward and background selection. Plan Theory 12(4):757

Zhang H, Chen J, Shan S, Cao F, Chen G, Zou Y, Huang M, Abou-Elwafa SF (2020) Proteomic profiling reveals differentially expressed proteins associated with amylose accumulation during rice grain filling. BMC Genomics 21(1):1051–1059

Zhang YM, Yu HX, Ye WW, Shan JX, Dong NQ, Guo T, Kan Y, Xiang YH, Zhang H, Yang YB, Li YC, Zhao HY, Lu ZQ, Guo SQ, Lei JJ, Liao B, Mu XR, Cao YJ, Yu JJ, Lin HX (2021) A rice QTL GS3.1 regulates grain size through metabolic-flux distribution between flavonoid and lignin metabolons without affecting stress tolerance. Commun Biol 4(1):1171

Genome-Assisted Breeding and Genome-Wide Association Studies for Rice Improvement

6

Lopamudra Singha, Prastuti Bhattacharyya, Parshant Kumar Sharma, Ritik Digamber Bisane, Vemula Anjula Krishna, Akansha Singh, Jai Prakash Lal, and Shravan Kumar Singh

Abstract

Numerous crops have benefited from the application of genetic and genomic techniques, including rice, which is one of the most significant staple cereals. A century ago, the introduction of artificial crossing methods for inbreeding rice species significantly expanded the scope of genetic variation accessible to breeders. In light of its diversified germplasm, minuscule and well-annotated genome, and major role in ensuring global food security, rice is a great model crop for genetic research. Comprehending the genetic characteristics of rice is essential for both genome-assisted breeding (GAB) and genome-wide association studies (GWAS). GWAS in rice is instrumental in unraveling the genetic architecture of various traits, aiding in crop improvement strategies, and contributing to global food security by enhancing rice productivity and resilience. To find the genetic variants, GWAS entails scanning the genomes of a significant number of individuals—often thousands or more. GAB empowers the integration of genomic tools with highly advanced phenotyping techniques and traditional breeding methods. Genomic-assisted selection (GAS) and GWAS serve as complementary tools in crop breeding. Genomic selection models are developed and improved upon by GWAS discoveries, including marker-trait relationships and candidate genes, increasing their precision and predictive capacity. By combining these methodologies, scientists and agricultural breeders can leverage the advantages of each method to enhance genomic prediction accuracy and expedite crop breeding initiatives.

L. Singha · P. Bhattacharyya · P. K. Sharma · R. D. Bisane · V. A. Krishna · A. Singh · J. P. Lal · S. K. Singh (✉)
Genetics and Plant Breeding, Institute of Agricultural Sciences, Banaras Hindu University, Varanasi, Uttar Pradesh, India

A. Singh et al. (eds.), *Climate-Smart Rice Breeding*,
https://doi.org/10.1007/978-981-97-7098-4_6

Keywords

GWAS · GAB · Rice · GAS · Genetic variation · SNP · Haplotypes

6.1 Introduction

Rice (*Oryza sativa* L.) is one of the most vital staple food crops globally, sustaining more than three billion people who rely on it for 20–80% of their daily calorie intake (Kumar et al. 2020). With the world's population projected to increase from 6 billion in 2000 to 9.6 billion by 2050, there is a pressing need to meet the growing demand for food grain. This underscores the critical role of rice in global food security and the challenge of ensuring an adequate supply to feed a rapidly expanding world population.

In response to the escalating demand for rice production, genetic enhancement has emerged as a pivotal strategy for augmenting rice productivity (Pingali 2012). Developing cultivars resistant to heat, drought, salinity, cold, and freezing stress using traditional and cutting-edge crop development techniques can aid in the fight against climate change (Singh et al. 2023). These advancements not only bolster food security by ensuring a more stable food supply, especially in regions where rice is a dietary staple but also positively impact farmers' livelihoods by improving crop yields and reducing input costs. In rice breeding, understanding the connections between different attributes is essential for crop yield and performance (Kumar et al. 2023). The adaptability of these improved varieties to diverse local conditions and their reduced reliance on chemical inputs makes them environmentally sustainable while promoting biodiversity conservation.

6.2 Rice Genome Structure and Complexity

Rice's small genome size and diploid nature make it an ideal candidate among cereals for launching genomic research initiatives (Vetriventhan et al. 2020). It stands as a model organism in plant biology and agricultural research, offering valuable insights and serving as a foundation for scientific exploration in these domains. The genome of rice, scientifically known as *Oryza sativa* with AA genome, comprises 12 chromosomes ($2n = 24$) and spans a total length of 430 Mb, corresponding to approximately 1500 cM (centi Morgans). The majority of the rice genome has been sequenced by two research groups: the Syngenta group (Goff et al. 2002) and the Chinese sequencing group (Yu et al. 2002). Examining these preliminary sequences to determine the quantity and functional classifications of anticipated genes reveals that rice possesses approximately 40,000–50,000 functional genes distributed among around 15,000 distinct gene families. This genetic complexity underpins rice's adaptability and resilience, reflecting its diverse ecological niches and agricultural significance. Understanding this intricate structure not only illuminates the plant's biology but also holds the key to enhancing crop yield and ensuring food security for millions worldwide. In the annals of scientific achievement, the rice genome sequencing stands as a monumental milestone, ushering humanity into a new era of agricultural understanding and biotechnological innovation.

6.3 Genetic Variation in Rice Population

Genetic variation in rice populations refers to the natural diversity within different rice plants, encompassing variations in their genes, traits, and genetic makeup. This diversity is a result of evolutionary processes and environmental adaptation, making rice an excellent model for studying genetic variation. A century ago, the introduction of artificial crossing methods for inbreeding rice species significantly expanded the scope of genetic variation accessible to breeders. By creating diverse populations through the controlled crossing of two or more superior lines, breeders dramatically increased the pool of genetic traits available for selection. Recognizing and broadening this genetic diversity stands as a pivotal aspect of rice breeding, enabling the development of novel varieties with enhanced characteristics. The analysis of natural genetic variation in rice has primarily concentrated on vital agronomic and quantitative traits such as flowering time, yield components, and stress tolerance. The groundbreaking sequencing of the rice genome, the first among cereals, has seamlessly integrated gene discovery and cloning into genetic research. This achievement has not only facilitated gene cloning but has also heightened interest in comprehending allelic diversity within genes and gene sets. The significance of sequencing lies in its ability to unravel the intricate genetic makeup of rice, providing invaluable insights into the diverse traits crucial for crop improvement. Through this comprehensive understanding, researchers can develop rice varieties with enhanced features, ensuring the continued advancement of agricultural practices and global food security. Studying genetic diversity, relationships, and population structures serves various purposes, such as identifying parent pairs to create offspring with superior traits and substantially higher yields than their parents. Genetic variation in rice encompasses a wide array of traits that significantly impact its adaptability and productivity. Variability in yield, quality, resistance to pests and diseases, and tolerance to environmental stresses such as drought and salinity are prominent aspects of genetic diversity in rice populations. Additionally, differences in morphological characteristics, nutritional content, and photoperiod sensitivity contribute to the overall genetic variation. Researchers and breeders leverage this diversity to develop new rice varieties with desirable traits, ensuring sustainable agriculture and global food security.

6.4 Genetic Insight of Rice

Plant scientists are faced with tremendous obstacles in their efforts to enhance crops for greater yield, improved quality, and more resilience to stress in light of the rapidly expanding population and changing environmental conditions. Numerous crops have benefited from the application of genetic and genomic techniques (Afzal et al. 2023) including rice (*Oryza sativa* L.), which is one of the most significant staple cereals. In light of its diversified germplasm, minuscule and well-annotated genome, and major role in ensuring global food security, rice is a great model crop for genetic research.

Comprehending the genetic characteristics of rice is essential for both GAB and GWAS. Breeding programs that use molecular marker technologies are more productive, can move desired genes more quickly between varieties, and can introduce novel genes from related or cultivated species (Majhi et al. 2022). Targeted breeding efforts are made possible by genetic discoveries that identify the precise genes linked to important traits. Furthermore, this information facilitates the creation of molecular markers, simplifying the process of choosing desired characteristics. Breeders can forecast and choose traits more quickly thanks to GAB, aided by genetic data. Furthermore, precision agriculture is made easier by this genetic insight, which raises yields and improves resource efficiency. Moreover, it is essential to conservation efforts because it maintains genetic variability. Crucially, this genetic information enables scientists to create climate-adaptive rice varieties, guaranteeing food security for all people even in the face of climatic difficulties. This in-depth understanding allows for precise genetic modifications, such as targeted breeding and genetic engineering, to develop crop varieties that exhibit improved traits like higher yields, resistance to diseases and pests, and tolerance to environmental stresses. Additionally, the integration of this knowledge facilitates the development of more resilient and nutritious crop varieties, ensuring sustainable agricultural practices and addressing global food security challenges.

6.5 Genome-Wide Association Studies (GWAS)

GWAS have revolutionized the field of crop genetics and breeding, including the study and improvement of rice, one of the world's most important staple crops. It is a powerful and systematic method for identifying genetic variants associated with rice's key agronomic traits and diseases (Yu et al. 2017). To find these genetic variants, GWAS entails scanning the genomes of a significant number of individuals—often thousands or more. The application of GWAS in rice has been a pivotal area in agricultural genetics.

To uncover genetic differences linked to numerous attributes like yield, disease resistance, grain quality, and tolerance to environmental challenges, comprehensive genome analyses of rice varieties are performed as part of GWAS. The major goal of GWAS is to find single nucleotide polymorphisms (SNPs) whose allele frequencies systematically fluctuate with phenotypic trait values. SNPs are the subject matter of numerous GWAS. These research efforts are structured to enable researchers to analyze hundreds or thousands of SNPs in a large number of subjects at once.

6.5.1 Experimentation

Understanding the genetic basis of variables essential for rice agriculture has been made easier because of the insights obtained from GWAS in rice. For example, these investigations have determined the genetic variants causing characteristics like grain size, cooking quality, drought or flooding tolerance, disease resistance against primary rice pathogens, and fertilizer responsiveness.

The experimentation process involved in conducting a GWAS typically follows a structured methodology:

6.5.1.1 Sample Selection

Sample selection is a crucial aspect of conducting a GWAS. The selected samples or individuals represent the diversity within a population and significantly impact the study's accuracy and reliability. Rigorous sample selection criteria contribute significantly to the accuracy of identifying genetic variants associated with specific traits in rice, ultimately benefiting breeding programs and crop improvement efforts.

In the context of GWAS for crops like rice, here are key considerations in sample selection:

(a) Genetic diversity.
(b) Population structure.
(c) Phenotypic variation.
(d) Sample size.
(e) Control population.
(f) Validation sets.
(g) Linkage disequilibrium (LD) structure.

6.5.1.2 Phenotype Measurement

In GWAS, measuring phenotypes accurately is crucial for identifying genetic variants associated with various traits. Some common phenotypic measurements such as days to flowering, maturity period, nutrient uptake, photosynthetic efficiency, plant height and architecture, biotic and abiotic stress tolerance, grain and cooking quality, yield attributing traits, etc., were conducted in rice GWAS. The genetic architecture of shoot growth dynamics was explored during the early and active tillering stages by applying a functional modeling and genome-wide association mapping approach on a diversity panel of ~360 rice accessions and found a gibberellic acid (GA_3) catabolic gene, OsGA2ox7, which could influence GA_3 levels to regulate vigor in the early tillering stage (Campbell et al. 2017). A GWAS was performed using 115 rice japonica accessions, including tropical and temperate genotypes, to expand the knowledge of the genetic bases affecting resistant starch (RS) and amylose content (AAC) & identified 11 associations for RS on seven chromosomes (chromosomes 1, 2, 3 5, 6, 8, and 11) and five associations for AAC on chromosome 6 (Biselli et al. 2019).

6.5.2 SNP Genotyping

This process aims to associate these genetic variations with particular traits or phenotypes of interest. The choice of genotyping method in rice GWAS depends on factors like the number of SNPs required, the size of the genome to be analyzed, budget constraints, throughput requirements, and the desired resolution of the study. Several methods used in SNP genotyping:

6.5.2.1 SNP Array Technology

Arrays are probes that target known SNP loci designed to detect and genotype thousands to millions of SNPs across the rice genome. Affymetrix and Illumina are popular platforms for SNP arrays in rice research. SNP probes designed from 801 rice varieties and produced a "RiceSNP50" array containing 51,478 evenly distributed markers, 68% of which were within genic regions on the Illumina Infinium platform (Chen et al. 2014). Cornell-IR LD Rice Array (C7AIR), a second-generation SNP array, was developed containing 7098 markers and designed to detect genome-wide polymorphisms within and between subpopulations of *O. sativa*, as well as *O. glaberrima, O. rufipogon* and *O. nivara* (Pingali 2012). They carried out a GWAS for plant height, employing the C7AIR across a diversity panel of 189 rice accessions and identified 20 QTLs contributing to plant height.

6.5.3 Restriction Fragment Length Polymorphism (RFLP)

Involves digesting genomic DNA with restriction enzymes and detecting SNPs by analyzing the resulting fragment sizes via gel electrophoresis.

6.5.4 Whole-Genome Sequencing (WGS)

Complete sequencing of the entire rice genome to identify SNPs, providing high-resolution data but typically more expensive and computationally intensive. After completing whole-genome resequencing on 50 accessions, scientists mapped the results against the *Oryza sativa ssp. japonica* Nipponbare genome and discovered 2,145,095 SNPs and 317,832 insertion-deletions (indels) (Tanaka et al. 2020). They also used the multiple genome browser TASUKE to look at the haplotypes of flowering genes and find new mutations in these genes. Using Japanese Landraces (JRC) as a reference, GWAS revealed two notable trends in leaf blade width: a strong peak at NAL1, a crucial gene for the regulation of leaf width, and a peak near HESO1, involved in flowering regulation, in heading date. A set of 853 diverse accessions was obtained with a whole genome sequence (WGS). Additionally, a group that will be receptive to GWAS was assembled based on a number of factors, including geographical, physiological, and molecular diversity (Prabhu et al. 2023).

6.5.5 Reduced Representation Sequencing (RRS)

Restriction-site Associated DNA sequencing (RAD-seq) or Genotyping-by-sequencing (GBS) sequence specific regions of the genome, reducing complexity and cost while identifying a subset of SNPs. With a set of 15,083 high-quality SNP markers, a GWAS was conducted along with GBS for seven agronomically important traits on a panel of 346 *O. rufipogon* accessions. The study revealed a total of 47 strong MTAs, with 19 SNPs located in or close to previously reported QTL/genic regions and 27 novel SNPs that were found to be significantly associated with various traits (Malik et al. 2022). The absence of superior alleles at 12 locations in the cultivated and wild rice samples at significant MTAs indicates a major potential for improvement by targeted introgression of these alleles into the cultivar.

6.5.6 High-Throughput Genotyping Technologies

TaqMan assays, Kompetitive Allele-Specific PCR (KASP), and Fluidigm SNP genotyping platforms are used for targeted SNP genotyping. In designing KASP primers, 596 SNPs were selected, which were categorized into eight subsets: quality control, indica-indica variation, highly polymorphic, functional genes, key genes targeting sites, gene cloned region, critical trait linked, and gap filling sites. Thirty-one KASP markers were found to be significantly ($P < 0.01$) linked with both ML and the LWR when the core KASP arrays were employed for association analysis with milled grain characteristics (Yang et al. 2019).

6.5.7 Data Quality Control

It is a critical step in GWAS to ensure the reliability and accuracy of the genetic and phenotypic data used for analysis. It enables scientific research to be transparent and reproducible. Researchers can strengthen the robustness and validity of the connections found in GWAS by minimizing mistakes and reducing biases through the implementation of stringent quality control methods. Some aspects of data quality control in GWAS are:

6.5.7.1 Sample Quality Control

Sample filtering excludes samples with high levels of missing data or high relatedness. Duplicated or Cryptic Samples prevent inflated acquaintances from finding and eliminating repetitions or samples that are closely linked.

6.5.7.2 Genotype Data

Call rate thresholds exclude SNPs or samples with low call rates to ensure data quality. Minor Allele Frequency (MAF) removes SNPs with low minor allele frequencies, as they might not provide sufficient statistical power for association analysis. Hardy Weinberg Equilibrium is used to test SNPs for deviations from Hardy Weinberg Equilibrium, which indicates genotyping errors or population stratification. Genotype Imputation improves coverage and accuracy of genotype data by imputing missing genotypes. Quality control was confirmed within the Illumina HiSeq software, and all reads passing the filtering specifications were mapped onto the reference genome IRGSP-1.0 (Guo et al. 2019).

6.5.7.3 Phenotype Data

Outlier detection identifies and removes outliers in phenotypic data that could skew association analysis. Phenotypic measurements are Normalized to correct for environmental variations or batch effects.

6.5.7.4 Batch Effects

Adjust for batch effects if genotyping or phenotyping was completed in separate batches to reduce their influence on association findings. The grain nutrients were analyzed in rice to provide quality control; a blank and a certified reference material (CRM; NIST 1567a wheat four) were included in every digestion batch during phenotyping (Cu et al. 2021).

6.5.7.5 Association Analysis

Leverage statistical approaches to examine the association between phenotypic features and genetic markers (SNPs) in the population, such as generalized or mixed linear models. This finds noteworthy connections between traits and markers (MTAs). In the study, a panel of 529 rice core varieties with 607,201 SNPs was used for the single-locus genome-wide association studies (SL-GWAS) method (MLM) and five multi-locus genome-wide association studies (ML-GWAS) approaches (mrMLM, FASTmrMLM, pLARmEB, pKWmEB, and ISIS EM-BLASSO) (Zhong et al. 2021).

6.5.7.6 Population Structure and Relatedness Correction (Kinship Analysis)

It reduces spurious associations and false positives. Principal component analysis (PCA) lowers the dimensionality of genetic data to identify population structure. It can be applied to association analysis to account for population stratification and pinpoint axes of genetic variation. Based on genetic similarities, tools such as hierarchical clustering or model-based clustering algorithms (e.g., STRUCTURE or ADMIXTURE) are available to locate subgroups or clusters within the population. By assessing anomalies from the predicted distribution of *P*-values, population stratification can be revealed using Manhattan plots or quantile–quantile (Q–Q) plots to visualize the results. Kinship estimation determines each member of the study population's pairwise relatedness coefficients (kinship matrix). Based on genetic markers, kinship calculates the degree of relatedness or genetic similarity between people. To account for population structure and relatedness, Linear Mixed Models (LMM) include both random effects (kinship matrix) and fixed effects (genetic markers linked to the trait). They achieve this by accounting for population stratification, which lowers false positives. GCTA (genome-wide complex trait analysis) software incorporates a genetic relationship matrix (GRM) into the statistical model by estimating genetic links using genome-wide SNP data, which aids in GWAS relatedness correction. A number of software programs, including PLINK, TASSEL, GCTA, EIGENSOFT, and others, include features for PCA, kinship estimates, and GWAS population structure and relatedness correction. Spurious MTAs were eliminated from the population structure, and an MLM implemented in GAPIT was adopted for association analysis for Mesocotyl Elongation in Rice (*Oryza sativa* L.) under multiple culture condition. 145 accessions' approximately 65 K SNP genotype was derived from a 90 K high-density SNP array. For the final genetic study, 56,456 SNPs dispersed across all 12 chromosomes were employed, having first been filtered using TASSEL 5.2.51 to exclude SNPs with minor allele frequency. When the paired accessions' kinship relatedness statistics were combined, they revealed that while the kinship relatedness among the Pop1 accessions was modest, a small number of accessions within Pop2 had rather substantial relatedness (Liu et al. 2019). 145 accessions were used for the final genetic study, and 56,456 SNPs dispersed across all 12 chromosomes were employed, having first been filtered using TASSEL 5.2.51 to exclude SNPs with minor allele frequency. When the paired accessions' kinship relatedness statistics were combined, they revealed that while the kinship relatedness among the Pop1 accessions was modest, a small number of accessions within Pop2 had rather substantial relatedness (Xu et al. 2020).

6.5.7.7 Multiple Testing Correction

It reduces the excessively high proportion of false positives that come from examining a large number of genetic markers throughout the entire genome. The trade-off between minimizing false positives and increasing the ability to identify genuine links should be taken into account when selecting a correction technique. Several methods can be utilized to address multiple testing issues in GWAS:

6.5.7.7.1 Permutation Testing

By comparing the distribution derived from permuted data with the actual test statistic, it provides a more precise calculation of significance and aids in evaluating the importance of observed relationships. Here, the genotypes are fixed but the phenotype labels (trait values) are randomly rearranged among the participants. An empirical null distribution of the test statistics is produced as a result.

6.5.7.7.2 False Discovery Rate (FDR) Correction

It allows for a balance between controlling false positives and maximizing power. It controls the expected proportion of false discoveries among all significant results rather than the family-wise error rate. Benjamini-Hochberg Procedure sets the threshold for establishing statistical significance by ranking the *P*-values from least to largest and comparing them to a critical value determined from the intended FDR level. For a range of characteristics and across several environments, GAS + de novo GWAS fared better in a tropical rice breeding population than six other models. Based on these findings, a two-part, extended breeding design suggested that can effectively incorporate novel variation into elite breeding populations, increasing genetic diversity and raising the possibility of long-term productivity increases. They performed the multiple-test correction using false discovery rate (FDR), then bin the SNPs on each chromosome into 500 Kb bins and output the lowest *P*-value SNP in each bin. This step was performed to group SNPs into GWAS peaks—as this was a breeding population with extensive linkage disequilibrium (LD), peaks were large, often spanning ~500 Kb (Spindel et al. 2016).

6.5.7.7.3 Bonferroni Correction

This rigorous approach modifies the significance threshold (α) by dividing the number of tests (number of markers tested) by the desired alpha level (usually 0.05). Although it keeps the family-wise error rate (FWER) constant, it may be unduly prudent, particularly in GWAS when many markers are examined. A total of 3,808,730 SNPs were employed from 150 types for GWAS. The EMMAX software was used to run a mixed model. P is less than 2.63×10^{-7} ($P = 1/n$), where n is the total number of markers utilized; this represents a rough Bonferroni correction of $-\log10(P) = 6.58$ (Zhang et al. 2019).

6.5.7.7.4 Adaptive False Discovery Rate (aFDR)

This adjusts the FDR control by considering the dependency structure between genetic markers and changing the FDR strategy accordingly.

6.5.7.7.5 Storey's *q*-Value

It estimates the proportion of false positives among significant tests by providing adjusted *P*-values (*q*-values), considering the dependency structure among tests.

While reducing false positives from testing a high number of genetic markers across the genome, applying the proper multiple testing correction methods in rice GWAS studies aids in reliably detecting significant marker-trait relationships. A multiple-testing correction strategy should be selected by weighing the trade-off between maintaining statistical power and enforcing strict control over false positives. Assess the results obtained after multiple testing corrections by replicating significant relationships in separate datasets or populations. In order to validate the initial GWAS results, lower the possibility of false positives, and increase the validity of genetic variants linked to certain traits that have been found, replication and validation are essential. Transforming GWAS findings into useful applications in agriculture and other domains requires this procedure.

6.5.7.8 Functional Annotation

It entails determining and comprehending the biological significance of genetic variations linked to particular phenotypes. It seeks to connect these polymorphisms to functional regions of the genome, biological pathways, and recognized genes. It helps bridge the gap between genetic variation and phenotypic traits by providing insights into the potential functional implications of identified genetic variants.

6.5.7.8.1 SNP Annotation

A targeted crop development strategy can be facilitated, and the genetic foundation of complex traits is simpler to comprehend with the help of effective SNP annotation in rice GWAS, which offers insights into the functional implications of genetic variants linked with certain features. Two sets of SNPs (groups I and II) were derived from 15 million polymorphisms according to annotated gene locations and possible influence on gene function. Then we obtained 572,511 non-synonymous SNPs and 1,801,421 SNPs located in the 5′ sequences of genes by performing GWAS using three types of SNPs (non-synonymous SNP, non-synonymous SNPs and SNPs within promoters and three million randomly selected SNPs from the entire set of SNPs (Zhao et al. 2018).

6.5.7.8.2 Genomic Location

Use genome browsers or bioinformatics tools to identify the exact genomic location of the relevant SNPs. Determine whether the SNPs are found in intergenic regions, coding regions, regulatory elements, or introns.

6.5.7.8.3 Functional Predictions

Taking use of bioinformatics databases and tools (e.g., HaploReg, RegulomeDB, ENCODE, GTEx) that offer data on transcription factor binding sites, chromatin states, regulatory elements, functional annotations, and epigenetic alterations connected to certain genomic areas. GWAS Databases are used to learn more about SNP correlations with different phenotypes and cross-reference them with the rice-specific GWAS databases that are currently available. Regulatory element analysis finds SNPs in regulatory areas (such as enhancers and promoters) that may have an impact on the expression or regulation of a gene. SIFT and PolyPhen determine the possible effects of SNPs on the structure and functionality of proteins. High-generation recombinant inbred lines (RILs) were created using Luohui 9 (indica) X RPY geng (japonica). Linkage mapping and RTM-GWAS techniques were used to examine the salt tolerance QTLs governing final survival rates based on the high-density bin map of this RIL population. A total of seven QTLs were identified on chromosomes 3, 4, 5, 6, and 8. qST-3.1, qST-5.1, qST-6.1, and qST-6.2 were novel salt tolerance QTLs in this study and their function were functionally verified by comparative analysis of parental genotype RILs (Kong et al. 2021).

6.5.7.9 Fine Mapping and Causality Determination

Translating GWAS connections into useful information for crop improvement or comprehending the genetic basis of complex traits requires precise mapping and causality determination. The identification of causative variations facilitates the development of targeted interventions, breeding plans, and gene editing techniques to modify particular features in crops such as rice. It indicates the exact genetic variations responsible for many of the noted correlations with particular characteristics. These procedures enable focusing on the genomic areas and identifying the causal variations affecting the desired trait. Fine mapping strives to compress the essential genomic areas correlated to a trait identified by initial GWAS to smaller intervals with higher resolution by locus refinement. Linkage Disequilibrium (LD) analysis explores other candidate genes or potential causative variants by examining the LD patterns encompassing the related SNPs. This can help uncover other nearby variants that may be highly linked to the identified SNP. The conditional analysis uses the lead SNP or the most frequently encountered variants as covariates in conditional analyses, such as stepwise regression or conditional association testing. This facilitates the discovery of further independent signals in the area. Imputation and sequencing aid in identifying uncommon or low minor allele frequency variants.

Causality determination mainly focuses on variants located in regulatory elements, coding regions, or those predicted to have functional effects on genes or regulatory elements by functional annotation and validates the effects of candidate variants. Set up statistical and computational techniques (e.g., Mendelian Randomization, colocalization analysis) to deduce causality from natural experiments or the integration of genetic association data with other molecular phenotypes (e.g., gene expression) using causal inference methods.

6.5.7.10 Validation and Functional Studies

In rice GWAS, functional research and validation are crucial for confirming and clarifying the biological significance of discovered genetic variations linked to particular phenotypes. The findings from these studies foster better knowledge of the genetic mechanisms underpinning key rice agronomic features, clarify the functional functions of discovered variations, and validate their significance. Some validation strategies like cross-environment validation, independent population validation, and meta-analysis are used to detect true associations. Functional Studies are used for gene editing techniques, gene expression analysis, phenotypic validation, pathway analysis and network studies, and transgenic approaches.

6.5.7.11 Functional Inference From GWAS

It entails evaluating the functional value and biological significance of genetic variations linked to particular phenotypes. Integrate data from the literature, pathway analysis, experimental validations, and functional annotations to fully understand the biological significance of discovered mutations.

6.5.7.12 Data Sharing Through Public Database

By rendering GWAS data accessible and transparent through publicly accessible databases, researchers may do meta-analyses and comparison studies and generate a new understanding of the genetic basis of significant rice attributes. Numerous databases facilitate data exchange in uniform formats, granting the scientific community access to the data for further scrutiny and comprehension. There are a few important open sources where researchers frequently exchange GWAS information for rice such as RiceVarMap, Gramene, National Center for Biotechnology

Information (NCBI), Rice SNP-Seek Database, International Rice Information System (IRIS), The European Variation Archive (EVA).

6.6 Application of GWAS in Rice Breeding

GWAS in rice is instrumental in unraveling the genetic architecture of various traits, aiding in crop improvement strategies, and contributing to global food security by enhancing rice productivity and resilience. By assisting in the establishment of hybrid crossings and directing the genetic improvement of inbred lines, GWAS may be a significant and useful method for breeding rice. It is possible to designate sequence variants that have been found to be linked to significant agronomic traits—such as disease resistance and the buildup of heavy metals—for direct application as molecular markers in rice breeding (Table 6.1).

Some substantial applications of GWAS in rice research include

1. Crop improvement and breeding.
2. Biodiversity and germplasm evaluation.
3. Trait mapping and gene discovery.
4. Stress tolerance and adaptation.
5. Functional genomics and pathway analysis.
6. Understanding genetic architecture.
7. Data integration and meta-analyses.

6.7 Expanding Breeding Horizons: From Traditional Breeding to GAB—Precision in Progress

Traditional methods of characterizing rice varieties, involving morphological, physiochemical, and enzymatic analyses, have proven ineffective in differentiating closely related groups. Additionally, these methods are time-consuming (Kaur et al. 2015). Breeding methods such as QTL/gene pyramiding and multi-parental applications have demonstrated effectiveness in developing resistant/tolerant crop varieties (Sandhu et al. 2021). In recent years, the development of high-quality genome sequences for crops like rice (*Oryza sativa*) has provided a wealth of genomic resources that significantly enhance breeding programs. The rich genetic diversity in the rice gene pool, combined with the availability of advanced genomic tools, offers opportunities to identify suitable donor plants without undesirable linkages and dissect the quantitative nature of associated traits. Genome Assisted Breeding (GAB) empowers the integration of genomic tools with highly advanced phenotyping techniques and traditional breeding methods. This synergy leverages molecular markers, streamlining the anticipation of phenotypic traits based on genetic information (Leng et al. 2017). The advancements in GAB have not only enabled a comprehensive understanding of the allelic variations responsible for crucial agronomic traits but have also facilitated their efficient integration into germplasm

Table 6.1 GWAS success stories in rice improvement

Sl. no.	Traits	Associated genes/loci/ QTLs	Impact on rice improvement	Authors contribution
1	Resistance to the root-knot nematode	NBS-LRR, Cf2/Cf5 resistance protein	Improved resistance against nematode	Hada et al. (2020)
2	Resistance to *Magnaporthe Oryzae*	LAFBRs	Increased resistance to field blast	Zhu et al. (2016)
3	Disease resistance	GA-related genes (OsXTH8, OsGAE1, EUI, G-alpha-(D1)	Enhanced resistance against foot rot	Meng et al. (2021)
4	Early root vigor	PTR5 and PTR6	Enhanced root vigor in DSR	Anandan et al. (2022)
5	Drought tolerance	qPH10.1, qEPN6.1, qPL9.1	Increased drought tolerance	Yi et al. (2023)
6	Cold tolerance	qLTG11-1	Increased germination at low temperature	Sales et al. (2017)
7	Chlorophyll content	OsSG1	Enhanced stay green character	Zhao et al. (2019)
8	Leaf anatomical traits and vein architecture	NARROW LEAF 1 (NAL1)	Improved yield	Narawatthana et al. (2023)
9	Appearance quality	LOC_Os01g66110, LOC_ Os01g66140, LOC_ Os07g44910, and LOC_Os02g14120	Enhanced grain aroma	Sachdeva et al. (2024)
10	Salt tolerance	OsSTL1, OsSTL2	Improved salt tolerance in japonica	Yuan et al. (2020)
11	Yield-related traits	(LOC_Os09g02830 and LOC_Os07g31450)	Increased thousand-grain weight	(Zhong et al. 2021)
12	Salt tolerance	qSES1.1, qSES1.3	Enhanced salt tolerance at the seeding stage	Phan et al. (2023)
13	Grain quality	qRS-I4	Increased resistant starch, slowly digestible starch, and rapidly digestible starch	Zhang et al. (2020)
14	Cold tolerance	Os07g0585500, Os07g0585700, Os07g0585900	Improved cold tolerance at the germination stage	Yang et al. (2020)
15	Salinity tolerance	OsFBA1, OsFBL7, and mTERF	Enhanced salinity tolerance during the vegetative stage	Yadav et al. (2021a)

(continued)

Table 6.1 (continued)

Sl. no.	Traits	Associated genes/loci/ QTLs	Impact on rice improvement	Authors contribution
16	Root growth	OsIAGLU	Influenced the expression of multi hormones for root growth	Zhao et al. (2021)
17	Agronomic traits	OsNAS3	Enhanced Zn and co concentration	Cu et al. (2021)
18	Clum length	Os09g0441900	Increased dense and erect panicle	Kim et al. (2022)
19	Rice Panicle architecture	OsGRRP and LOC_Os03g03480	Increased panicle length	Bai et al. (2021)
20	Agronomic traits	Epidermal patterning factor-like protein 8 (EPFL8)/regulator of awn elongation 2 (RAE2)/grain number, grain Length and awn development 1 (GAD1)	Improved yield	Suganami et al. (2023)

enhancement and cultivar development processes. By screening diverse genotypes under various conditions and assessing critical biometric traits, such as root architecture, plant height, and yield-related characteristics, GAB allows for the identification of genotypes exhibiting improved yield, better root traits, enhanced nutrient uptake and utilization efficiency, and superior plant type traits. The synergy between traditional breeding methods and genomic technologies in GAB has significantly accelerated the development of crops with enhanced resilience, productivity, and adaptability to diverse environmental conditions, marking a pivotal advancement in modern agriculture.

6.8 Revolutionizing Traditional Breeding: Integration of Molecular Markers

The International Rice Genome Sequencing Project (IRGSP) has successfully generated and meticulously analyzed a comprehensive rice genome sequence intricately linked to the genetic map. This method of map-based sequencing has proven invaluable in pinpointing numerous genes pivotal to a variety of agronomic traits in rice. In addition to gene identification, this process has also unveiled significant information on SNPs and SSRs. These findings are poised to expedite advancements in rice production, holding promise for future improvements in agricultural practices.

Simple sequence repeats (SSRs) or microsatellites, derived from conserved noncoding regions, are abundant, hypervariable, and multiallelic, offering an excellent source of polymorphism for DNA fingerprinting and varietal identification

(Varshney et al. 2005). With applications ranging from cultivar identification to gene detection and QTL analysis, SSR markers prove indispensable, providing a versatile tool for revealing multiple alleles in the rice genome (Kaur et al. 2015).

However, most markers lacked localization in target gene regions, hampering their impact on plant breeding. In contrast, SNPs have gained popularity due to their abundance and high-throughput detection methods, revolutionizing modern breeding programs.

Single nucleotide polymorphisms (SNPs) are the most common types of genetic markers used in genome-assisted breeding. Being prevalent and easily identifiable, SNPs paved the way for haplotype studies. **Haplotypes**, groups of co-inherited SNPs, provided a more comprehensive view of genetic variations. As genomics advanced, the integration of SNP and haplotype information facilitated more precise trait identification. Taking hybrid rice as an example, haplotyping can be instrumental in identifying parental lines (maintainers and restorers) of rice hybrids before conducting test crosses to generate new hybrids. This approach demonstrates the practical application of haplotypes in genetic research and crop improvement (Katara et al. 2021).

The enhancement of complex traits required efficient breeding methods to identify and leverage previously unexplained trait variations linked to numerous small-effect QTLs. As a result, breeding methods such as genomic selection (GS), which capitalize on genome-wide marker information, gained increased relevance for ongoing population improvement and accelerated genetic gain rate (Varshney et al. 2021).

Modern breeding approaches, such as **marker-assisted selection (MAS)**, **marker-assisted backcrossing (MABC)**, **marker-assisted recurrent selection (MARS)**, and **genomic selection (GS)**, have significantly advanced the development of improved crop varieties (Mir et al. 2021). MARS involves repeated cycles of selection and recombination guided by molecular markers to improve complex traits and is particularly effective for enhancing quantitative traits through multiple rounds of selection. GS presents a novel alternative in breeding methods, enabling harnessing within-family genetic variation and predicting genetic gain using genome-wide markers. These techniques collectively form the foundation for GAB, enabling breeders to make informed selections and accelerate the development of crop varieties with targeted traits. GAB represents a paradigm shift, leveraging genomic information to enhance breeding precision and efficiency.

6.9 GAB Methods

Genomic-assisted breeding in rice involves an iterative process, integrating genomic information with conventional breeding practices, to accelerate the development of high-yielding, stress-tolerant, and nutritionally superior rice varieties (Fig. 6.1).

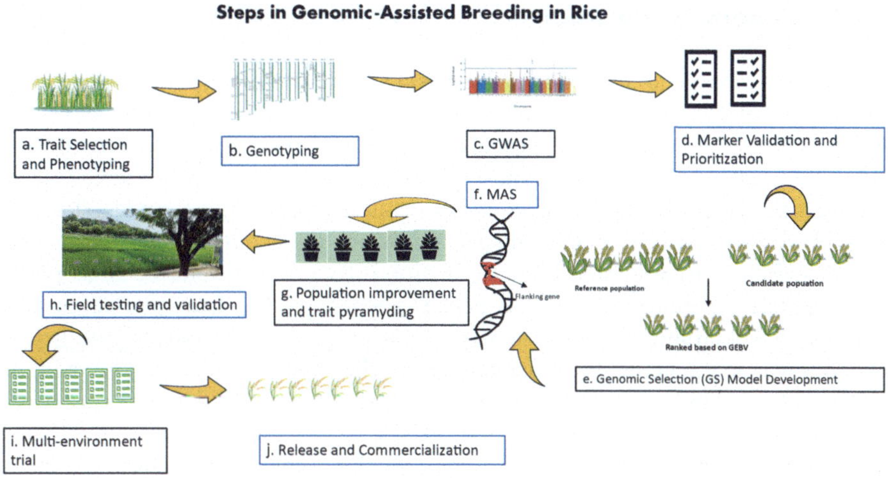

Fig. 6.1 Procedure for GAB

6.9.1 Advancements Paving the Way for GAB

Recent progress in germplasm enhancement, high-throughput genotyping and phenotyping technologies, marker-trait associations, and the exploration of natural variations, together, have set the stage for GAB in crop improvement. The key steps of GAB to comprehend how these advancements enhance breeding precision and efficiency are as follows:

6.9.2 Germplasm Enhancement

Crop germplasm, vital for genetic improvement, presents challenges in utilization despite extensive collection. Artificial methods, such as recombinant inbred lines and evolutionary breeding, are employed to create novel germplasm. Although rich, genetic diversity in natural populations poses challenges in genotyping and phenotyping. The core collection strategy (Goff et al. 2002) involves curating a minimal set of germplasms to uphold the maximum genetic diversity found in the entire collection, significantly improving the efficiency of genetic resource utilization. Traditionally, core collections were formulated based on factors like geographical distribution and pedigrees. However, with recent advancements in plant genomics, there is a paradigm shift in the establishment and utilization of crop core collections, ushering in a new era of precision and effectiveness (Kaur et al. 2015). In addition, it is crucial to recognize that epigenetic changes are enhancing germplasm diversity, and the latest technological advancements in identifying such variations offer a novel avenue for harnessing and preserving the genetic diversity present in crop germplasm through the utilization of epigenetic modifications (Khan et al. 2024).

6.9.3 Genome Acquisition

Advances in next-generation sequencing (NGS) and whole-genome resequencing are instrumental in discovering diagnostic markers in plants, crucial for developing informative genome-wide markers (Lai et al. 2010). Genome complexity reduction methods, such as genotype-by-sequencing, have played a crucial role in cost-effective genotyping of large numbers of recombinant progenies (Elshire et al. 2011). Recent long-read sequencers enhance the acquisition of end-to-end chromosome information and improve haplotype resolution, referred to as platinum standard reference genome sequences (psRfs). psRfs data acquisition is a vital step in obtaining a comprehensive view of genetic diversity at the genus level, known as the genus-level pangenome or the super-pangenome (Khan et al. 2020). The availability of psRfs facilitates the cataloging of structural variations (SVs), encompassing insertion/deletion (InDEL), copy number variation (CNV), and presence-absence variation in any crop (Varshney et al. 2021). These advancements mark significant strides in understanding and utilizing genetic diversity for crop improvement.

6.9.4 Whole-Genome Exploration and High-Throughput Phenotyping

To access genetic diversity is another essential step to associate genomic data with significant phenotypic variations. Artificial Intelligence (AI) plays a pivotal role by enhancing data processing and interpretation, facilitating the extraction of meaningful patterns from vast datasets. Additionally, remote sensing technologies, such as drones and satellites, contribute to non-invasive and timely data acquisition. Physiological breeding, integrating High-Throughput Phenotyping Platforms (HTPPs) and potentially combined with genomic selection, proves advantageous for refining breeding methodologies and introduces more precise strategies for enhancing stress tolerance and optimizing yield in crop breeding programs (Reynolds and Langridge 2016).

6.9.5 The Marker-Trait

Marker-trait association approach allows breeders to identify and select individuals with desired traits more efficiently, where genetic markers are linked to specific traits of interest in crops. By understanding the correlation between genetic markers and phenotypic traits, GAB enables the development of improved crop varieties with targeted characteristics, such as enhanced yield, resistance to diseases, or tolerance to environmental stresses. This precision in trait selection accelerates the breeding process and uses approaches like Genome-Wide Association studies, SNP and haplotype information, QTL mapping, linkage equilibrium, genomic selection, and others. Moreover, the integration of **bioinformatics** plays a pivotal role in the

Table 6.2 Success stories of incorporation of genome-assisted breeding for improvement of traits in rice

Sl. no.	Targeted trait	Associated genes/loci	Impact on rice improvement	Authors contribution
1.	Disease resistance	Pi-2 and Pi-54	Improved resistance to blast disease	Ratna Madhavi et al. (2016)
2.	Disease resistance	Xa38	Improved resistance to bacterial blight of rice	Yugander et al. (2018)
3.	Disease resistance	OsSWEET14	Improved resistance to bacterial blight of rice	Zeng et al. (2020)
4	Insect pest resistance	Bph3, Bph17	Enhanced resistance against Brown plant hopper infestation	Yadav et al. (2021b)
5	Insect pest resistance	Gm4	Enhanced resistance against gall midge infestation	Yadav et al. (2021b)
6	Salt tolerance	Saltol	Improved salinity tolerance	Nair and Shylaraj (2021)
7.	Submergence tolerance	Sub1	Enhanced submergence tolerance for growing cultivars in flood-prone areas	Nair and Shylaraj (2021)
8.	Drought tolerance	qDTY1.1	Improved drought tolerance to enhance productivity	Dhawan et al. (2021)
9.	Grain yield	$qGY_{6.1}$ and $qGY_{10.1}$	Improved grain yield	Sandhu et al. (2021)
10.	Fertility restoration	Rf3	Conversion of partial restorer to complete restorer to allow direct employment hybrid breeding	Ponnuswamy et al. (2020)

realm of genomic data analysis, providing essential tools for unraveling insights from both genomic and post-genomic data by DNA sequence analysis, marker discovery, and analyzing information. Through the synergy of computational and biological sciences, bioinformatics serves as a critical bridge, enhancing our ability to interpret and derive meaningful conclusions from the vast and complex datasets generated in genomics (Table 6.2).

6.10 Integrating Genomic-Assisted Selection (GAS) and GWAS in Rice Breeding Programs

6.10.1 Complementary Roles of GAS and GWAS

Using a range of sequence and construction approaches, several rice genomes have been produced to varying degrees of quality over the last two decades, from draft genomes to relatively close high-reference sequencing (Zaghum et al. 2022). GAS and GWAS serve as complementary tools in crop breeding, with GAS accelerating the selection process and GWAS providing insights into the genetic architecture and underlying biology of traits. Integrating these approaches maximizes their combined potential for more efficient and targeted crop improvement programs. GWAS informs breeding priorities with the help of focused breeding efforts to introduce advantageous alleles into breeding populations. GWAS data prioritize genomic areas or genes for additional functional validation. GAS models are developed and improved upon by GWAS discoveries, including marker-trait relationships and candidate genes, increasing their precision and predictive capacity. GWAS directs Marker development by developing SNP panels or markers for incorporation into GAS models with the help of marker information obtained from GWAS, guaranteeing the inclusion of markers associated with significant features.

6.10.2 Strategies for Integrating GAS and GWAS Data

Effective GAS and GWAS data integration necessitates meticulous marker selection, model building, and teamwork. By combining these methodologies, scientists and agricultural breeders can leverage the advantages of each method to enhance genomic prediction accuracy and expedite crop breeding initiatives (Fig. 6.2). Before adding markers found by GWAS to GAS models, confirm their utility and dependability by confirming their connection with the desired attributes.

6.10.3 Accelerating Genetic Gain in Rice Through Integrated Approaches

As one of the important concepts in conventional quantitative genetics and breeding, genetic gain can be defined as the increase in performance achieved annually through artificial selection (Xu et al. 2017). Accelerating genetic gain in rice

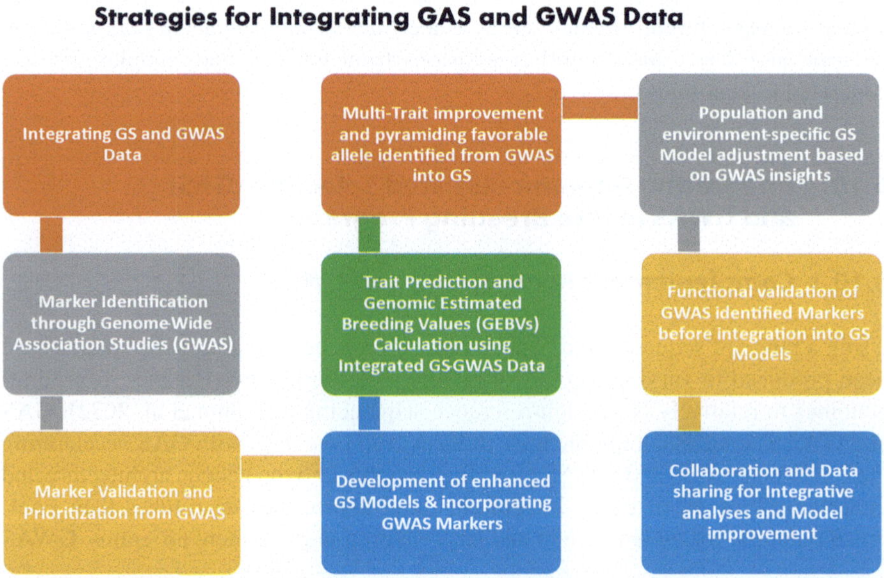

Fig. 6.2 Various strategies for integrating GAS and GWAS data

involves maximizing the rate of progress in breeding for improved varieties. Using multiple breeding techniques and technologies together, integrated approaches can greatly increase the effectiveness and pace of rice improvement. Deploy GWAS in rice breeding programs to determine more promising lines early in the breeding cycle by using genomic information to predict breeding values. Incorporate GWAS marker data or known candidate genes into GAS models to improve precision and effectively select for unique characteristics. Through the integration of these methodologies, breeding programs for rice can attain greater genetic gains in a shorter amount of time, resulting in the generation of superior rice varieties with improved quality, yield, and resilience to changing circumstances in agriculture.

6.11 Future Prospects and Challenges

Cooperation between researchers, breeders, legislators, and stakeholders is essential to overcome some obstacles and fully utilize GAS and GWAS in rice breeding. Technological developments, computational techniques, and data integration will further propel the use of these genetic methodologies to speed up rice improvement. Here are some key challenges:

- Create sophisticated GAS prediction models that include gene-gene interactions, epigenetics, and multi-omics data to improve the precision of genomic predictions.
- Provide unified platforms that combine GAS and GWAS data to enable the easy conversion of GWAS insights into GAS-assisted breeding initiatives.
- Address issues with missing heritability and complicated phenotypes by investigating rare variations and non-additive genetic effects that may go unnoticed by existing models.
- Breeders will be able to choose several traits in a single selection cycle for multiple trait prediction if GAS models are improved to predict numerous complicated features simultaneously.
- Perform GWAS and GAS research tailored to individual populations while considering a variety of rice germplasms to identify distinct alleles and genomic variants crucial for climate adaptation.
- Adopt measures to address genetic heterogeneity and population structure issues so that results can be accurately predicted and applied to a variety of rice populations.
- Functional genomics techniques are utilized to validate GWAS results to enhance trait predictions and clarify the functional significance of discovered genetic variations.
- Discuss moral issues pertaining to intellectual property rights, data privacy, and fair access to cutting-edge breeding technology and their advantages.
- Engage in bioinformatics and data analytics capacity building and computational resources to efficiently manage and understand large-scale genomic datasets.

References

Afzal M, Hindawi SES, Alghamdi SS, Migdadi HH, Khan MA, Hasnain MU, Sohaib M (2023) Potential breeding strategies for improving salt tolerance in crop plants. J Plant Growth Regul 42:3365–3387

Anandan A, Panda S, Sabarinathan S, Travis AJ, Norton GJ, Price AH (2022) Superior haplotypes for early root vigor traits in rice under dry direct seeded low nitrogen condition through genome wide association mapping. Front Plant Sci 13:911775

Bai S, Hong J, Li L, Su S, Li Z, Wang W, Zhang F, Liang W, Zhang D (2021) Dissection of the genetic basis of rice panicle architecture using a genome-wide association study. Rice 14:1–12

Biselli C, Volante A, Desiderio F, Tondelli A, Gianinetti A, Finocchiaro F, Taddei F, Gazza L, Sgrulletta D, Cattivelli L, Valè G (2019) GWAS for starch-related parameters in japonica rice (Oryza sativa L.). Plan Theory 8:292

Campbell MT, Du Q, Liu K, Brien CJ, Berger B, Zhang C, Walia H (2017) A comprehensive image-based phenomic analysis reveals the complex genetic architecture of shoot growth dynamics in rice (Oryza sativa). Plant Genome 10:2016–2007

Chen H, Xie W, He H, Yu H, Chen W, Li J, Yu R, Yao Y, Zhang W, He Y, Tang X (2014) A high-density SNP genotyping array for rice biology and molecular breeding. Mol Plant 7:541–553

Cu ST, Warnock NI, Pasuquin J, Dingkuhn M, Stangoulis J (2021) A high-resolution genome-wide association study of the grain ionome and agronomic traits in rice Oryza sativa subsp. Indica Sci Rep 11:19230

Dhawan G, Kumar A, Dwivedi P, Gopala Krishnan S, Pal M, Vinod KK, Singh AK (2021) Introgression of qDTY1.1 governing reproductive stage drought tolerance into an elite basmati rice variety "Pusa basmati 1" through marker assisted backcross breeding. Agronomy 11:202

Elshire RJ, Glaubitz JC, Sun Q, Poland JA, Kawamoto K, Buckler ES, Mitchell SE (2011) A robust, simple genotyping-by-sequencing (GBS) approach for high diversity species. PLoS One 6:e19379

Goff SA, Ricke D, Lan TH, Presting G, Wang R, Dunn M, Glazebrook J, Sessions A, Oeller P, Varma H, Hadley D (2002) A draft sequence of the rice genome (Oryza sativa L. ssp. japonica). Science 296:92–100

Guo T, Yang J, Li D, Sun K, Luo L, Xiao W, Wang J, Liu Y, Wang S, Wang H, Chen Z (2019) Integrating GWAS, QTL, mapping and RNA-seq to identify candidate genes for seed vigor in rice (Oryza sativa L.). Mol Breed 39:1–16

Hada A, Dutta TK, Singh N, Singh B, Rai V, Singh NK, Rao U (2020) A genome-wide association study in Indian wild rice accessions for resistance to the root-knot nematode Meloidogyne graminicola. PLoS One 15:e0239085

Katara JL, Parameswaran C, Devanna B, Verma RL, Anil Kumar C, Patra B, Samantaray S (2021) Genomics assisted breeding: the need and current perspective for rice improvement in India. Oryza Int J Rice 58:61–68

Kaur S, Panesar PS, Bera MB, Kaur V (2015) Simple sequence repeat markers in genetic divergence and marker-assisted selection of rice cultivars: a review. Crit Rev Food Sci Nutr 55:41–49

Khan AW, Garg V, Roorkiwal M, Golicz AA, Edwards D, Varshney RK (2020) Super-pangenome by integrating the wild side of a species for accelerated crop improvement. Trends Plant Sci 25:148–158

Khan AA, Iqbal B, Jalal A, Khan KA, Al-Andal A, Khan I, Suboktagin S, Qayum A, Elboughdiri N (2024) Advanced molecular approaches for improving crop yield and quality: a review. J Plant Growth Regul 43:2091

Kim KW, Nawade B, Nam J, Chu SH, Ha J, Park YJ (2022) Development of an inclusive 580K SNP array and its application for genomic selection and genome-wide association studies in rice. Front Plant Sci 13:1036177

Kong W, Zhang C, Zhang S, Qiang Y, Zhang Y, Zhong H, Li Y (2021) Uncovering the novel QTLs and candidate genes of salt tolerance in rice with linkage mapping, RTM-GWAS, and RNA-seq. Rice 14:1–12

Kumar K, Kaur P, Kishore A, Vikal Y, Singh K, Neelam K (2020) Recent advances in genomics-assisted breeding of brown planthopper (Nilaparvata lugens) resistance in rice (Oryza sativa). Plant Breed 139:1052–1066

Kumar S, Singh A, Singh SK (2023) Deciding selection indices using correlation and path coefficient analysis in Rice (Oryza sativa L.) genotypes. Bullet Natl Res Centre

Lai J, Li R, Xu X, Jin W, Xu M, Zhao H, Xiang Z, Song W, Ying K, Zhang M, Jiao Y, Ni P, Zhang J, Li D, Guo X, Ye K, Jian M, Wang B, Zheng H, Liang H (2010) Genome-wide patterns of genetic variation among elite maize inbred lines. Nat Genet 42:1027–1030

Leng PF, Lübberstedt T, Xu ML (2017) Genomics-assisted breeding—a revolutionary strategy for crop improvement. J Integr Agric 16:2674–2685

Liu H, Zhan J, Li J, Lu X, Liu J, Wang Y, Zhao Q, Ye G (2019) Genome-wide association study (GWAS) for mesocotyl elongation in rice (Oryza sativa L.) under multiple culture conditions. Genes 11:49

Majhi PK, Singh SK, Anandan A, Khaire AR, Korada M, Habde SV, Singh DK, Bhagvan AP, Jayasudha S, Singh A (2022) Parental polymorphism survey for evaluation and selection of contrasting parents for drought tolerance in rice (Oryza sativa L.) by using SSR markers. Int J Environ Clim 12:2507–2519

Malik P, Huang M, Neelam K, Bhatia D, Kaur R, Yadav B, Singh J, Sneller C, Singh K (2022) Genotyping-by-sequencing based investigation of population structure and genome wide association studies for seven agronomically important traits in a set of 346 Oryza rufipogon accessions. Rice 15:1–23

Meng B, Wang T, Luo Y, Xu D, Li L, Diao Y, Gao Z, Hu Z, Zheng X (2021) Genome-wide association study identified novel candidate loci/genes affecting lodging resistance in rice. Genes 12:718

Mir RR, Kumar A, Pandey MK, Isobe SN (2021) Achieving nutritional security and food safety through genomics-based breeding of crops. Front Nutr 8:638845

Nair MM, Shylaraj KS (2021) Introgression of dual abiotic stress tolerance QTLs (Saltol QTL and Sub1 gene) into Rice (Oryza sativa L.) variety Aiswarya through marker assisted backcross breeding. Physiol Mol Biol Plants 27:497–514

Narawatthana S, Phansenee Y, Thammasamisorn BO, Vejchasarn P (2023) Multi-model genome-wide association studies of leaf anatomical traits and vein architecture in rice. Front Plant Sci 14:1107718

Phan NTH, Van Pham C, Tang HT, Van Nguyen L, Nguyen LV, Bertin P (2023) Integration of genome-wide association studies reveal loci associated with salt tolerance score of rice at the seedling stage. J Appl Genet 64:603–614

Pingali PL (2012) Green revolution: impacts, limits, and the path ahead. Proc Natl Acad Sci U S A 109:12302–12308

Ponnuswamy R, Singh AK, Raman MS, Subbarao LV, Neeraja CN (2020) Conversion of partial restorer Swarna into restorer by transferring fertility restorer Rf gene (s) through marker assisted back cross breeding (MABB) in rice. Sci Rep 10:1101

Prabhu BM, Gowda HR, Abhishree R, Sheshshayee MS (2023) Arranging a set of accessions with whole genome sequence amenable for GWAS studies in rice. Mysore J Agric Sci 57(2)

Ratna Madhavi K, Rambabu R, Abhilash Kumar V, Vijay Kumar S, Aruna J, Ramesh S, Sundaram RM, Laha GS, Sheshu Madhav M, Ravindra Babu V, Prasad MS (2016) Marker assisted introgression of blast (Pi-2 and Pi-54) genes in to the genetic background of elite, bacterial blight resistant indica rice variety, improved samba Mahsuri. Euphytica 212:331–342

Reynolds M, Langridge P (2016) Physiological breeding. Curr Opin Plant Biol 31:162–171

Sachdeva S, Singh R, Maurya A, Singh VK, Singh UM, Kumar A, Singh GP (2024) Multi-model genome-wide association studies for appearance quality in rice. Front Plant Sci 14:1304388

Sales E, Viruel J, Domingo C, Marqués L (2017) Genome wide association analysis of cold tolerance at germination in temperate japonica rice (Oryza sativa L.) varieties. PLoS One 12(8):e0183416

Sandhu N, Yadav S, Catolos M, Cruz MTS, Kumar A (2021) Developing climate-resilient, direct-seeded, adapted multiple-stress-tolerant rice applying genomics-assisted breeding. Front Plant Sci 12:637488

Singh A, Rami E, Upadhyay P, Gangawane AK (2023) The impact of climate change on crop production and combat strategies. In: Climate change and sustainable development. CRC Press, pp 103–115

Spindel JE, Begum H, Akdemir D, Collard B, Redoña E, Jannink JL, McCouch S (2016) Genome-wide prediction models that incorporate de novo GWAS are a powerful new tool for tropical rice improvement. Heredity 116:395–408

Suganami M, Kojima S, Wang F, Yoshida H, Miura K, Morinaka Y, Watanabe M, Matsuda T, Yamamoto E, Matsuoka M (2023) Effective use of legacy data in a genome-wide association studies improves the credibility of quantitative trait loci detection in rice. Plant Physiol 191:1561–1573

Tanaka N, Shenton M, Kawahara Y, Kumagai M, Sakai H, Kanamori H, Yonemaru JI, Fukuoka S, Sugimoto K, Ishimoto M, Wu J (2020) Investigation of the genetic diversity of a rice core collection of Japanese landraces using whole-genome sequencing. Plant Cell Physiol 61:2087–2096

Varshney RK, Graner A, Sorrells ME (2005) Genomics-assisted breeding for crop improvement. Trends Plant Sci 10:621–630

Varshney RK, Bohra A, Yu J, Graner A, Zhang Q, Sorrells ME (2021) Designing future crops: genomics-assisted breeding comes of age. Trends Plant Sci 26:631–649

Vetriventhan M, Azevedo VC, Upadhyaya HD, Nirmalakumari A, Kane-Potaka J, Anitha S, Tonapi VA (2020) Genetic and genomic resources, and breeding for accelerating improvement of small millets: current status and future interventions. Nucleus 63:217–239

Xu Y, Li P, Zou C, Lu Y, Xie C, Zhang X, Prasanna BM, Olsen MS (2017) Enhancing genetic gain in the era of molecular breeding. J Exp Bot 68:2641–2666

Xu X, Ye J, Yang Y, Zhang M, Xu Q, Feng Y, Yuan X, Yu H, Wang Y, Yang Y, Wei X (2020) Genome-wide association study of rice rooting ability at the seedling stage. Rice 13:1–15

Yadav AK, Kumar A, Grover N, Ellur RK, Bollinedi H, Krishnan SG, Bhowmick PK, Vinod KK, Nagarajan M, Singh AK (2021a) Genome-wide association study reveals marker–trait associations for early vegetative stage salinity tolerance in rice. Plan Theory 10:559

Yadav S, Sandhu N, Dixit S, Singh VK, Catolos M, Mazumder RR, Kumar A (2021b) Genomics-assisted breeding for successful development of multiple-stress-tolerant, climate-smart rice for southern and southeastern Asia. Plant Genome 14:e20074

Yang G, Chen S, Chen L, Sun K, Huang C, Zhou D, Huang Y, Wang J, Liu Y, Wang H, Chen Z (2019) Development of a core SNP arrays based on the KASP method for molecular breeding of rice. Rice 12:1–18

Yang J, Yang M, Su L, Zhou D, Huang C, Wang H, Guo T, Chen Z (2020) Genome-wide association study reveals novel genetic loci contributing to cold tolerance at the germination stage in Indica rice. Plant Sci 301:110669

Yi Y, Hassan MA, Cheng X, Li Y, Liu H, Fang W, Zhu Q, Wang S (2023) QTL mapping and analysis for drought tolerance in rice by genome-wide association study. Front Plant Sci 14:1223782

Yu J, Hu S, Wang J, Wong GKS, Li S, Liu B, Deng Y, Dai L, Zhou Y, Zhang X, Cao M (2002) A draft sequence of the rice genome (Oryza sativa L. ssp. indica). Science 296:79–92

Yu J, Zao W, He Q, Kim TS, Park YJ (2017) Genome-wide association study and gene set analysis for understanding candidate genes involved in salt tolerance at the rice seedling stage. Mol Gen Genomics 292:1391–1403

Yuan J, Wang X, Zhao Y, Khan NU, Zhao Z, Zhang Y, Wen X, Tang F, Wang F, Li Z (2020) Genetic basis and identification of candidate genes for salt tolerance in rice by GWAS. Sci Rep 10:9958

Yugander A, Sundaram RM, Singh K, Ladhalakshmi D, Subba Rao LV, Madhav MS, Laha GS (2018) Incorporation of the novel bacterial blight resistance gene Xa38 into the genetic background of elite rice variety improved samba Mahsuri. PLoS One 13:e0198260

Zaghum MJ, Ali K, Teng S (2022) Integrated genetic and omics approaches for the regulation of nutritional activities in rice (Oryza sativa L.). Agriculture 12:1757

Zeng X, Luo Y, Vu NTQ, Shen S, Xia K, Zhang M (2020) CRISPR/Cas9-mediated mutation of OsSWEET14 in rice cv. Zhonghua11 confers resistance to Xanthomonas oryzae pv. Oryzae without yield penalty. BMC Plant Biol 20:1–11

Zhang P, Zhong K, Zhong Z, Tong H (2019) Genome-wide association study of important agronomic traits within a core collection of rice (Oryza sativa L.). BMC Plant Biol 19:1–12

Zhang N, Wang M, Fu J, Shen Y, Ding Y, Wu D, Shu X, Song W (2020) Identifying genes for resistant starch, slowly digestible starch, and rapidly digestible starch in rice using genome-wide association studies. Genes Genomics 42:1227–1238

Zhao Y, Zhao W, Jiang C, Wang X, Xiong H, Todorovska EG, Yin Z, Chen Y, Wang X, Xie J, Pan Y (2018) Genetic architecture and candidate genes for deep-sowing tolerance in rice revealed by non-syn GWAS. Front Plant Sci 9:332

Zhao Y, Qiang C, Wang X, Chen Y, Deng J, Jiang C, Sun X, Chen H, Li J, Piao W, Zhu X (2019) New alleles for chlorophyll content and stay-green traits revealed by a genome wide association study in rice (Oryza sativa). Sci Rep 9:2541

Zhao J, Yang B, Li W, Sun S, Peng L, Feng D, Li L, Di H, He Y, Wang Z (2021) A genome-wide association study reveals that the glucosyltransferase OsIAGLU regulates root growth in rice. J Exp Bot 72:1119–1134

Zhong H, Liu S, Sun T, Kong W, Deng X, Peng Z, Li Y (2021) Multi-locus genome-wide association studies for five yield-related traits in rice. BMC Plant Biol 21:364

Zhu D, Kang H, Li Z, Liu M, Zhu X, Wang Y, Wang D, Wang Z, Liu W, Wang GL (2016) A genome-wide association study of field resistance to Magnaporthe oryzae in rice. Rice 9:1–9

Haplotype-Assisted Breeding in Rice

7

U. S. Sree Vathsa Sagar, R. Pushpam, M. Raveendran, Vikas Kulkarni, and B. Rajanna

Abstract

Breeding assisted by haplotype is currently one of the promising crop improvement methods. A haplotype is a specific DNA markers combination or nucleotides that are jointly inherited from polymorphic regions, such as polymorphic Single Nucleotide Polymorphisms (SNPs), that have a low probability of current recombination between them, i.e., substantial linkage disequilibrium (LD). Rice is the most important staple food crop grown all over the world; extensive work has been done and is currently going on, with all these available huge data of genome sequence of rice, sequencing-based trait mapping data, and next-generation sequencing (NGS) technologies equipped with precise phenotyping were useful in the identification of haplotypes indirectly. Still, the expense and duration of haplotyping are increased since it necessitates certain statistical conclusions from genotypic data. On the other hand, third-generation sequencing systems, including from Pacific Biosciences (PacBio) and Oxford Nanopore Technologies (ONT), may create lengthy reads that can be used to directly iden-

U. S. Sree Vathsa Sagar (✉)
Centre for Plant Breeding and Genetics, Tamil Nadu Agricultural University, Coimbatore, Tamil Nadu, India

R. Pushpam
Department of Forage, CPBG, TNAU, Coimbatore, Tamil Nadu, India

M. Raveendran
Director of Research, Tamil Nadu Agricultural University, Coimbatore, Tamil Nadu, India

V. Kulkarni
AICRP on Sunflower, MARS, Raichur, Karnataka, India

B. Rajanna
AICRP on Groundnut, MARS, Raichur, Karnataka, India

A. Singh et al. (eds.), *Climate-Smart Rice Breeding*,
https://doi.org/10.1007/978-981-97-7098-4_7

tify haplotypes. The technique of generating haplotypes from sequencing data via estimation of haplotype is known as phasing. In recent times, numerous phasing techniques have been created. When compared to single SNP markers, the usage of these haplotypes had been shown to significantly improve the accuracy of prediction of quantitative traits that are low-heritable in rice. It is also more effective in identifying candidate genes and genomic regions regulating rice traits of interest, like many quantitative and qualitative traits of rice. This illustrates how effective gene identification and GS are. Using haplotype markers can handle complicated features more effectively than individual SNP markers. Haplotype-assisted breeding could be used as an effective advanced tool to develop genotypes that can prevail in current changing climatic conditions, which might help tackle the adverse effects of many biotic and abiotic stresses.

Keywords

Rice · Breeding · Haplotype · SNP and haplotyping

7.1 Introduction

Rice is the staple diet for major countries all over the world. Especially in Asian countries, it is the sole source of the carbohydrate. Currently, it provides food for almost 50% of the total population. The projected total amount of rice used in 2022–2023 is 522.0 million tons (FAO 2023). This data illustrates the high demand for the crop and the ongoing intensive production system. While previous researchers have found numerous genes controlling significant rice characteristics, the best haplotype combinations for creating the most desirable variety are still unknown. Numerous important genes that regulate rice grain yield and associated characteristics have had their functions characterized. Several platforms for high-throughput OMICs have been created and have helped to functionally characterize several important genes governing important rice features (Li et al. 2018). About 2296 genes have been cloned and functionally validated for grain yield (189), disease resistance (221), growth and development (513), fertility (174), nutrient-use efficiency (207), insect resistance (31), inflorescence and flowering time (276), phytohormones (472), seed quality (63), and responsiveness to stress (367) (Wing et al. 2018). Similarly, yield-attributing two genes, DEP1 and Gn1a, were also mutated through genome editing techniques. Further assessment of the genetically edited lines revealed that, compared to other naturally occurring high-yield alleles, three DEP1 mutant alleles and one mutant Gn1a allele confer a higher yield (Huang et al. 2018).

7.2 Haplotypes

A haplotype consists of allele combinations on the same chromosome that are inherited jointly, have a limited chance of contemporary recombination and have a variety of polymorphisms (Stram 2017). Putting it differently, a haplotype is characterized by a group of adjacent genetic structural changes, like polymorphic SNPs, that exhibit a high degree of linkage disequilibrium (LD) among them. As seen in Fig. 7.1, a haplotype consisting of haploid sequences in more than two inherited SNPs shows polymorphism. The trade-offs associated with traditional introgression techniques for integrating "compound" loci into disparate genetic backgrounds can be minimized by carefully assembling haplotypes.

The resequencing of several varied germplasm lines would be helpful in locating and analyzing allelic/haplotype variants to aid in the exploitation of genetic diversity. This would contribute to selecting the rare and novel alleles as donors for the important yield-contributing traits or any trait of interest in rice crop improvement (Bhat et al. 2021). Important genes might be identified to obtain allelic diversity from the rice gene bank collections. For instance, the GW2 locus' haplotype diversity was examined in 93 indica and aromatic rice germplasm samples, where four novel SNPs were identified based on the GW2 locus' resequencing data (Dixit et al. 2013). Singh et al. (2017) examined the variation in SNP haplotype in the primary gene for grain size (GS3 gene) by using a sample of 160 wild rice genotypes. Six minor haplotypes (GS3-H2, GS3-H5, GS3-H3, GS3-H6, H7GS3-H8, and GS3-) and three major haplotypes (GS3-H1, GS3-H9 and GS3-H4) were identified by the investigation.

The haplotypes are identified or assigned by using either of these three methods (Huang et al. 2007):

- Using the chromosomal segment's haplotype diversity directly.

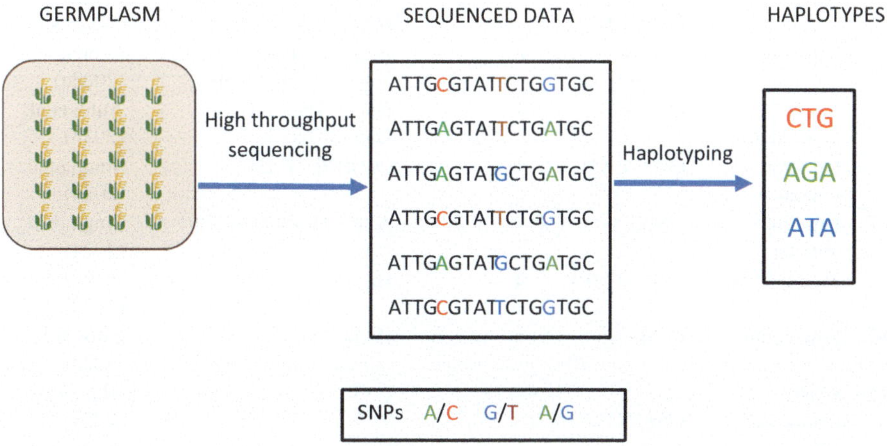

Fig. 7.1 Development of haplotypes

- The LD pair that shows no evidence of past recombination between the markers inherited together is used to calculate r^2, a measure of LD.
- By organizing SNPs into groups using sliding-windows with varying or set lengths.

It has been simpler to identify the haplotype for a given crop species because sequencing data from many sources is now available. The concept of haplotype assembly was defined by Bevan et al. (2017) through the utilization of whole genome sequencing data. Combined with phenotypic information of breeding lines and germplasm, this allows us to examine and confirm the phenotypic effects of these "component" haplotypes. Considering this idea, Abbai et al. (2019) used datasets obtained by complete-genome resequencing on a large scale in integration with haplo-pheno assessment to identify useful haplotypes for future rice breeding, while Sinha et al. (2020) used a similar methodology in pigeon pea. Haplotype mapping in numerous plant species has been accomplished by using high-density SNP information acquired from many genotypes using array-based or NGS-based techniques. As shown in Table 7.1 below, these haplotypes have also been used in rice breeding and research for various purposes.

Table 7.1 Haplotype markers for different traits for GWAS studies

SI no.	Trait	Gene/LOC ID	Haplotype group	Haplotypes	Reference
1	Salt tolerance	SKC1/ OsHKT8	11	HapE, HapK-4, HapH, HapA, HapD, HapF, HapJ	Li et al. (2023a)
2	Salt tolerance	GS3	9	HapA, HapB, HapD, HapG, HapE	Li et al. (2023a)
3	Salt tolerance	qSE3/ OsHAK21	10	HapA, HapE, HapH, HapB, HapF-1, HapF-2, HapC, HapG, HapI	Li et al. (2023a)
4	Salt tolerance	RST1/ OsARF18	7	HapC-1, HapC-3, HapA, HapB	Li et al. (2023a)
5	Salt tolerance	qSTG4/ OsPAO3	6	HapA-1, HapA-2, HapB, HapC, HapD, HapF	Li et al. (2023a)
6	Anaerobic germination	OsTPP7	18	Hap-1, Hap-2, Hap-3, Hap-4, Hap-5	Aung et al. (2023)
7	Drought tolerance	OsDREB1F	1	OsDREB1F-H1	Singh et al. (2024)
8	Drought tolerance	ASR3	91	ASR3-H1—ASR3-H91	Singh et al. (2024)
9	Zinc content	OsNRAMP7	4	Hap1	Joshi et al. (2023)
10	Zinc content	OsMT1g	4	Hap4	Joshi et al. (2023)
11	Blast resistance	*LOC_ OS12G39700*	1	H33	Alam et al. (2024)
12	Blight resistance	*LOC_ Os05g35490*	1	H49	Alam et al. (2024)

7.3 Advanced Sequencing Technologies for Easy Haplotype Identification

Rather than finding the causative mutation, association or linkage mapping uses single genetic markers or variants to ascertain their connection(s) among the trait(s) of interest. This helps to identify or clarify the effect of genetic variation in DNA sequence on plant traits. The desired phenotypes are then chosen using the DNA markers with trait association as surrogates. Genetic studies now lack sufficient information on the haplotypes governing the significant traits, making it impossible to accurately determine ancestry reconstruction, chromosome rearrangements, allele-specific expression, and selective sweep detection. On the other hand, haplotype identification and its consequences in genetic research have been greatly impacted by the availability of high-throughput sequencing technology (Bhat et al. 2016). Although second-generation sequencing methods can handle 150 bp short reads, these short reads typically only include one variation. As a result, particular statistical conclusions from the population genotyping data are needed to produce the haplotypes, which are derived indirectly from the data. Nevertheless, this adds to the time and expense of creating the haplotypes (Maestri et al. 2020; Delaneau et al. 2019). On the other hand, lengthy reads from TGS systems, such as PacBio and ONT, allow for the direct construction of haplotypes.

Platforms for long-read sequencing allow direct study of DNA molecules, which is more efficient than second-generation sequencing techniques (Amarasinghe et al. 2020). To boost the effectiveness of haplotype identification, the long-read sequencing data is adjusted using a technique called "phasing." "Phasing" is the process of using haplotype estimation by developing haplotypes from the sequence data. It is crucial to understanding sequence-specific differences, including compound heterozygosity, the impact of methylation, and specific allele expression.

7.4 Phasing Methods for Haplotype Construction

Numerous techniques for phasing long-read sequencing data to create haplotypes include (Garg 2021):

- **Reference-based phasing**, which includes polyploid, single-cell, and molecular haplotyping.
- **De novo assembly** of the genome (including polyploid and diploid assemblies).
- **Strain-resolved assembly of metagenomes** (read and contig binning, de-novo reassembly and SNP variant-based assembly).

HapCut2, WhatsHap-polyphase, HapTree, Falcon phase, SDip, Hifiasm, MetaMaps, POLYTE, ProxiMeta and DESMAN are just a few examples of the computational tools that can be used in conjunction with these haplotype analysis methods from long-read sequencing data to find rare and do novo variants with greater efficiency and precision (Garg 2021). By combining several bioinformatics tools and phasing with long-read sequencing technology in haplotype construction, we have thus been able to leverage the possibilities of these methods for sequencing completely.

7.5 Haplo-Tagging

A new method of sequencing to find haplotypes quickly. This approach involves molecular barcoding of large DNA molecules before sequencing, which preserves the connected variations and long-range information. The individual short reads are then connected by the shared barcode to create the initial haplotype (Wang et al. 2019). Larger plant and animal populations will be able to be haplotyped, and tens of thousands of samples will be sequenced and systematically used to find haplotypes in plant species that are models and non-models. However, a number of problems, such as the need for specific sequencing primers, the inefficiency of the current methods, and their limited scalability, prohibit the commercial use of haplo-tagging in genetic studies. Although the effectiveness of haplo-tagging technology in identifying haplotypes within the genome is not yet proven in plants, it has recently been proven in two species of butterflies (Meier et al. 2020).

The aforementioned results indicate a possible use for haplo-tagging in identifying the haplotype alleles controlling various phenotypes for a given trait of interest. In light of this, the haplo-tagging methodology may be a useful method for locating the alleles of superior haplotype in a variety of germplasm and plant populations for its use in breeding to create enhanced agricultural varieties. This technique will be suitable for completely realizing the haplotype-based breeding potential for rice and other plant species' crop improvement.

7.6 Functional Haplotype-GWAS (FH-GWAS)

FH-GWAS analysis is the most sophisticated technique for identifying the best haplotypes for desired attributes. Haplotypes with non-consecutive SNPs can be identified using a variety of techniques; these haplotypes will provide much more detailed data than those with consecutive SNPs (Laramie et al. 2007). Furthermore, these techniques have a significant computing burden, thus restricting their application in association research. FH-GWAS is a different and more efficient approach that had been developed to take out the superior haplotype alleles for the characteristic of interest, mitigating the drawbacks of the haplotype-based GWAS. Given the significance of epistasis in complex trait regulation, SNPs' epistatic effects are taken into account for trait identification in FH-GWAS. Therefore, in FH-GWAS, the SNPs with mid thresholds for the main effects are chosen initially. Then an identified chromosomal region of a defined size is searched for SNPs of non-consecutive combinations (epistatic effects with significance) and/or consecutive. Thus this method, only combines SNPs into a functional haplotype those which are genuinely contributing to the haplotype effects via epistatic and/or additive effects, this technique prevents duplicated SNPs (with high LD) from integrating into a haplotype. FH-GWAS is therefore more potent and effective than SNP and haplotype-based methods in this regard (Maldonado et al. 2019; Knüppel et al. 2012).

Steps involved in SNP identification using FH-GWAS are (Fig. 7.2) (Bhat et al. 2021):

- It involves the use of individual SNP-marker-based GWAS analysis.
- This makes it possible to find candidate SNPs (SNP trait association) and surrounding SNPs in a chromosomal region that are closely associated with a specified range.
- The functional haplotype is created by combining SNPs in this particular region that have real contributions to the haplotype and/or additive and/or epistatic effects.

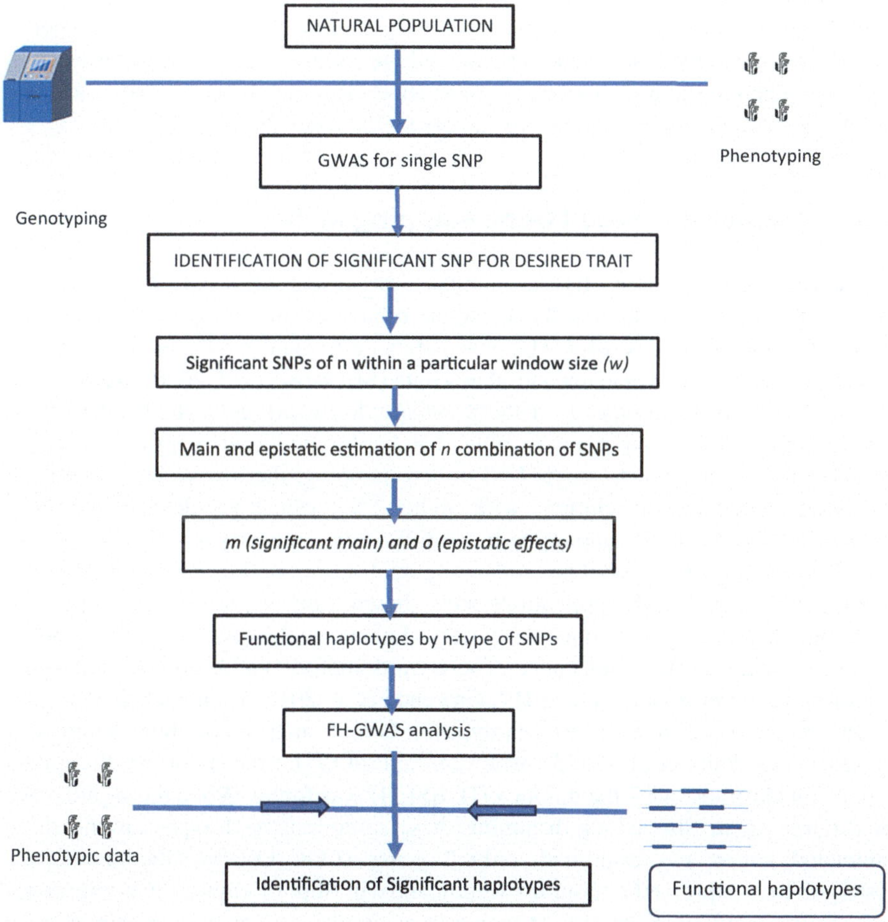

Fig. 7.2 Functional haplotype-GWAS (FH-GWAS)

- Again, the GWAS process involves integrating functional haplotypes with phenotypic data, which eventually helps identify significant haplotypes linked to the desired characteristic.

7.7 Status in Other Crops

Around 7.4 million genotypes are preserved in 1700 seed reservoirs worldwide, providing breeders with abundant naturally occurring variation (Varshney et al. 2018). Climate data and gene bank passport data may be utilized to discover genotypes harboring significant haplotypes that can be incorporated into breeding programs, acting as surrogates for abiotic stress. Novel and superior genes from the Crop Wild Relatives (CWR) and landraces stored in gene banks must be found to improve the genetic benefits made possible by adding fresh sources of genetic diversity (Gur and Zamir 2004). Numerous instances exist whereby NGS-based methods are effectively employed to pinpoint DNA variants linked to desirable characteristics. For example, there are chances to find unique variations for many genes through genotype-phenotypic associations in 3000 Rice Genome (Li et al. 2014) and the 3000 Chickpea Genome Sequencing Initiative (Varshney 2016).

7.8 Haplotype-Based Gene Mapping in Rice

Haplotypes are favored by a number of researchers while performing GWAS. Haplotype blocks have been shown to have greater mapping accuracy and power for the discovery of QTLs/genes than individual SNPs in recent GWA research based on empirical and simulated data (Srivastava et al. 2020). Reasons for the superiority of haplotypes over the SNPs are discussed under the heading 10.1, considering that it's important to note that haplotype-based mapping of maize revealed that, compared to single SNP-based mapping, there were less significant candidate genes and correlations with higher PVE values for drought tolerance (Yuan and Biswas 2019). Haplotype-based GWAS was recently used by Wang et al. (2017) to enhance rice's cooking and eating qualities as well as its protein content. Haplotype-based GWAS applications have shown a lot of promise for improving crops and discovering new traits in a variety of plant species, including Arabidopsis, soybean, wheat, barley, and maize. These traits include yield, quality, and stress tolerance (Contreras-Soto et al. 2017; Gawenda et al. 2015; Yuan et al. 2019).

In a rice collection with 391 genotypes, a GWAS analysis of shoot length was conducted by Yang et al. (2023), who reported 24 QTLs for shoot length, among which, on chromosome 1 the unique QTL qSL-1f was found. When the significance of differences in Shoot Length among these important haplotypes among these annotated genes was examined, only 2 genes (LOC_Os01g68500 and LOC_Os01g68460) expressed alterations significantly in Shoot Length. Fifty-four genes have been identified in the qSL-1f region. It is simple to identify the potential gene behind qSL-1f in addition to performing gene expression analysis. Additional knockout transgenic studies verified that LOC_Os01g68500 was the causative gene for Shoot Length, LOC_Os01g68500 will encode a protein containing a DUF538 (Domain of uncertain function), and that the DUF538 protein gene affects the development of rice seedlings. These findings demonstrated that locating the candidate gene and its favorable haplotype/allele using gene-based haplotype analysis with GWAS in conjunction is a successful strategy.

7.9 Haplotype-Based Breeding (HBB)

The breeding of higher-yield potential crop cultivars that are stress-tolerant is one of the main issues faced by breeders today. A powerful method for using haplotypes in GS has been proposed to boost prediction ability efficiency and accuracy since most crop species have complex genomes. This is because complete haplotype maps will allow for more accurate identification and use of genomic areas in populations associated with a particular trait with prominent LD structures.

Haplotype-assisted breeding can be implemented using two approaches (Bevan et al. 2017):

Retrospective approach.
Prospective approach.

7.9.1 Retrospective Approach

Plant breeders have selected the advantageous haplotypes through long-term selection that produce the desired phenotypes for the traits of interest. Therefore, by sequencing an elite gene pool using the genome resequencing approach, these beneficial haplotypes could be detected in these elite crop germplasm. It is also feasible to create molecular markers that identify beneficial haplotypes. The best desirable haplotype combination governing the specific phenotype can then be selected utilizing all of these markers. In addition, these haplotype-related markers identify lines that have distinct recombination in relevant chromosomal blocks, which can be used to discriminate between advantageous and disadvantageous genetic variation.

7.9.2 Prospective Approach

Instead of focusing exclusively on the best breeding pools, this approach enables the identification of a greater range of genetic variation in haplotypes by resequencing a substantial collection of ancestral and wild germplasm belonging to a crop species. By identifying the novel haplotypes incorporated in the natural germplasm widely, we may use genome-wide haplotypes to find new, desired, and superior haplotypes.

7.10 Haplotyping in Rice for Crop Improvement of Different Traits

7.10.1 Haplotypes Identified for Yield-Related Traits and Grain Quality in Rice

All breeders will primarily focus on yield as the most significant characteristic and grain quality is also becoming a very important attribute, especially in rice. Abbai et al. (2019), in their haplotype study of 120 functionally defined genes linked to grain yield (87 genes) and grain quality (33 genes) of rice, amply demonstrated the existence of wide variances in the 3 K RG panel. Using available datasets, the spatiotemporal and developmental stage-specific transcriptome analysis revealed intriguing findings, including the fact that *OsSPL14* had approximately 15 haplotypes, *OsSS1* had two haplogroups, and *OsMADS1* had only one haplotype among the 3 K RG panel. Additionally, the target genes' expression patterns showed positive co-expression of *OsWAK120* with *Ghd7* and *LTPL144* with *OsNAS3*. An investigation of 120 genes using candidate gene-based association found that ten important grain yield and grain quality parameters were significantly correlated with 21 genes that had been cloned previously. Crucially, superior/desired haplotypes were identified after phenotypic validation for every target traits viz., plant height (SD1-H8), number of productive tillers (MOC1-H9 & IPA1-H14), flowering time (Ghd7-H14, OsVIL3-H14 & TRX1-H9), panicle length (DEP3-H2, DEP1-H2 & SP1-H3), number of primary branches per panicle (LAX1-H5, LP-H13, OSH1-H4), grain yield per plant (PHD1-H14, AGO7-H15 & ROC5-H2), size of grains (GS5-H4 & GW2-H2), amylose content (RSR1-H8), and Fe and Zn concentration of grains (OsNAS3-H2). The efficient haplotype-based breeding method's introduction of these superior haplotypes holds promise for creating future rice varieties that will meet food and nutritional demands in the face of declining cultivable land, a rapidly growing population and changing climate conditions.

7.10.2 Haplotypes Identified for Salinity Stress

Crop growth and productivity are greatly threatened by salinity stress. Hence, it is crucial to produce salt-tolerant crops to ensure food security. Li et al. (2023b) considered five rice genes that were salt-tolerant by performing haplotype analysis. The primary haplotype groups were categorized with the help of 38 accessions with reference genomes and 4726 imputed rice. Three markers, with the exception of "SKC1 E1C," were created at position +551. Notably, certain haplotypes within the elite group HapK were the only ones with these two marker variants. The 11 markers that were created as a result of this work will be very helpful in rice breeding for marker-assisted genetic salt tolerance enhancement.

7.10.3 Haplotypes Identified for Drought Stress

Drought is one of the main factors affecting rice growth and development, causing a negative impact on rice yield. This issue may be successfully resolved by analyzing rice's drought tolerance mechanism, producing superior germplasm that was drought-tolerant, and thus discovering additional drought-tolerant genes. Wang et al. (2023) found that 15, 33, 17, and 13 SNPs were connected to plant height, plant number of panicles, number of grains per panicle, and plant yield per plant, respectively. Every gene within the QTL region has been functionally annotated, and haplotype block structure analysis was used to further screen for potential genes associated with drought resistance. Expectedly, they successfully reduced the number of candidate genes to choose from for target qualities across a wide LD region using functional annotation and haplotype analysis. Eleven known drought-tolerant genes were identified in 42 QTLs, and consequently, haplotype analysis resulted in 4 candidate genes (LOC_Os03g48890, LOC_Os04g35114, LOC_Os06g38950 and LOC_Os11g45924). Singh et al. (2024) conducted an analysis of the haplotypes of the 16 related genes in 399 genotypes to determine which of the drought-responsive 95 candidate genes' superior haplotypes. Superior haplotypes for 7 of these 16 genes, which were linked to higher single plant yield under drought conditions, were found by haplo-pheno analysis. The number of haplotypes found differed greatly. Further validation of results indicated that a superior haplotype for the DREB transcription factor (OsDREB1C) was present in all the drought-tolerant rice varieties, which could be used in haplotype-assisted breeding for drought tolerance.

7.11 Comparison Between the Haplotype and Individual Markers

Variations occurring among the polygenic traits are characterized by the presence of SNPs, insertion—changes in copy number and deletions at specific genomic regions. Nowadays, the majority of plant breeders are tagging unique genetic variations that underlie distinct phenotypes using SNP markers and into best crop cultivars these variations were introduced. On the other hand, compared to individual SNP markers, haplotype markers are better at addressing polygenic traits. For instance, using haplotypes has greatly enhanced the accuracy of low-heritable quantitative trait prediction, even when compared with individual SNP markers. Additionally, it has been shown that using haplotypes in gene mapping investigations is a more effective way to find genomic regions and potential genes that control desired attributes (Joshi et al. 2023; Garg 2021).

According to this most recent data, the haplotype-based strategy may enhance GS models' prediction capabilities as well as the accuracy with which genomic loci are found in GWAS.

7.11.1 Reasons for the Superiority of Haplotype Over SNPs

- Since all new mutations in the beginning remain at low frequency and thus a significant portion of them might perish before reaching notable frequency, it is anticipated that most SNPs in commercial chips are older mutations.
- Meuwissen et al. (2014) proposed that the accuracy of genomic predictions might be increased by constructing the association matrix with haplotypes rather than single SNPs.
- Because SNPs are biallelic, they have a low polymorphism information content (PIC) value. This limits the capacity to detect multi-allelic QTLs LD value and SNPs. However, haplotype blocks are typically "multi-allelic," which means that they could more accurately represent LD with multi-allelic QTLs than single SNPs.
- Unlike G_{SNP}, G_{HAPH} is able to distinguish between identical by state (IBS) and identical by descent (IBD). This is because long-shared haplotype blocks likely originated from common ancestors.
- Haplotype-based research is predicted to better disclose the intricate mechanism of causative haplotypes and control false positives when compared to individual SNPs.
- Haplotype variants are more common than SNPs, suggesting recombination and recurring mutation events inside and between the haplotype's genes (Qian et al. 2017).
- The visualization at SNPs locus for epistatic interactions between them can be achieved by haplotype-based analysis (Clark 2004).
- Evaluate how adjacent amino acids biologically affect a protein's structure (Clark 2004).
- Offer a more useful system for analyzing an allelic sequence present at a specific locus than a single marker (Akey et al. 2001).

7.12 Conclusion

Growing population and climate change has made us to think of increasing the production, with limited resources like area, thus there is requirement for innovative breeding methods which could be utilized for the obtaining the new improved cultivars that can tackle the current problems like climate change and this can affect the plants in many ways like it may reduce the plant efficiency as it may include high temperatures, drought, pest resurgence, disease resistance by pathogens, and many other biotic and abiotic stress. This will obviously reduce the plant productivity and plant inherent genetic potential, so either to have genotypes that can overcome these situations and yield normally or gets least affected by the environment (low G × E effect), advanced breeding tools need to be employed, especially in staple crops like rice, which is the crop important at international level as it being a source of food for a majority of people around the globe. Haplotype-assisted breeding is an efficient method for developing new genotypes for climate resilience, which reduces the effort to select genes for desired traits. Even mapping of genes can also be achieved. Haplotype is already a good method for improving traits, and it was employed to improve many traits that are favorable for climate-resilient genotypes. However, many haplotypes have been earlier mapped for very important traits in rice, which are listed in this chapter. These are currently used extensively for precise genotype selection with desirable trait combinations like drought, salinity, quality, and yield. Haplotypes are far superior to normal SNP markers in both efficiency and accuracy in mapping genes or genomic selection. Considering these aspects, although there are many haplotypes mapped for a few traits, extensive work is needed to map haplotypes in other traits of rice that will aid in selection of the genotypes and lay the foundation for further genetic improvement in rice.

References

Abbai R et al (2019) Haplotype analysis of key genes governing grain yield and quality traits across 3K RG panel reveals scope for the development of tailormade rice with enhanced genetic gains. Plant Biotechnol J 17:1612–1622

Akey J, Jin L, Xiong M (2001) Haplotypes vs single marker linkage disequilibrium tests: what do we gain? Euro J Hum Genet 9:291–300

Alam S, Sundaram KT, Singh UM, Srinivas Prasad M, Laha GS, Sinha P, Singh VK (2024) Superior haplotypes towards the development of blast and bacterial blight-resistant rice. Front Plant Sci 15:1272326

Amarasinghe SL et al (2020) Opportunities and challenges in long-read sequencing data analysis. Genome Biol 21:1–16

Aung KM, Oo WH, Maung TZ, Min MH, Nam J, Kim KW, Nawade B, Chu SH, Park YJ (2023) Genomic landscape of the OsTPP7 gene in its haplotype diversity and association with anaerobic germination tolerance in rice. Front Plant Sci 25(14):1225445

Bevan MW, Uauy C, Wulff BB, Zhou J, Krasileva K, Clark MD (2017) Genomic innovation for crop improvement. Nature 543:346

Bhat JA et al (2016) Genomic selection in the era of next generation sequencing for complex traits in plant breeding. Front Genet 7:221

Bhat JA, Yu D, Bohra A, Ganie SA, Varshney RK (2021) Features and applications of haplotypes in crop breeding. Commun Biol 4(1):1266

Clark AG (2004) The role of haplotypes in candidate gene studies. Genet Epidemiol 27:321–333

Contreras-Soto RI et al (2017) A genome-wide association study for agronomic traits in soybean using SNP markers and SNP-based haplotype analysis. PLoS One 12:e0171105

Delaneau O et al (2019) Accurate, scalable and integrative haplotype estimation. Nat Commun 10:1–10

Dixit N, Dokku P, Mithra SA, Parida S, Singh A, Singh N, Mohapatra T (2013) Haplotype structure in grain weight gene GW2 and its association with grain characteristics in rice. Euphytica 192:55–61

FAO (2023) The Food and Agricultural Organisation: www.fao.org. Accessed 29 Nov 2023

Garg S (2021) Computational methods for chromosome-scale haplotype reconstruction. Genome Biol 22:1–24

Gawenda I, Thorwarth P, Günther T, Ordon F, Schmid KJ (2015) Genome-wide association studies in elite varieties of German winter barley using single-marker and haplotype-based methods. Plant Breed 134:28–39

Gur A, Zamir D (2004) Unused genetic variation can lift yield barriers in plant breeding. PLoS Biol 2:e245

Huang BE, Amos CI, Lin DY (2007) Detecting haplotype effects in genome wide association studies. Genet Epidemiol 31:803–812

Huang L, Zhang R, Huang G, Li Y, Melaku G, Zhang S, Chen H et al (2018) Developing superior alleles of yield genes in rice by artificial mutagenesis using the CRISPR/Cas9 system. Crop J 6:475–481

Joshi G, Soe YP, Palanog A, Hore TK, Nha CT, Calayugan MI, Inabangan-Asilo MA, Amparado A, Pandey ID, Cruz PC, Hernandez JE (2023) Meta-QTL s and haplotypes for efficient zinc biofortification of rice. Plant Genome 16(4):e20315

Knüppel S et al (2012) Multi locus stepwise regression: a haplotype-based algorithm for finding genetic associations applied to atopic dermatitis. BMC Med Genet 13:8

Laramie JM, Wilk JB, DeStefano AL, Myers RH (2007) HaploBuild: an algorithm to construct non-contiguous associated haplotypes in family based genetic studies. Bioinformatics 23:2190–2192

Li JY, Wang J, Zeigler RS (2014) The 3,000 rice genomes project: new opportunities and challenges for future rice research. GigaScience 3:8

Li Y, Xiao J, Chen L, Huang X, Cheng Z, Han B, Zhang Q et al (2018) Rice functional genomics research: past decade and future. Mol Plant 11:359–380

Li P, Li Z, Yang Y (2023a) Haplotype analysis and marker development of five salt-tolerant–related genes in rice (*Oryza sativa* L.). Front Plant Sci 14:1259462

Li P, Li Z, Liu X, Zhang H, Zhang S, Liu F, Li N, Yang Y, Xie K, Ding H, Yao F (2023b) Haplotype analysis and marker development of five salt-tolerant–related genes in rice (*Oryza sativa* L.). Front Plant Sci 14:1259462

Maestri S et al (2020) A long-read sequencing approach for direct haplotype phasing in clinical settings. Int J Mol Sci 21:9177

Maldonado C, Mora F, Scapim CA, Coan M (2019) Genome-wide haplotype based association analysis of key traits of plant lodging and architecture of maize identifies major determinants for leaf angle: hap LA4. PLoS One 14(3):e0212925

Meier JI et al (2020) Haplotype tagging reveals parallel formation of hybrid races in two butterfly species. Proc Natl Acad Sci 118(25):e2015005118. bioRxiv. https://doi.org/10.1073/pnas.2015005118

Meuwissen TH, Odegard J, Andersen-Ranberg I, Grindflek E (2014) On the distance of genetic relationships and the accuracy of genomic prediction in pig breeding. Genet Sel Evol 46:1–8

Qian L et al (2017) Exploring and harnessing haplotype diversity to improve yield stability in crops. Front Plant Sci 8:1534

Singh N, Singh B, Rai V, Sidhu S, Singh AK, Singh NK (2017) Evolutionary insights based on SNP haplotypes of red pericarp, grain size and starch synthase genes in wild and cultivated rice. Front Plant Sci 8:972

Singh P, Sundaram KT, Vinukonda VP, Venkateshwarlu C, Paul PJ, Pahi B, Gurjar A, Singh UM, Kalia S, Kumar A, Singh VK (2024) Superior haplotypes of key drought-responsive genes reveal opportunities for the development of climate-resilient rice varieties. Commun Biol 7(1):89

Sinha P et al (2020) Superior haplotypes for haplotype-based breeding for drought tolerance in pigeonpea (*Cajanus cajan* L.). Plant Biotechnol J 18:2482–2490

Srivastava RK et al (2020) Genome-wide association studies and genomic selection in pearl millet: advances and prospects. Front Genet 10:1389

Stram DO (2017) Multi-SNP haplotype analysis methods for association analysis. In statistical human genetics. Methods Mol Biol 1666:485–504

Varshney RK (2016) Exciting journey of 10 years from genomes to fields and markets: some success stories of genomics assisted breeding in chickpea, pigeon pea and groundnut. Plant Sci 242:98–107

Varshney RK, Singh VK, Kumar A, Powell W, Sorrells ME (2018) Can genomics deliver climate-change ready crops? Curr Opin Plant Biol 45:205–211

Wang X et al (2017) Genome-wide and gene-based association mapping for rice eating and cooking characteristics and protein content. Sci Rep 7:1–10

Wang O et al (2019) Efficient and unique cobarcoding of second-generation sequencing reads from long DNA molecules enabling cost-effective and accurate sequencing, haplotyping, and de novo assembly. Genome Res 29:798–808

Wang N, Gao Z, Zhang W, Qian Y, Bai D, Zhao X, Bao Y, Zheng Z, Wang X, Li J et al (2023) Genome-wide association analysis reveals the gene loci of yield traits under drought stress at the rice reproductive stage. Agronomy 13(8):2096

Wing RA, Purugganan MD, Zhang Q (2018) The rice genome revolution: from an ancient grain to green super Rice. Nat Rev Genet 19:505–517

Yang T, Dong J, Zhao J, Zhang L, Zhou L, Yang W, Ma Y, Wang J, Fu H, Chen J, Li W (2023) Genome-wide association mapping combined with gene-based haplotype analysis identify a novel gene for shoot length in rice (*Oryza sativa* L.). Theor Appl Genet 136(12):251

Yuan X, Biswas S (2019) Bivariate logistic Bayesian LASSO for detecting rare haplotype association with two correlated phenotypes. Genet Epidemiol 43:996–1017

Yuan Y et al (2019) Genome-wide association mapping and genomic prediction analyses reveal the genetic architecture of grain yield and flowering time under drought and heat stress conditions in maize. Front Plant Sci 9:1919

Genomic Selection for Phenotype Prediction in Rice

8

Sakthi Anand Muthazhagu Kuppuraj,
Bharathi Raja Ramadoss, Karthikeyan Adhimoolam,
Vengadessan Vedachalam, Tamilzharasi Murugesan,
and Anandhan Tamilselvan

Abstract

Rice (*Oryza sativa* L.) is one of the world's most critical staple crops, serving as a primary source of food for the large population of the world. To meet the future demand of a growing population under the adverse effects of global warming, the need for innovative and efficient breeding strategies to enhance rice yield, quality, and resilience to biotic and abiotic stresses has become important. Genomic selection (GS) has emerged as a transformative approach in rice breeding, revolutionizing traditional plant breeding methods. This chapter provides an in-depth information of the application of genomic selection for phenotype prediction in rice breeding programs. It explains the principles, methodologies, and recent advancements in GS, highlighting its potential to accelerate rice breeding efforts. It underscores the potential of GS to revolutionize rice breeding by reducing the

S. A. Muthazhagu Kuppuraj · V. Vedachalam · T. Murugesan
Department of Plant Breeding and Genetics, Pandit Jawaharlal Nehru College of Agriculture and Research Institute (PAJANCOA & RI), Karaikal, Puducherry (U.T), India

B. R. Ramadoss
Centre for Plant Breeding and Genetics (CPBG), Tamil Nadu Agricultural University (TNAU), Coimbatore, Tamil Nadu, India

Expresstec, Junction City, KS, USA

K. Adhimoolam
Subtropical Horticulture Research Institute, Jeju National University, Jeju City, Jeju, South Korea

A. Tamilselvan (✉)
Department of Plant Breeding and Genetics, Pandit Jawaharlal Nehru College of Agriculture and Research Institute (PAJANCOA & RI), Karaikal, Puducherry (U.T), India

Centre for Plant Breeding and Genetics (CPBG), Tamil Nadu Agricultural University (TNAU), Coimbatore, Tamil Nadu, India
e-mail: anandhan@pajancoa.ac.in

A. Singh et al. (eds.), *Climate-Smart Rice Breeding*,
https://doi.org/10.1007/978-981-97-7098-4_8

breeding cycle, increasing selection accuracy, and facilitating the development of rice varieties with enhanced yield, quality, and resilience. This chapter further provides a critical assessment of the challenges and opportunities associated with implementing GS in rice breeding programs. It explores the molecular tools and genomic resources available for rice, including high-density single-nucleotide polymorphism (SNP) markers, whole-genome sequencing data, and the rapidly expanding knowledge of rice genomics. This chapter also outlines the key principles of GS, elucidating how it influences genomic information to predict the performance of rice individuals based on their genetic makeup. It discusses the various statistical models and algorithms used in GS, highlighting their strengths and limitations. Furthermore, it explores the integration of multi-omics data, such as transcriptomics and epigenomics, to enhance the accuracy of phenotype prediction in rice. Finally, it outlines prospects and research directions, emphasizing the need for collaborative efforts between plant breeders, geneticists, and bioinformaticians to fully harness the power of genomic selection in securing global rice production and food security.

Keywords

Rice breeding · Genomic selection (GS) · Phenotype prediction · Breeding cycle · SNP markers · Whole-genome sequencing · Multi-omics data · Epigenomics

8.1 Introduction

Rice (*Oryza sativa* L.) stands as a cornerstone of global agriculture, playing a pivotal role in providing sustenance for a substantial portion of the world's population. Rice ranks as the world's third most vital cereal crop, following maize and wheat. It is cultivated extensively in 115 countries with a total rice cultivation area of 166 million hectares, yielding a total production of 517.60 million tonnes (FAO 2023). In India, approximately 43% of the overall food grain production originates from rice, representing around 46% of the total cereal production in the nation. India's share in the global rice area and production amounts to about 26.7% and 23.5%, respectively (Bal et al. 2023). Population growth and changing dietary habits are driving an ever-increasing demand for rice production, placing substantial pressure on the agricultural sector. Recent years have seen a dramatic rise in the world population, raising concerns about food shortages. Soil salinization remains a critical global agricultural challenge. Soil salinity represents a prominent issue on a global scale within dry and semi-arid regions, where the limited precipitation fails to leach the salts that have accumulated in the soil, thereby threatening agricultural productivity (Singh et al. 2023). According to Mukhopadhyay et al. (2021), approximately 33% of irrigated agricultural land and 20% of all cultivated land worldwide are currently impacted by soil salinity, with projections indicating an increase by 2050. The frequency and intensity of floods and droughts have thus experienced a significant rise, whereas the duration and timing of these occurrences have displayed a greater level of unpredictability (Duan et al. 2022). This alarming trend poses grave threats to food security and crop productivity worldwide. To address the food security threats posed by climate change, researchers are working to develop flood-tolerant rice varieties. This involves leveraging traditional flood-resilient cultivars in breeding programs to boost rice yields and resilience to flooding in vulnerable regions (Michael et al. 2023; Mwakyusa et al. 2023). To meet these demands and ensure food security, there is an urgent need for innovative and efficient breeding strategies that can boost rice yield, improve quality, and enhance resilience to both biotic and abiotic stresses.

If fully integrated into a breeding strategy, genomic prediction can be a powerful tool for increasing rates of genetic gain for quantitative traits. Similar to other crops, rice is a subject of great interest for genomic prediction, with several research addressing different aspects of its use, from the more theoretical to the more practical. Genomic selection (GS) revolutionizes traditional breeding methods by leveraging genomic information to predict the genetic merit of individuals for various traits of interest in crops and livestock. This approach utilizes genetic markers distributed across the genome, such as single-nucleotide polymorphisms (SNPs), to create genomic profiles of individuals, which are then analyzed using statistical models like genomic prediction models or genomic estimated breeding values (GEBVs). By training these models on populations with both genotypic and phenotypic data, breeders can accurately predict the genetic potential of individuals for traits like yield, disease resistance, or quality characteristics. This allows breeders to identify superior candidates at an earlier stage, accelerating genetic progress (Merrick et al. 2022). Genomic selection has been widely adopted in various crops and livestock species, offering benefits such as increased selection accuracy, targeting of complex traits, and contributing to the sustainable improvement of agricultural productivity and food security. GS enables the acceleration of genetic gain by increasing selection intensity and shortening the selection cycle through genomic predictions that approximate future pure line performances in an earlier generation when a large number of entries can be genotyped. GS has emerged as a revolutionary approach in rice breeding, offering the promise of transforming traditional plant breeding methods. The foundation of GS in rice breeding hinges on a wealth of molecular tools and genomic resources that have become available over the years. High-density single-nucleotide polymorphism (HDSNP) markers, whole-genome sequencing data, and a growing understanding of rice genomics have paved the way for the practical implementation of GS. This chapter will explore the significance of these resources in facilitating the application of GS and how they have revolutionized the field. This comprehensive chapter explores the application of GS for phenotype prediction in rice breeding programs, providing a detailed exploration of its principles, methodologies, and recent advancements.

8.2 Genomic Selection: A Transformative Approach

Genomic selection (GS) is a method for plant breeding that allows the selection of lines that were not tested and phenotyped at a chosen location; however, their performance is predicted based on genomic estimated breeding values. Genomic selection (GS), introduced in 2001 by Meuwissen and his team, offers an innovative approach distinct from traditional marker-assisted selection (MAS), which is recognized as an efficient approach for optimizing breeding programs by accelerating genetic gain without necessitating significant expansions in program size, largely due to shortened breeding cycles. The efficacy of genomic prediction models that incorporate multi-environmental data, particularly in accounting for genotype-by-environment interactions (G × E), can significantly enhance predictive accuracy for complex traits in plant breeding endeavors (Nguyen et al. 2023). GS holds great promise for enhancing the genetic gain per selection within a breeding program over a given timeframe, ultimately improving breeding efficiency. In a GS breeding framework, a multitude of DNA markers spread across the genome are employed to forecast which individuals within a breeding population hold the greatest potential as parents for the subsequent generation of offspring. Building a training population set (TRS) for genomic prediction models involves phenotypic and genome-wide marker data. Since predictions are dependent on markers or individual impacts calculated on the TRS, choosing the TRS is essential for the effectiveness of GS.

Traditional breeding has been effective in achieving phenotypic selection-based genetic improvement in crops. However, recent advances in genomics have revealed a number of underlying genes and quantitative trait loci (QTLs) that confer tolerance to these particular environments and are utilized in marker-assisted selection (MAS) (Singh et al. 2022a; Majhi et al. 2022). Incorporating environmental factors and pedigree data into genomic prediction models can improve accuracy for complex traits with low heritability, as it captures important nonadditive genetic effects. This allows breeders to make more informed decisions compared to MAS, which only considers a limited number of markers (Sinha et al. 2023). Additionally, machine learning techniques like neural networks are increasingly being applied to genomic prediction problems, with ongoing efforts to develop robust cross-validation procedures to assess model performance across different prediction scenarios. These advanced statistical methods enable more precise prediction of breeding values than the typical linear models used in MAS (Sigala et al. 2023). According to Jeon et al. (2023), genomic prediction models that integrate environmental factors and genetic data, such as pedigree information, have proven more effective than traditional phenotype-based selection for improving accuracy in plant breeding. This approach enhances the ability to select for complex, low-heritability traits by capturing important nonadditive genetic effects.

Genomic selection has also aided in the enhancement of rice by helping to choose parents for hybridization based on breeding qualities. Plant breeders aiming to enhance the accuracy of phenomic selection (PS) must carefully weigh the costs associated with adding more phenotype trials to their program. In this evaluation, it becomes crucial to determine whether the expense of conducting additional phenotype trials exceeds the cost of essential genotyping required for implementing GS. According to Sinha et al. (2023), molecular markers are essential for determining an individual's genetic value by capturing the effects of diverse trait loci across the genome. Research has emphasized the benefits of using dense markers to improve the accuracy and efficiency of genomic selection in breeding programs.

High-density markers ensure more comprehensive coverage of trait loci, increasing the chances of identifying individuals with superior genetic potential for desired traits (Bohar et al. 2020). By incorporating these dense markers into genomic selection, breeders can make more informed decisions about trait selection and genetic improvement in plant breeding. Alternatively, genotyping should substantially contribute to the program's genetic gain by improving accuracy, surpassing the advantages gained from additional phenotyping. GS is particularly important in specific circumstances. Firstly, it is most advantageous when there is a limited availability of seeds. Additionally, GS proves especially beneficial in situations where the cost associated with phenotyping is significantly high. Moreover, its effectiveness is heightened when GS attains exceptionally high accuracy in predicting traits. Lastly, GS is particularly valuable when there is a feasibility of intensifying the selection process substantially. This intensification can be accomplished through the implementation of rapid generation advancement techniques.

8.3 Factors Influencing the Rice Genomic Selection Efficiency

The rapid advancements in GS methodologies over the past 3 years have underscored its effectiveness in enhancing breeding programs and boosting genetic gains across various crops, including rice (Montesinos-López et al. 2023). The continuous evolution and successful application of GS techniques in breeding strategies have firmly established it as a valuable tool for modern plant breeding, offering substantial benefits in terms of increased genetic gain and selection efficiency (Anilkumar et al. 2022). Along with trait heritability and the amount of time needed to complete the selection cycle, the selection response is the most important variable in the breeder's equation; raising the selection response can enhance the genetic gain of the desired trait. To optimize the advantages of a genomic selection-assisted breeding program, it is essential to take into account all the factors that impact the effectiveness of genomic selection (Table 8.1).

8.4 Molecular Tools and Genomic Resources for Rice

Global rice breeding programs have led to the development of various genomic resources, including marker systems like restriction fragment length polymorphism (RFLP), amplified fragment length polymorphism (AFLP), and SNP markers. These markers have greatly aided functional genomic research due to the availability of sequence information and the use of next-generation sequencing (NGS) platforms and bioinformatics tools. This has expedited the identification of genes and their impact on traits. One important breeding tool is MAS, which assists in tracking and tagging markers to genes of interest. While MAS has successfully integrated major genes, or QTLs, controlling key rice traits, it has limitations in pinpointing combinations of genes with minor effects that give rise to new traits. According to

Table 8.1 Factors influencing genomic selection in rice

Component	Contributors
Population	1. Population size, selection proportion, and selection method are crucial factors in breeding programs 2. Novel germplasm introduction is an important aspect of diversifying the genetic pool 3. Population structure plays a key role in shaping breeding strategies 4. Population type is a fundamental consideration in breeding programs
Trait heritability	1. Heritability of traits and its method of estimation, number of genes involved in the trait's regulation, field management, interactions with the environment, selective mating design, and phenotyping accuracy
GS model	1. A single-model or multi-model approach is a critical decision in the analysis 2. Model used is an essential factor in genetic effect estimation 3. Accuracy of model estimates is a key consideration in the analysis 4. Genetic effects are estimated as part of the modeling process
Markers used	1. Target genes, marker density, marker distribution, marker types, and linkage disequilibrium (LD) between markers and quantitative trait loci (QTL)

Fujino et al. (2019), the limitation of MAS is the difficulty in elucidating the gene combinations that produce novel phenotypes in rice. Seck et al. (2023) highlighted the need for a clear separation between pre-breeding and breeding stages in rice programs. Pre-breeding should focus on introgressing favorable alleles into elite germplasm, while breeding activities should implement population improvement through recurrent selection to increase the frequency of favorable alleles for quantitative traits. Additionally, Merrick et al. (2022) reviewed the importance of leveraging phenotypic and genotypic data, identifying optimized training populations, and utilizing genomic selection models that incorporate multi-trait, genotype-by-environment, and multi-trait, multi-environment data to increase prediction accuracy in plant breeding programs.

SNP-based markers have emerged as a valuable alternative, enabling the selection of high-performing offspring based on breeding values and streamlining the breeding process. Advanced high-throughput DNA marker-based technologies and modern statistical or data science approaches have driven the emergence of genome selection in rice. With these methods, genetic factors linked to complicated quantitative traits can be studied in well-defined statistical models, leading to accurate model effects. GS can be viewed as an evolved version of MAS, employing genome-wide dense DNA markers to predict genetic values with high precision, facilitating subsequent selection. While there have been significant measures in the field of rice molecular genetics and genomics, newer and potentially quicker breeding methods are under development. The molecular tools and resources have revolutionized rice research by enabling scientists to study the rice genome, identify genes associated with specific traits, and develop improved rice varieties through techniques like genome selection. This wealth of information has accelerated the pace of rice breeding and contributed to the development of high-yielding, disease-resistant, and nutritionally enhanced rice varieties to meet global food demand. It is particularly well-suited for quantitative traits influenced by numerous genes with small individual effects.

8.5 Genomic Selection for Hybrid Genotypes

According to Bartholomé et al. (2022), researchers have explored the relationship between prediction ability and prediction accuracy in genomic selection studies. The importance of understanding the factors that influence the accuracy of genomic prediction models, such as the size of the training population, the number of markers used, and the relatedness between training and validation populations, is recommended. Recent research has emphasized the significance of optimizing training population design and leveraging high-density molecular markers to enhance prediction accuracy in GS for various traits, like growth and yield in crops such as rice (Dekkers et al. 2021). The correlation between genomic predictions and observed phenotypes expresses prediction ability. It is more important to measure the effects of GS-based indirect selection than to predict the genotypic correlation based on prediction accuracy. Predictions are generated for a group of individuals that have been genotyped but not phenotyped as potential candidates for selection. Werner et al. (2020) highlighted the importance of using appropriate validation strategies to accurately assess the predictive ability of GS models in plant breeding. Overestimation of predictive ability can occur when the same genotypes used to estimate marker effects are also used for prediction, even if different phenotypes are used. Prediction accuracy can be influenced by factors such as population structure, family relationships, and the composition of the training and validation sets. According to Selga et al. (2022), random cross-validation, while useful for model comparison, may not accurately represent the prediction accuracy expected in a practical breeding program; it is suggested to use cross-validation strategies that mimic the actual breeding scenario, such as predicting the performance of individuals from earlier selection cycles using a training population from later cycles.

8.6 Effect of Trait Genetic Architecture on Genomic Selection and Association Mapping in Rice

Genome-wide association studies (GWAS) have been extensively utilized to identify genes or quantitative trait loci (QTLs) associated with important breeding traits like yield, flowering time, and plant height, particularly in rice. Biswas et al. (2023) applied GS to develop improved irrigated rice varieties for favorable ecosystems in Bangladesh and found that the accuracy of GS varied depending on the trait, population, and prediction models used.

While obtaining genomic data has become high-throughput and low-cost, collecting phenotypic data still requires significant effort and time. In past breeding programs, researchers and breeders have conducted extensive phenotypic surveys and accumulated the results as legacy data (Suganami et al. 2023). In rice, MAS has been effectively applied to introduce major genes and/or QTLs with significant effects that control abiotic stresses like submergence, salinity, and drought into new rice varieties. The sequencing data from advanced technologies are now being combined with high-throughput phenotyping (HTP), enabling GWAS and SNP identification within complex genetic architectures (Uffelmann et al. 2021).

Rice, which is frequently bred in massive, interconnected pedigree schemes and is prone to profound population structure, will likely face similar restrictions. Furthermore, rice frequently combines with other related species, resulting in many cultivars having introgression from other subpopulations. Accounting for subpopulation structures can produce more realistic estimates of prediction accuracy, avoiding the inflated results that arise when population structure is not properly addressed. Genomic prediction model accuracy is influenced by key factors, such as training population size, marker count, and relatedness between training and validation sets. Understanding these factors is essential for optimizing genomic prediction models and enhancing breeding efficiency.

A study conducted in 2020 by Syngenta prompted participants to forecast the yield performance of potential breeding crosses by utilizing historical data. The study involved the amalgamation of random forest and optimization models for the detection of G × E interaction, resulting in a higher level of accuracy in comparison to other contemporary models. The utilization of genotype to phenotype models is a common practice for analyzing sparsely collected phenotypic traits such as plant height and germination rate. These traits are typically manually gathered, which could potentially introduce biases and escalate experimental expenditures. Conversely, the adoption of high-throughput phenotyping enables the simultaneous monitoring of hundreds or even thousands of plants in both field and greenhouse settings. This approach facilitates nondestructive measurements of traits like biomass and plant height (Danilevicz et al. 2022). However, making predictions within a particular subpopulation revealed minimal impact of the structure on prediction accuracy. Yet rice breeding using GS and GWAS indicates that GS can produce more precise predictions of breeding line performance than pedigree data alone and that GWAS results may impact GS results.

8.7 Effect of Prediction Methods on Predictive Accuracy

The selection techniques used to identify the genetic components controlling the expression and distribution of phenotypes serve as a foundation for genomic selection techniques. Prior to implementing a specific GS model in a breeding program, it is crucial to assess the model's efficacy for the traits being studied in order to ensure a better fit of the model and maximize genetic gain. Rice, being a prominent crop model for genetic manipulation, gene replication, and functional examination, has offered profound insights into various aspects such as phytohormones, nutrient utilization efficiency, responses to environmental stresses, defense mechanisms, signaling in biological interactions, flowering influenced by light duration, as well as the regulation of reproductive capabilities (Chen et al. 2022). The effectiveness of predictive models based on genomics that integrate data from multiple environments, especially in addressing genotype–environment interactions (G × E), is crucial for improving the accuracy of predictions for intricate traits in the field of plant breeding. Schrauf et al. (2020) demonstrated the significance of marker density in determining the superiority of models incorporating interactions among markers. Enhancing the precision of genomic prediction models, with a specific focus on managing G × E interactions, is a key area of interest in current research endeavors. However, using the models that capture the dominance and epistasis effects is recommended for GS-assisted hybrid rice breeding programs. Validating predictability is essential to demonstrating the true potential of GS in plant breeding.

The precision of the assessments is a crucial factor influenced by various aspects, including the demographic attributes of the population for enhancement (such as the number of breeding individuals and the turnover rate), the genomic composition (in terms of allelic diversity and the extent of linkage disequilibrium), and the genetic architecture of the specific traits being targeted (considering the distribution of gene effects and heritability). For instance, a study on the genomic prediction of resistance to brown planthopper (BPH) in rice reported an accuracy ranging from 0.32 to 0.51 depending on the genetic relatedness between the training and testing populations and the statistical model used (Zhou et al. 2024). Another study evaluating multi-environment genomic selection in elite rice breeding lines achieved prediction accuracies of 0.38–0.49 for grain yield across different environments (Bhandari et al. 2019).

8.8 Accuracy Predictions and Advancements in Genomic Selection for Rice Breeding

Individual genotype acquires genomic estimated breeding values (GEBV) through five iterations of the procedure, with each fold acting as the validation fold once. The primary metric for assessing accuracy will be the mean Pearson correlation between the predicted GEBV and the observed phenotypic data in the validation population. GS efforts were performed for various traits in rice using different statistical models, software packages, and next-generation sequencing (NGS) marker genotyping platforms. The list of software used for GS in breeding programs is presented in Table 8.2. Emerging sequencing methodologies have transformed the landscape of genomic research in rice, facilitating the construction of top-notch rice genomes and the creation of tools for functional genomic research. The most distinctly employed statistical approaches encompassed four genomic selection methods: ridge regression best linear unbiased prediction (RR-BLUP), Bayesian lasso (BL), reproducing kernel Hilbert spaces (RKHS), and random forest (RF), and two non-genomic selection methods: multiple linear regression (MLR) and pedigree best linear unbiased prediction (PED). Gu et al. (2022) suggested that the selection of these four genomic methods will be based on their proven effectiveness in

Table 8.2 The list of software used for GS in breeding programs

Software	Description	Utilized for	Reference
rrBLUP	Ridge regression best linear unbiased prediction	Statistical method for genomic prediction incorporating marker information	Xu et al. (2021)
BLR	Bayesian linear regression	Utilizes Bayesian statistical methods to provide a framework for modeling and prediction	Zheng and Zhang (2023)
BGLR	Bayesian generalized linear regression	Employs Bayesian methods for generalized linear regression, allowing for flexible modeling of relationships among variables	Asai et al. (2021)
solGS	Web-based tool for genomic selection	Provides users with tools for marker-assisted breeding and prediction of phenotypic traits	Budhlakoti et al. (2022)
GVCBLUP	Genomic prediction of additive and dominance effects	Method for genomic prediction that accounts for both additive and dominance genetic effects in breeding value estimation	Budhlakoti et al. (2022)
GenoMatrix	Pedigree-based genomic prediction	Utilizes pedigree information in combination with genomic data for predicting breeding values, enhancing the accuracy of genomic selection	Tong and Nikoloski (2021)
ShinyGPAS	Interactive genomic prediction	An interactive tool for genomic prediction, offering users a user-friendly interface and interactive features for conducting genomic selection analysis	Thorwarth (2019)
GVCHAP	Genomic prediction using haplotypes and SNPs	Utilizes haplotype and single-nucleotide polymorphism (SNP) data for genomic prediction, enhancing the accuracy of breeding value estimation	Prakapenka et al. (2020)
SeqBreed	Python tool for genomic prediction (extended version)	Developed in Python for conducting genomic prediction analysis, providing users with a flexible and customizable framework for genomic selection studies	Pérez-Enciso et al. (2020)
BWGS	Genomic selection in wheat breeding program	Specifically designed for GS in wheat breeding programs, offering tailored features and analysis options for wheat genetics research	Charmet et al. (2020)

accurately forecasting genomic estimated breeding values (GEBV) across various crop types. The utilization of genetic data, encompassing high-throughput genotyping tools, bulk segregant analysis, and genome-wide association studies, has hastened marker-assisted selection, genomic selection, and genome editing in rice breeding programs.

8.9 G × E Interaction vs. GS in Rice Breeding

In recent times, there has been a development of multiple models aimed at addressing genotype–environment interaction (GEI). Variations in how genes are expressed in different environments contribute to GEI, leading to the manifestation of intricate characteristics. The G × E interaction in rice breeding refers to how genotypes respond differently to various environments, impacting traits differently. On the other hand, GS involves using genomic information to predict the performance of genotypes in different environments. According to Yang et al. (2024), wild relatives of rice are a valuable genetic resource for genomics and breeding research. Recent studies have identified promising yield QTLs in wild species like *Oryza rufipogon*, *Oryza nivara*, and landraces. Integrating cutting-edge technologies like multi-omics, gene editing, and high-throughput phenotyping can enable the next breakthrough in rice yield. Strategies to bring out hidden genes from distant rice gene pools and induce variation are crucial. Integrating G × E modeling in genomic prediction enhances predictive abilities, especially for untested lines in diverse environments. However, predicting performance in new environments, particularly for less heritable traits like yield, remains challenging. While incorporating environmental covariates can improve accuracy, it may not always yield better predictions across all environments. These findings highlight the importance of G × E interactions and the effective use of genomic tools to optimize rice breeding strategies. Recent advances such as molecular marker technology, MAS, and genomic selection have facilitated the discovery and stacking of yield genes from wild species (Mehandi et al. 2021; Siddiq and Vemireddy 2021). Yang et al. (2024) highlighted the significant contributions of genome sequencing, functional gene mining, and genomics-based innovations such as genome-wide molecular navigation breeding, molecular module design breeding, and genomic selection. Additionally, metabolomics research has provided crucial insights into the metabolic products of rice, shedding light on its adaptive responses and environmental resilience (Cerioli et al. 2022). Furthermore, Iqbal et al. (2021) highlighted the growing use of multi-omics analysis to explore molecular mechanisms in rice, showing increased integration of various omics data for a comprehensive understanding of its biological processes.

8.10 Strategies to Optimize Genomic Selection in Rice Breeding

Researchers repeatedly move many genes responsible for disease and insect resistance between different varietal backgrounds. A significant portion of these genes exhibits dominant or recessive traits, making the transfer process time-consuming. The integration of molecular markers can streamline this process, resulting in both time and cost savings when transferring these genes across different varietal backgrounds. Molecular markers closely linked to the target gene can essentially serve as "tags," enabling the indirect selection of the desired gene at an early stage.

In the context of rice breeding, the integration of MAS with genomic selection offers several notable advantages. It significantly enhances the efficiency of breeding programs by expediting the identification and transfer of specific genes, contributing to more rapid crop improvement. This combination also results in cost reduction, as the use of molecular markers streamlines the gene transfer process, conserving valuable resources and ultimately reducing expenses. Molecular markers enable precise selection of target genes, resulting in more reliable breeding outcomes.

However, there are some noteworthy disadvantages to consider. Implementing MAS with genomic selection may require an initial investment in infrastructure and training, which can pose a barrier for certain breeding programs, particularly those with limited resources. The effectiveness of MAS is contingent on the linkage between the molecular marker and the target gene, and this linkage may not always be as tight as desired, impacting the accuracy of selection. Additionally, MAS is most effective for monogenic traits, and its utility may be limited for complex, polygenic traits where multiple genes are involved.

Epistatic interactions play a crucial role in understanding how plants, including rice, respond to environmental stressors and how they grow and produce grain. These interactions involve the combined effects of multiple genes and can greatly influence the expression of traits. While conventional breeding and MAS have been used to improve these traits, the projected outcomes have struggled to keep up with the increasing demands of the rising human population. To address this challenge, researchers have been exploring alternative approaches. Nguyen et al. (2023) employed a combination of genotypic and phenotypic selection methods to identify new rice lines with improved traits. This study showcased the potential to enhance rice breeding outcomes by considering both genetic markers and observed phenotypic characteristics. Cerioli et al. (2022) analyzed the genetic diversity and trait associations in elite rice genotypes across multiple locations. Their findings provided insights into the relationships among genes, traits, and environments, which can further inform breeding strategies. GS has shown promise in enhancing the efficiency and effectiveness of rice breeding efforts.

Overall, the exploration of epistatic interactions, the integration of genotypic and phenotypic selection, and the application of genomic selection techniques collectively contribute to advancing rice breeding by improving predictions and selecting desired traits. Without needing to compile phenotype data for these individuals, the GS may reliably choose individuals with better breeding values and perhaps correct all genetic variance. This has made it possible for the breeding cycle to be shortened and for early-generation breeding material to be selected and intercrossed quickly. Since the advent of GS, the cost of genotyping has decreased significantly, but the cost of phenotyping is rising as a result of increasing labor and land-use costs, which has improved the usefulness of GS in crop improvement. Recent research indicates that GS has the potential to revolutionize crop breeding, with many authors concluding that GS can achieve an annual genetic gain significantly higher than that of conventional breeding methods. This will hasten the supply of crop varieties with agronomically improved traits like enhanced yield, quality, and biotic and abiotic stress tolerance, as well as broaden the genetic evaluation of germplasm in crop development programs (Table 8.3).

Table 8.3 Application of GS in identifying agronomically useful traits in rice

S. no.	Statistical methodology	Target trait	Plant material	Reference
1.	**BLUP**: Best linear unbiased prediction			
(i)	2D BLUP-HAT (BLUP with haplotype allelic effects)	Yield, grain weight, and grain number	210 F_9 recombinant inbred lines (RILs)	Wang et al. (2020)
2.	**GBLUP**: Genomic best linear unbiased prediction			
(i)	GBLUP	Yield and related traits	Combinations of 1495 hybrid rice	Cui et al. (2020)
(ii)				
(iii)				
(iv)		Plant height and protein content	413 accessions in a structured diversity panel	Yu et al. (2022)
3.	**LASSO**: Least absolute shrinkage and selection operator			
(i)	LASSO and GBLUP	Yield, tillers, and grain number	240 RILs and 360 derived crosses	Xu et al. (2021)
4.	**EBL**: Empirical best linear unbiased			
(i)	EBL, GBLUP C, and CB methods (R software)	Days to heading (DTH)	Untested genotypes (under complex environments)	Jarquin et al. (2020)
5.	**Combined analysis**			
(i)	GBLUP, BayesA, and BayesC	Rice blast (0.15–0.72) *prediction accuracy enclosed in ()	161 African accessions and 162 USDA accessions	Huang et al. (2019)
(ii)		Protein and nutritional content	Rice *indica* accessions	Wang et al. (2020)
(iii)	GBLUP and LASSO	Grain yield (0.54), grain number (0.62), plant height (0.58), 1000 grain weight (0.54) *prediction accuracy enclosed in ()	1495 hybrids derived from incomplete NC II design and 100 hybrids derived from half diallel crosses	Cui et al. (2020)
(iv)	GBLUP and RKHS (reproducing kernel Hilbert space)	Arsenic content in flag leaf and cargo grain in rice	225 accessions and 95 elite lines	Frouin et al. (2019)
(v)	RRBLUP GBS	Grain yield	Tropical rice breeding lines	

8.11 Important Findings for Genomic Prediction in Rice

The sequencing and resequencing of the rice genome have significantly deepened our understanding and identification of functional genes, greatly benefiting breeding programs (Yang et al. 2024). GWAS have significantly contributed to genomic selection by improving our understanding of crucial yield-related traits in rice. Sachdeva et al. (2024) utilized both single-locus and multi-locus models to identify numerous quantitative trait nucleotides (QTNs) and candidate genes associated with key agronomic traits like plant height, days to flowering, and plot yield. Similarly, Sanchez et al. (2023) explored the phenotypic variation in head rice and chalky grain percentages across a diverse array of rice accessions. They used GWAS to identify SNP markers linked to these traits and pinpointed potential candidate genes related to the discovered SNPs. These findings demonstrate the power of GWAS in identifying genetic markers and candidate genes that are essential for GS, enabling more effective breeding programs aimed at improving rice yield and quality traits. These discoveries are instrumental for genomics-assisted breeding. Recent advancements in functional genomics, proteomics, and metabolomics are being integrated with genomic data to unravel the molecular foundations of complex traits like yield. This integrated approach is instrumental in developing new high-yielding rice varieties. Additionally, machine learning models that incorporate genetic, environmental, and historical data have demonstrated potential in forecasting the performance of both modern and traditional rice varieties under future climate scenarios, aiding in the formulation of effective breeding strategies. Modern genomics-based breeding techniques, including genome-wide molecular navigation breeding, molecular module design breeding, and genomic selection, are increasingly being adopted to expedite genetic improvements in rice (Singh et al. 2024). These approaches leverage genomic information more efficiently than traditional breeding methods, offering a promising path forward for rice cultivation (Table 8.4).

Table 8.4 Key advancements and findings in rice genomic selection

Finding	Importance	Example	Relevant citation
Expanded rice genome sequencing	Advancements in rice genome sequencing and resequencing have greatly improved our understanding and revealed numerous functional genes	OsSEH1 gene regulates the phenylpropane pathway in response to cold stress	Bartholomé et al. (2022), Gu et al. (2023)
Genome-wide association studies (GWAS)	GWAS, employing single-locus and multi-locus models, have pinpointed many quantitative trait nucleotides (QTNs) and candidate genes linked to vital yield-related traits	Numerous loci associated with antibiosis to BPH have been identified from a study of 1520 rice varieties	Zhou et al. (2024)
Genomics-based breeding approaches	Genomic methods such as molecular navigation, module design breeding, and genomic selection are speeding up rice genetic improvement	Resequencing of 33 wild rice species identified different alleles of the OsCENH3 gene	Kaur et al. (2023)
Integration of multi-omics	Functional genomics, proteomics, and metabolomics are merging with genomics to enhance our grasp of complex traits	Discovery of the multi-effect gene DEP1, the low-temperature-responsive gene COLD1, and the high-temperature-responsive gene TT1	Gu et al. (2022), Hu et al. (2021)
Climate-resilient rice varieties	Identifying superior haplotypes of key drought-responsive genes has opened new opportunities for creating climate-resilient rice varieties	PAPH1 plays a crucial role in helping rice manage drought stress	Bing et al. (2022)

8.12 Challenges and Opportunities in GS Implementation

With today's sequencing advancement, we can gather whole-genome sequences for a wide range of genotypes. The 3000 Rice Genomes Project (2014) has achieved considerable advancements in the sequencing of rice genomes. Recent research indicates that this project has effectively sequenced more than 3000 rice accessions, establishing a comprehensive database for rice breeding and scientific investigation (Bernal-Gallardo and de Folter 2024). The lack of assessment of genetic gain achieved by hybrid rice breeding programs presents a significant challenge in evaluating the efficiency of breeding strategies. Quantifying the progress made through these programs, particularly based on quantitative traits, is essential for understanding the impact of breeding efforts. One approach to evaluating genetic gain in breeding programs involves assessing progress based on quantitative traits and the need to establish appropriate marker sets and reliable, cost-effective genotyping platforms for target germplasm. This is crucial for accurately measuring genetic gain and informing breeding decisions. Recent developments have focused on optimizing GS in hybrid crops to maximize genetic gain. Yang et al. (2024) have developed optimal designs for GS in hybrid crops, demonstrating the predictive ability of GS with accuracies ranging from 0.13 to 0.78 for key traits across different market classes and yield trials. Despite the demonstrated potential, the widespread implementation of GS in rice breeding programs has been limited thus far, as highlighted by Cerioli et al. (2022). However, integrating GS with other cutting-edge technologies such as gene editing and high-throughput phenotyping can further enhance the effectiveness of rice breeding programs. As genotyping costs continue to decrease, GS is poised to become an increasingly important tool for rice breeders to meet the growing global demand for this staple crop. Through the integration of GS and other innovative technologies, breeders can make significant strides in improving the genetic gain of hybrid rice breeding programs and addressing the challenges associated with meeting the demands for high-yielding and resilient rice varieties. The GBLUP method, which presumes uniform genetic variance among all markers and utilizes the genomic relationship matrix to forecast phenotypes without estimating marker effects, proves to be robust and efficient, particularly for highly polygenic traits (Xu et al. 2021). Furthermore, GBLUP is used to evaluate the significance of markers. However, the increased quantity of genome-wide markers may pose a problem for genomic selection based on whole-genome sequencing.

8.13 Enhancing Current Rice Breeding Initiatives with Genomic Selection

Genomics-assisted breeding programs should therefore aim to maximize selection gain per unit of time compared to other breeding strategies, mainly phenotypic selection (PS) or MAS. Rice, being a crop that is strictly self-pollinated, relies on a variety of male sterility systems for commercial hybrid breeding. Within a male sterility system, a male sterile line is unable to produce seeds through self-fertilization, making it suitable as the female progenitor for creating hybrids by pollination from a normally fertile line. It has been a standard practice to generate new hybrids through test crosses involving one or a few male sterile lines and multiple superior rice varieties or lines. The most successful hybrids are typically determined based on their performance in experimental fields. The expense of phenotyping is generally high in hybrid breeding due to the extensive labor and land requirements, making the development of hybrid rice challenging. Encouragingly, genomic prediction of hybrid rice shows promise, as the phenotypes of hybrids can be forecasted prior to test crosses and evaluation, leading to a significant reduction in costs and labor (Cui et al. 2020). In breeding rice, extensive pedigree schemes have typically been used due to its primarily inbred nature. Researchers have combined phenotypic selection and GS in various ways within pedigree schemes. In order to save time and money by eliminating underperforming lines early in the pedigree line fixation process, one design uses GEBVs to carry lines until the end of fixation. These experimental breeding initiatives have mostly consisted of modifications to recurrent selection systems. Recent studies on genomic selection in rice have shown significant advancements and promising results, enhancing the efficiency and effectiveness of rice breeding programs (Table 8.5).

In a study conducted by Spindel et al. (2015), parents are chosen and crossed to generate an F_1 population. After fixing 20,000 F_1 lines over the course of 7–8

Table 8.5 Recent studies on genomic selection in rice

Study title	Gene/trait of interest	References
Increasing selection differential and decreasing cycle time for accelerated genetic improvement	Yield improvement using GS and MAS	Biswas et al. (2023)
Genomic selection in Rice breeding: Factors affecting predictive performance and strategies for improvement	Yield growth, multiple traits, genotype-by-environment interactions, and functional markers	Xu et al. (2021)
Genomic selection for enhanced grain quality in Rice: Insights and challenges	Grain quality	Huang et al. (2019)
Genomic selection for enhanced nutritional content in Rice: Opportunities and challenges	Nutritional content	Wang et al. (2020)
Comparison of various phenotypic and genomics-assisted reciprocal recurrent selection breeding schemes for hybrid performance	Hybrid performance with updated testers increased hybrid genetic gain with phenotypic RRS	Fritsche-Neto et al. (2024)
Biofortification of Rice to alleviate micronutrient deficiencies	Zn, Fe, protein, and provitamin A	Senguttuvel et al. (2023)
Consumer preferences for specific nutritional, physical, eating, and cooking characteristics of Rice	Grain width, length, color (translucence or whiteness), chalkiness, cooking time (gelatinization temperature and viscosity), ability to remain soft after cooking (gel consistency), and textural properties (amylose content and aroma)	Patra et al. (2022)
Genetic basis underlying grain quality traits in Rice	Quantitative trait loci (QTL) controlling grain appearance, chalkiness, protein content, and grain length	

generations with families selected for heritable features, about 25% of the pedigree lines are eventually chosen for enrollment in the observational yield experiment (OYT). If resources allow, selection can be carried out using GEBVs during fixation at two or more generations. In this case, it is suggested to apply GEBVs to the F_3 and F_6 generations. Furthermore, the fixed lines are selected from the F_8 that advance to the OYT based on GEBVs. The breeding population was further improved by cycling back the top lines, which were advanced to the OYT based on GEBV, into the crossing block. The lines with the best phenotypic performance from the OYT move on to the repeated yield trials (RYT), while the lines with the best performance from the RYT go on to the multi-environment trials (MET). For the subsequent generation of recurrent selection based on GEBV, lines from the MET were chosen as parents. For example, in a rice breeding process, F_1 lines were bred over 7–8 generations, selecting 25% for the observational yield trial based on heritable traits. GEBVs guided selection at F_3, F_6, and F_8-OYT stages, with top-performing lines cycled back for continual improvement. Replicated yield trials and multi-environment trials followed, with GEBVs used to choose parents for recurrent selection. Models were constructed at each stage using 300 genotyped and phenotyped individuals for training (Fig. 8.1).

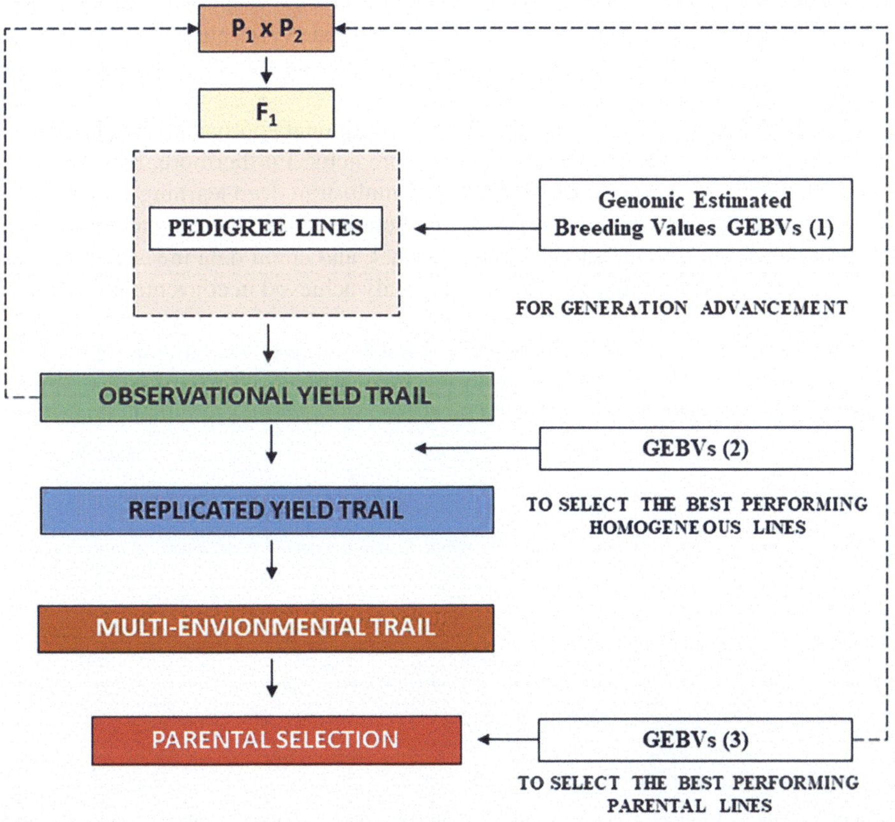

Fig. 8.1 A rice breeding process using genomic selection

8.14 Correlation Between Traits and Consideration for Genomic Prediction

Genomic-enabled prediction is a pivotal element contributing to the success of GS. However, according to Sterkenburg and Grünwald (2021), it is important to acknowledge that no universally effective model can perform optimally for all datasets, as stated by the "No Free Lunch (NFL)" theorem according to James Le (i.e., no single machine learning algorithm works best for every problem). Consequently, there is a diverse array of statistical and machine learning models available for genomic prediction. In cases where there is multi-trait data available, it is preferable to employ models capable of considering the correlations between different phenotypic traits. These models become particularly valuable when there is a moderate to substantial degree of correlation among the traits, as they tend to enhance prediction accuracy. According to Montesinos et al. (2019), the inability to estimate a variance–covariance matrix indicating the degree of correlation between characteristics is a drawback of multi-trait deep learning models over traditional multi-trait linear models. When the data is large enough, they can outperform linear multi-trait models because they can capture both linear and nonlinear patterns in the data that traditional multi-trait linear models are unable to capture. However, because they jointly model all the response variables, this helps increase the precision of the parameter estimates (weights and intercepts). These models have a significant disadvantage: they require a laborious and complex altering procedure.

The multi-trait models are compatible with traditional approaches like Bayesian ridge regression and the best linear unbiased predictor. Furthermore, they are also adaptable to deep learning frameworks. Using multi-trait deep learning frameworks is very helpful because they let you make prediction models that can handle a range of outcomes, such as continuous, binary, ordinal, and count data measured on different scales. This flexibility is a feature not easily achieved in conventional statistical models.

If a dominant male sterility system is not made publicly available to rice breeders, implementing a recurrent selection program in rice will likely be extremely challenging. Focusing on the utilization of GS at an early stage in initial yield trials, particularly in circumstances with restricted seed availability, may enable breeders to circumvent the need to enhance precision by increasing the number of phenotype replicates or focusing on traits that present challenges in phenotyping due to their time-consuming, costly, or labor-intensive nature. This represents a viable approach for integrating the advantages of GS into rice breeding initiatives.

8.15 Phenomic Selection vs. Genomic Selection in Rice

Phenomic selection (PS), a new technique in plant breeding, suggests switching from molecular markers to near-infrared (NIR) spectra to figure out how different individuals are related. A genomics-based approach closely related to PS is referred to as genomic-like omics-based (GLOB) selection, where molecular markers are replaced by endophenotypes such as metabolites or transcript levels (Robert et al. 2022). Phenomics, particularly through near-infrared spectroscopy (NIRS), often provides more cost-effective information compared to other omics techniques. Despite the extensive historical application of NIRS in breeding, primarily for assessing end-product quality traits, its utilization for predicting and selecting other traits is a relatively new effort. Since the important work on PS, multiple publications have championed the adoption of spectral acquisition, encompassing methods like NIRS and hyperspectral imaging, in the realm of plant breeding, offering a glimpse of the possibilities it holds. PS offers exciting possibilities to enhance breeding efficiency, potentially leading to accelerated genetic gain and crop improvement. Nevertheless, the successful implementation of PS is contingent on several factors, including the choice of technology, data preprocessing, and robust statistical modeling. The utilization of both phenomic and genomic selection to increase prediction accuracy in rice breeding under diverse environments is illustrated in Fig. 8.2. These considerations are pivotal in realizing the full potential of PS, aligning it with, and perhaps surpassing, the established GS approach.

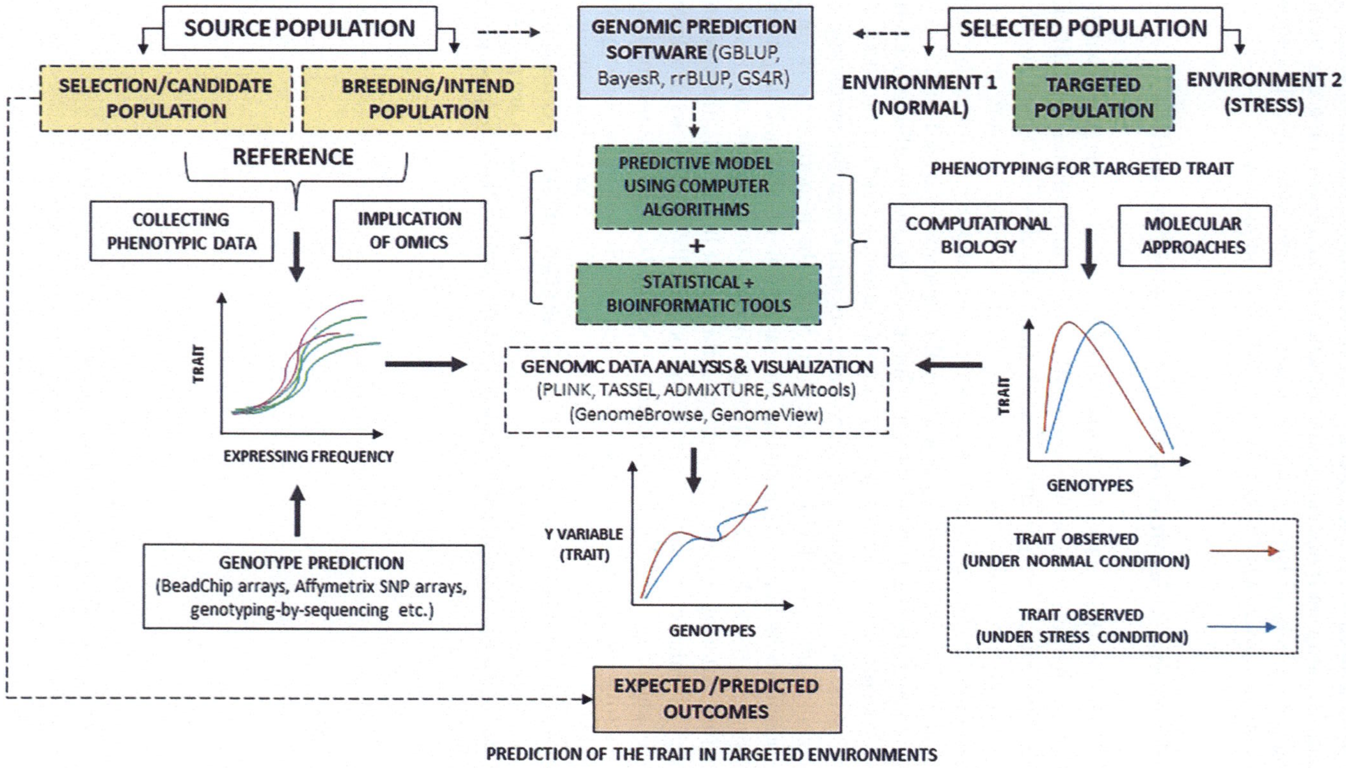

Fig. 8.2 Enhancing selection precision in diverse environments through GLOB selection

8.16 Breeding Strategies for Genomic Selection in Rice

Genomic selection, utilizing genomic prediction models, stands as a cutting-edge approach in plant breeding, streamlining the identification of candidate individuals and hastening genetic progress. This contrasts with the recent practices in hybrid rice breeding strategies, which focus on evaluating a set number of doubled haploid (DH) lines from various crosses in nurseries. It is important to note that traits selected in the nursery are assumed to have less correlation with the main target trait, grain yield. After the nursery phase, the first stage of grain yield selection is carried out using GS. Subsequently, in four distinct ways, grain yield performance in the field is assessed, each with its respective allocation of testing resources. In stages 2 and 3, phenotypic selection is based on general combining ability (GCA), and selection occurs after the respective field trials (Singh et al. 2022b). Nursery selection involves evaluating a smaller number of candidates for grain yield, which can decrease the selection intensity. According to Singh et al. (2022b), this approach is often used in conjunction with other breeding methods to optimize the efficiency of breeding programs, the potential of using 2Gs (grain number and grain weight), and plant architecture to break the grain yield ceiling in rice breeding. Nursery selection is crucial since high-yielding hybrids with poor disease resistance, lodging, or a lack of adaptation to particular growing conditions will never make it onto the market. It makes sense to screen these qualities in a nursery prior to GS and grain yield phenotyping since scoring these traits is easy, inexpensive, and found 28.5% higher annual selection gains for GS rapid compared with PS standard. Biswas et al. (2024) support this approach that involves selecting for desirable traits like plant height, days to flowering, and grain quality highlighting the importance of integrating these traits into breeding programs to achieve higher yields. This can be an effective approach in this context, especially when combined with other breeding methods like GS and MAS.

The "GS-only" strategy employs only genomic selection without any additional field tests, minimizing the length of the breeding cycle. A final set of five lines is chosen in all these breeding strategies to evaluate their potential for commercial use in creating new hybrid combinations. It is important to highlight that, for the sake of comparison, standard hybrid breeding strategies that do not incorporate genomic selection, known as PSstandard and PSrapid, have also been optimized (Fig. 8.3).

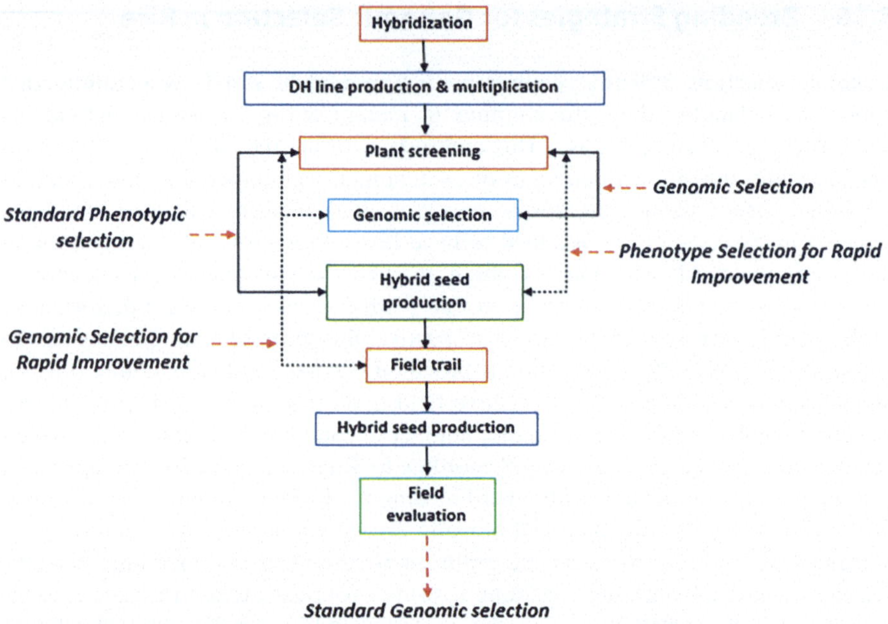

Fig. 8.3 Genomic selection flowchart for hybrid rice breeding techniques

8.17 Perspective of Genomic Selection in Rice

The perspective on GS is overwhelmingly positive due to its transformative potential in agricultural breeding programs offering a paradigm shift from traditional breeding methods by harnessing genomic information to accelerate genetic progress and enhance breeding efficiency. By leveraging genetic markers distributed throughout the genome, GS enables breeders to predict the genetic merit of individuals for various traits with greater accuracy and efficiency. This not only expedites the breeding process but also allows for the targeting of complex traits that are difficult to measure directly, such as drought tolerance or disease resistance. Moreover, GS facilitates the selection of superior breeding candidates at an earlier stage, reducing generation intervals and ultimately leading to faster genetic improvement in crops and livestock. As a result, GS has been widely adopted across different agricultural sectors, offering immense potential to address global challenges such as food security, climate change resilience, and sustainable agricultural production. However, challenges such as the cost of genotyping, data management, and the need for robust statistical models remain, highlighting the importance of continued research and innovation to fully realize the benefits of genomic selection in agriculture. Overall, the perspective on genomic selection is one of optimism and excitement, recognizing its pivotal role in shaping the future of agricultural breeding and fostering resilience and productivity in global food systems.

8.18 Conclusion

Extensive DNA marker genotyping has enabled more precise "genomic" assessments of breeding animals, surpassing the accuracy of earlier evaluations that relied on kinship information found in pedigrees. Numerous empirical findings substantiate this improvement, and the enthusiasm for these technologies remains unabated. Nonetheless, there are several scenarios where the implementation of genomic selection strategies faces challenges, whether due to technical constraints, such as the unavailability of a suitably large reference population, or budgetary limitations, encompassing the investment required for equipment and expertise as well as the per-unit cost of genotyping. However, the adoption of GS in rice breeding is not without its challenges. Data quality, model selection, and the establishment of extensive training populations are critical considerations. The adoption of GS holds great promise for ensuring a sustainable and productive future for rice cultivation, and this chapter aims to shed light on the path forward in harnessing its potential to meet the world's rice production demands.

Author Contribution Conceptualization: Sakthi Anand Muthazhagu Kuppuraj, Anandhan Tamilselvan, Bharathi Raja Ramadoss.

Methodology: Sakthi Anand Muthazhagu Kuppuraj, Karthikeyan Adhimoolam, Vengadessan Vedachalam.

Validation: Sakthi Anand Muthazhagu Kuppuraj, Tamilzharasi Murugesan, Anandhan Tamilselvan.

Formal Analysis: Karthikeyan Adhimoolam, Anandhan Tamilselvan, Sakthi Anand Muthazhagu Kuppuraj.

Resources: Sakthi Anand Muthazhagu Kuppuraj, Bharathi Raja Ramadoss, Anandhan Tamilselvan.

Data Curation: Sakthi Anand Muthazhagu Kuppuraj, Anandhan Tamilselvan.

Writing—Original Draft Preparation: Sakthi Anand Muthazhagu Kuppuraj, Anandhan Tamilselvan.

Writing—Review & Editing: Sakthi Anand Muthazhagu Kuppuraj, Anandhan Tamilselvan.

Visualization: Sakthi Anand Muthazhagu Kuppuraj, Anandhan Tamilselvan.

Supervision: Sakthi Anand Muthazhagu Kuppuraj, Vengadessan Vedachalam, Anandhan Tamilselvan.

References

Anilkumar C, Sah RP, Muhammed AT, Sunitha NC, Behera S, Marndi BC, Sharma TR, Singh AK (2022) Genomic selection in rice: current status and future prospects. In: Elias AA, Goel S (eds) Genomic selection in plants: a guide for breeders, 1st edn. CRC Press, Boca Raton, pp 68–82. https://doi.org/10.1201/9781003214991-4

Asai H, Saito K, Kawamura K (2021) Application of a Bayesian approach to quantify the impact of nitrogen fertilizer on upland rice yield in sub-Saharan Africa. Field Crop Res 272:108284

Bal SK, Sattar A, Nidhi CMAS, Subba Rao AVM, Manikandan N, Banerjee S, Choudhary JL, More VG, Singh CB, Sandhu SS, Singh VK (2023) Critical weather limits for paddy rice under diverse ecosystems of India. Front Plant Sci 14:1226064. https://doi.org/10.3389/fpls.2023.1226064

Bartholomé J, Prakash PT, Cobb JN (2022) Genomic prediction: progress and perspectives for rice improvement. In: Ahmadi N, Bartholomé J (eds) Genomic prediction of complex traits. Methods in molecular biology, vol 2467. Humana, New York, NY. https://doi.org/10.1007/978-1-0716-2205-6_21

Bernal-Gallardo JJ, de Folter S (2024) Plant genome information facilitates plant functional genomics. Planta 259:117. https://doi.org/10.1007/s00425-024-04397-z

Bhandari A, Bartholomé J, Cao-Hamadoun T-V, Kumari N, Frouin J, Kumar A et al (2019) Selection of trait-specific markers and multi-environment models improve genomic predictive ability in rice. PLoS One 14(5):e0208871. https://doi.org/10.1371/journal.pone.0208871

Bing H, Cui D, Ma X, Cao G, Zhang H, Koh HJ, Han L (2022) Evidence for evolution and selection of drought-resistant genes based on high-throughput resequencing in weedy rice. J Exp Bot 73(7):1949–1962. https://doi.org/10.1093/jxb/erab515

Biswas PS, Ahmed MME, Afrin W, Rahman A, Shalahuddin AKM, Islam R, Akter F, Syed MA, Sarker MRA, Ifterkharuddaula KM, Islam MR (2023) Enhancing genetic gain through the application of genomic selection in developing irrigated rice for the favorable ecosystem in Bangladesh. Front Genet 14:1083221. https://doi.org/10.3389/fgene.2023.1083221

Biswas PS, Santelices R, Mendoza R, Lopena V, Arbelaez JD, Manigbas NL, Cobb JN, Collard B, Islam MR (2024) Assessment of efficiency of breeding methods in accelerating genetic gain in rice. Agronomy 14(3):566. https://doi.org/10.3390/agronomy14030566

Bohar R, Chitkineni A, Varshney RK (2020) Genetic molecular markers to accelerate genetic gains in crops. BioTechniques 69(3):158–160. https://doi.org/10.2144/btn-2020-0066

Budhlakoti N, Kushwaha AK, Rai A, Chaturvedi KK, Kumar A, Pradhan AK, Kumar U et al (2022) Genomic selection: a tool for accelerating the efficiency of molecular breeding for development of climate-resilient crops. Front Genet 13:832153. https://doi.org/10.3389/fgene.2022.832153

Cerioli T, Hernandez CO, Angira B, McCouch SR, Robbins KR, Famoso AN (2022) Development and validation of an optimized marker set for genomic selection in southern U.S. rice breeding programs. Plant Genome 15(3):e20219. https://doi.org/10.1002/tpg2.20219

Charmet G, Tran LG, Auzanneau J, Rincent R, Bouchet S (2020) BWGS: a R package for genomic selection and its application to a wheat breeding programme. PLoS One 15:e0222733. https://doi.org/10.1371/journal.pone.0222733

Chen R, Deng Y, Ding Y, Guo J, Qiu J, Wang B, Wang C et al (2022) Rice functional genomics: decades' efforts and roads ahead. Sci China Life Sci. https://doi.org/10.1007/s11427-021-1982-5

Cui Y, Li R, Li G, Zhang F, Zhu T et al (2020) Hybrid breeding of rice via genomic selection. Plant Biotechnol J 18(1):57–67

Danilevicz MF, Gill M, Anderson R, Batley J, Bennamoun M, Bayer PE, Edwards D (2022) Plant genotype to phenotype prediction using machine learning. Front Genet 13:822173. https://doi.org/10.3389/fgene.2022.822173

Dekkers JCM, Su H, Cheng J (2021) Predicting the accuracy of genomic predictions. Genet Sel Evol 53:55. https://doi.org/10.1186/s12711-021-00647-w

Duan W, Zou S, Christidis N, Schaller N, Chen Y, Sahu N et al (2022) Changes in temporal inequality of precipitation extremes over China due to anthropogenic forcings. NPJ Clim Atmos Sci 5(1):33. https://doi.org/10.1038/s41612-022-00255-5

FAO (2023) Food and agriculture organization. In: FAO cereal supply and demand. Available at: https://www.fao.org/worldfoodsituation/csdb/en/ (Accessed 12 July 2023)

Fritsche-Neto R, Ali J, De Asis EJ et al (2024) Improving hybrid rice breeding programs via stochastic simulations: number of parents, number of hybrids, tester update, and genomic prediction of hybrid performance. Theor Appl Genet 137(3):3. https://doi.org/10.1007/s00122-023-04508-6

Frouin J, Labeyrie A, Boisnard A, Sacchi GA, Ahmadi N et al (2019) Genomic prediction offers the most effective marker assisted breeding approach for ability to prevent arsenic accumulation in rice grains. PLoS One 14(6):e0217516. https://doi.org/10.1371/journal.pone.0217516

Fujino K, Hirayama Y, Kaji R (2019) Marker-assisted selection in rice breeding programs in Hokkaido. Breed Sci 69(3):383–392. https://doi.org/10.1270/jsbbs.19062

Gu H, Liang S, Zhao J (2022) Novel sequencing and genomic technologies revolutionized rice genomic study and breeding. Agronomy 12(1):218. https://doi.org/10.3390/agronomy12010218

Gu S, Zhuang J, Zhang Z, Chen W, Xu H, Zhao M, Ma D (2023) Multi-omics approach reveals the contribution of OsSEH1 to rice cold tolerance. Front Plant Sci 13:1110724

Hu Y, Song S, Weng X, You A, Xing Y (2021) The heading-date gene Ghd7 inhibits seed germination by modulating the balance between abscisic acid and gibberellins. Crop J 9(2):297–304

Huang M, Balimponya EG, Mgonja EM, McHale LK, Luzi-Kihupi A et al (2019) Use of genomic selection in breeding rice (Oryza sativa L.) for resistance to rice blast (*Magnaporthe oryzae*). Mol Breed 39(8):114

Iqbal Z, Iqbal MS, Khan MIR, Ansari MI (2021) Toward integrated multi-omics intervention: Rice trait improvement and stress management. Front Plant Sci 12:741419. https://doi.org/10.3389/fpls.2021.741419

Jarquin D, Kajiya-Kanegae H, Taishen C, Yabe S, Persa R, Yu J, Nakagawa H, Yamasaki M, Iwata H (2020) Coupling day length data and genomic prediction tools for predicting time-related traits under complex scenarios. Sci Rep 10(1):13382. bioRxiv: 703488. https://doi.org/10.1101/703488

Jeon D, Kang Y, Lee S, Choi S, Sung Y, Lee TH, Kim C (2023) Digitalizing breeding in plants: a new trend of next-generation breeding based on genomic prediction. Front Plant Sci 14:1092584. https://doi.org/10.3389/fpls.2023.1092584

Kaur K, Neelam K, Singh J, Malik P, Singh K (2023) Uncovering natural allelic and structural variants of OsCENH3 gene by targeted resequencing and in silico mining in genus Oryza. Sci Rep 13(1):830

Majhi PK, Singh SK, Anandan A, Khaire AR, Korada M, Habde SV et al (2022) Parental polymorphism survey for evaluation and selection of contrasting parents for drought tolerance in rice (Oryza sativa L.) by using SSR markers. Int J Environ Clim 12(11):2507–2519

Mehandi S, Yadav A, Maurya R, Mishra SP, Quatadah SM, Praveen N, Dwivedi N (2021) Current scenario of breeding approaches in rice. In: Goyal AK (ed) Cereal grains, vol 1. IntechOpen. https://doi.org/10.5772/intechopen.98744

Merrick LF, Herr AW, Sandhu KS, Lozada DN, Carter AH (2022) Optimizing plant breeding programs for genomic selection. Agronomy 12(3):714. https://doi.org/10.3390/agronomy12030714

Michael PS, Mwakyusa L, Sanga HG, Shitindi MJ, Kwaslema DR, Herzog M, Meliyo JL, Massawe BHJ (2023) Floods stress in lowland rice production: experiences of rice farmers in Kilombero and lower-Rufiji floodplains, Tanzania. Front Sustain Food Syst 7:1206754. https://doi.org/10.3389/fsufs.2023.1206754

Montesinos-López OA, Montesinos-López A, Luna-Vázquez FJ, Toledo FH, Pérez-Rodríguez P et al (2019) An R package for Bayesian analysis of multi-environment and multi-trait multi-environment data for genome-based prediction. G3 Genes Genomes Genet 9:1355–1369. https://doi.org/10.1534/g3.119.400126

Montesinos-López OA, Kismiantini, Montesinos-López A (2023) Two simple methods to improve the accuracy of the genomic selection methodology. BMC Genomics 24:220. https://doi.org/10.1186/s12864-023-09294-5

Mukhopadhyay R, Sarkar B, Jat HS, Sharma PC, Bolan NS (2021) Soil salinity under climate change: challenges for sustainable agriculture and food security. J Environ Manag 280:111736

Mwakyusa L, Dixit S, Herzog M, Heredia MC, Madege RR, Kilasi NL (2023) Flood-tolerant rice for enhanced production and livelihood of smallholder farmers of Africa. Front Sustain Food Syst 7:1244460. https://doi.org/10.3389/fsufs.2023.1244460

Nguyen VH, Morantte RIZ, Lopena V et al (2023) Multi-environment genomic selection in rice elite breeding lines. Rice 16:7. https://doi.org/10.1186/s12284-023-00623-6

Patra B, Majhi PK, Tripathy SK, Tripathy SP, Khan A, Behera PP, Das S, Ahamad A (2022) Genomic-assisted breeding tools for grain and nutritional quality improvement in rice. Int J Environ Clim 12:10–24. https://doi.org/10.9734/ijecc/2022/v12i130609

Pérez-Enciso M, Ramírez-Ayala LC, Zingaretti LM (2020) SeqBreed: a python tool to evaluate genomic prediction in complex scenarios. Genet Sel Evol 52:7. https://doi.org/10.1186/s12711-020-0530-2

Prakapenka D, Wang C, Liang Z, Tan C et al (2020) GVCHAP: a computing pipeline for genomic prediction and variance component estimation using haplotypes and SNP markers. Front Genet 11:282. https://doi.org/10.3389/fgene.2020.00282

Robert P, Brault C, Rincent R, Segura V (2022) Phenomic selection: a new and efficient alternative to genomic selection. Methods Mol Biol (Clifton, NJ) 2467:397–420. https://doi.org/10.1007/978-1-0716-2205-6_14

Sachdeva S, Singh R, Maurya A et al (2024) New insights into QTNs and potential candidate genes governing rice yield via a multi-model genome-wide association study. BMC Plant Biol 24:124. https://doi.org/10.1186/s12870-024-04810-5

Sanchez DL, Samonte SOP, Wilson LT (2023) Genetic architecture of head rice and rice chalky grain percentages using genome-wide association studies. Front Plant Sci 14:1274823. https://doi.org/10.3389/fpls.2023.1274823

Schrauf MF, Martini JW, Simianer H, de Los CG, Cantet R, Freudenthal J et al (2020) Phantom epistasis in genomic selection: on the predictive ability of epistatic models. G3: Genes Genomes Genet 10:3137–3145. https://doi.org/10.1534/g3.120.401300

Seck F, Covarrubias-Pazaran G, Gueye T et al (2023) Realized genetic gain in rice: achievements from breeding programs. Rice 16:61. https://doi.org/10.1186/s12284-023-00677-6

Selga C, Reslow F, Pérez-Rodríguez P, Ortiz R (2022) The power of genomic estimated breeding values for selection when using a finite population size in genetic improvement of tetraploid potato. G3 (Bethesda) 12(1):jkab362. https://doi.org/10.1093/g3journal/jkab362

Senguttuvel P, Padmavathi G, Jasmine C, Rao DS, Neeraja CN, Jaldhani V, Beulah P, Gobinath R, Kumar JA, Prasad SVS, Rao LV, Hariprasad AS, Sruthi K, Shivani D, Sundaram RM, Govindaraj M (2023) Rice biofortification: breeding and genomic approaches for genetic enhancement of grain zinc and iron contents. Front Plant Sci 14:1138408. https://doi.org/10.3389/fpls.2023.1138408

Siddiq EA, Vemireddy LR (2021) Advances in genetics and breeding of rice: an overview. In: Ali J, Wani SH (eds) Rice improvement. Springer, Cham. https://doi.org/10.1007/978-3-030-66530-2_1

Sigala RE, Lagou V, Shmeliov A, Atito S, Kouchaki S, Awais M, Prokopenko I, Mahdi A, Demirkan A (2023) Machine learning to advance human genome-wide association studies. Genes (Basel) 15(1):34. https://doi.org/10.3390/genes15010034

Singh SK, Majhi PK, Anandan A, Korada M, Habde SV, Khaire AR et al (2022a) Microsatellites based parental polymorphism survey for moisture stress in rice (Oryza sativa L.) between the parental genotypes HUR-1309 and CR Dhan 801. Emergent Life Sci Res 8:248–258

Singh G, Kaur N, Khanna R, Kaur R, Gudi S, Kaur R, Sidhu N, Vikal Y, Mangat GS (2022b) 2Gs and plant architecture: breaking grain yield ceiling through breeding approaches for next wave of revolution in rice (Oryza sativa L.). Crit Rev Biotechnol 44:139–162

Singh A, Rami E, Upadhyay P, Gangawane AK (2023) The impact of climate change on crop production and combat strategies. In: Climate change and sustainable development. CRC Press, pp 103–115

Singh P, Sundaram KT, Vinukonda VP et al (2024) Superior haplotypes of key drought-responsive genes reveal opportunities for the development of climate-resilient rice varieties. Commun Biol 7:89. https://doi.org/10.1038/s42003-024-05769-7

Sinha D, Maurya AK, Abdi G, Majeed M, Agarwal R, Mukherjee R, Ganguly S, Aziz R, Bhatia M, Majgaonkar A, Seal S, Das M, Banerjee S, Chowdhury S, Adeyemi SB, Chen JT (2023) Integrated genomic selection for accelerating breeding programs of climate-smart cereals. Genes (Basel) 14(7):1484. https://doi.org/10.3390/genes14071484

Spindel J, Begum H, Akdemir D et al (2015) Genomic selection and association mapping in rice (Oryza sativa): effect of trait genetic architecture, training population composition, marker number and statistical model on accuracy of rice genomic selection in elite, tropical rice breeding lines. PLoS Genet 11:e1004982. https://doi.org/10.1371/journal.pgen.1004982

Sterkenburg TF, Grünwald PD (2021) The no-free-lunch theorems of supervised learning, vol 199. SpringerLink, p 9979

Suganami M, Kojima S, Wang F, Yoshida H, Miura K, Morinaka Y, Watanabe M, Matsuda T, Yamamoto E, Matsuoka M (2023) Effective use of legacy data in a genome-wide association studies improves the credibility of quantitative trait loci detection in rice. Plant Physiol 191(3):1561–1573. https://doi.org/10.1093/plphys/kiad018

Thorwarth P (2019) Genomic prediction in crops: advantages and drawbacks. In: Advances in breeding techniques for cereal crops. Burleigh Dodds Science Publishing, pp 537–558

Tong H, Nikoloski Z (2021) Machine learning approaches for crop improvement: leveraging phenotypic and genotypic big data. J Plant Physiol 257:153354

Uffelmann E, Huang QQ, Munung NS, De Vries J, Okada Y, Martin AR, Martin HC, Lappalainen T, Posthuma D (2021) Genome-wide association studies. Nat Rev Methods Primer 1:59. https://doi.org/10.1038/s43586-021-00056-9

Wang S, Xu Y, Qu H, Cui Y, Li R, Chater JM, Yu L, Zhou R, Ma R, Huang Y, Qiao Y, Hu X, Xie W, Jia Z (2020) Boosting predictabilities of agronomic traits in rice using bivariate genomic selection. Brief Bioinform 22(3):bbaa103. https://doi.org/10.1093/bib/bbaa103

Werner CR, Gaynor RC, Gorjanc G, Hickey JM, Kox T, Abbadi A, Leckband G, Snowdon RJ, Stahl A (2020) How population structure impacts genomic selection accuracy in cross-validation: implications for practical breeding. Front Plant Sci 11:592977. https://doi.org/10.3389/fpls.2020.592977

Xu Y, Ma K, Zhao Y, Wang X, Zhou K, Yu G, Li C, Li P, Yang Z, Xu C, Xu S (2021) Genomic selection: a breakthrough technology in rice breeding. Crop J 9:305–311. https://doi.org/10.1016/j.cj.2021.03.008

Yang X, Yu S, Yan S, Wang H, Fang W, Chen Y, Ma X, Han L (2024) Progress in rice breeding based on genomic research. Genes 15(5):564. https://doi.org/10.3390/genes15050564

Yu P, Ye C, Li L, Yin H, Zhao J, Wang Y, Zhang Z, Li W, Long Y, Hu X, Xiao J, Jia G, Tian B (2022) Genome-wide association study and genomic prediction for yield and grain quality traits of hybrid rice. Mol Breed 42(4):16. https://doi.org/10.1007/s11032-022-01289-6

Zheng J, Zhang S (2023) Improving rice phenology simulations based on the Bayesian model averaging method. Eur J Agron 142:126646

Zhou C, Jiang W, Guo J, Zhu L, Liu L, Liu S, Chen R, Du B, Huang J (2024) Genome-wide association study and genomic prediction for resistance to brown planthopper in rice. Front Plant Sci 15:1373081. https://doi.org/10.3389/fpls.2024.1373081

Genome Editing for Trait-Specific Improvement in Rice

9

Ambothi Rathnasamy Sakthi, Shobica Priya Ramasamy, Shubham Rajaram Salunkhe, Ameena Premnath, Anandhan Tamilselvan, and Bharathi Raja Ramadoss

Abstract

Rapid advances in genome editing technology have revolutionized agriculture by giving precise tools for trait-specific crop improvement. As a staple food for a substantial fraction of the world's population, rice is among the most important crops. This comprehensive chapter explores the complex landscape of genome editing in rice, elucidating the scientific principles of cutting-edge techniques, including CRISPR-Cas9, TALENs and ZFNs and their potential for revolutionizing pest/disease resistance, abiotic stress tolerance, nutritional enrichment, yield optimization and quality enhancement. Furthermore, genome editing facilitates the development of rice varieties that thrive in diverse environmental conditions, displaying increased tolerance to drought, salinity and temperature extremes. The nutritional landscape of rice is also in position for transformation,

A. R. Sakthi · S. R. Salunkhe · A. Premnath
Department of Plant Biotechnology, Centre for Plant Molecular Biology and Biotechnology (CPMB&B), Tamil Nadu Agricultural University (TNAU), Coimbatore, Tamil Nadu, India

S. P. Ramasamy
Department of Genetics and Plant Breeding, Centre for Plant Breeding and Genetics (CPBG), Tamil Nadu Agricultural University (TNAU), Coimbatore, Tamil Nadu, India

A. Tamilselvan
Department of Genetics and Plant Breeding, Centre for Plant Breeding and Genetics (CPBG), Tamil Nadu Agricultural University (TNAU), Coimbatore, Tamil Nadu, India

Department of Plant Breeding and Genetics, Pandit Jawaharlal Nehru College of Agriculture and Research Institute, Karaikal, Puducherry, India

B. R. Ramadoss (✉)
Centre for Plant Breeding and Genetics (CPBG), Tamil Nadu Agricultural University (TNAU), Coimbatore, Tamil Nadu, India

Plenty Unlimited Inc., Laramie, WY, USA

A. Singh et al. (eds.), *Climate-Smart Rice Breeding*,
https://doi.org/10.1007/978-981-97-7098-4_9

as genome editing enables the elevation of essential vitamins, minerals and amino acids, addressing global malnutrition challenges. Beyond productivity, genome editing tailors rice to consumers' preferences by enhancing grain quality attributes like aroma, texture and cooking properties. This chapter also provides comprehensive knowledge of how genome editing transforms rice agriculture and contributes to global food security by addressing the scientific, ethical, and regulatory components.

Keywords

Genome editing · Rice · CRISPR-Cas9 · Trait-specific improvement · Genetic transformation

9.1 Introduction

Rice, a fundamental dietary component for over half of the world's population, is cultivated in more than 100 countries, with Asia contributing to 90% of the global production. The primary sustenance for many people, rice holds significance due to its inherent nutritional properties and high energy content (Burlando and Cornara 2014). Rice grains are consumed as cooked rice, while a smaller proportion serves as ingredients in various food and non-food processes. Beyond being a valuable source of fiber, energy, minerals, proteins, vitamins, antioxidants and other biomolecules, rice components may collaborate synergistically to confer health benefits (Ramadoss et al. 2019; Goufo and Trindade 2014).

Rice production has increased from 35.66 million tonnes in 1961–1962 to 129.47 million tons (Indiastat 2022). By 2050, the world's population is expected to reach approximately 10 billion (Farooq et al. 2023; Singh et al. 2023). To meet the requirements of an ever-growing population, the productivity of rice has to be increased three-fold in another 20 years. In the face of changing climatic conditions, plants are continually subjected to biotic and abiotic stresses, emphasizing the necessity for developing novel crop varieties (Singh et al. 2023). The high-yielding varieties should possess the ability to withstand biotic and abiotic stress factors. Initially, crop varieties were developed by employing traditional plant breeding techniques, and marker-assisted selection techniques played a crucial role in expediting crop improvement (Singh et al. 2022; Majhi et al. 2022). Also, using genetically modified crops has become prevalent, addressing concerns such as pest control, herbicide tolerance and other desirable traits. However, the production of GM crops has sparked controversy primarily due to regulatory/ethical issues and a limited understanding among the public. Genome engineering techniques precisely altering the target DNA sequence open up a wide scope for the genetic enhancement of crops.

The discovery of molecular scissors, viz. restriction enzymes in the late 1970s, which are part of the bacterial immune system, marked a pivotal moment that spurred the onset of recombinant DNA technology. This breakthrough discovery led to numerous advancements in molecular biology and genetics for developing plants possessing novel traits. The ability to transfer genes from unrelated sources into economically significant crop plants so they can develop resistance against biotic and abiotic stresses is the most significant advancement in plant biotechnology (Dhaliwal and Uchimiya 1999). The availability of genome sequences has led to the identification of genes responsible for the positive/negative regulation of key traits. Due to regulatory issues, the commercial release of genetically modified food crops was hindered and the researchers developed gene-editing technology by introducing a DNA segment with homologous arms at both ends into a cell and the integration of a desired sequence of DNA into the host genome through homologous recombination (Capecchi 1989). Subsequently, precise gene editing was employed to generate new genetic variations, enhancing resistance to biotic and abiotic stresses while elevating yield and quality characteristics. Initially, meganucleases, Zinc finger nucleases (ZFNs) and transcription activator-like effector nuclease (TALENs) were used to generate variation in targeted regions and later, CRISPR-Cas9 was widely used for gene editing in various crops (Premnath et al. 2022).

CRISPR (Clustered Regularly Interspaced Short Palindromic Repeats)—Cas9 (CRISPR-Cas9) is a breakthrough genome engineering technology that facilitates the introduction of desired mutations at precise locations in the genome within a remarkably short time frame and with minimal side effects. The CRISPR-Cas9 method surpasses other genome-editing techniques, such as transcriptional activator-like effector nucleases (TALENs) and zinc finger nucleases, owing to its adaptability and simplicity for a wide range of crop species (Cong et al. 2013; Miao et al. 2013; Ma et al. 2015). In addition to precise targeted manipulation of DNA, CRISPR technology eliminates the usage of transgenes and vector-free gene-edited plants can be obtained with desirable genetic changes (He and Zhao 2020). This chapter deals with the overall improvement of CRISPR–Cas9 technology and the deployment of this technology for trait-specific improvement in rice.

9.2 Gene Editing Tools

In response to the growing demand for food production, various genome-editing tools and techniques have been increasingly utilized in plants. These advancements aim to address challenges and enhance crop productivity to meet future food needs (Zhang et al. 2017). The prominent gene-editing techniques are engineered endonucleases/mega-nucleases (EMNs), zinc-finger nucleases (ZFNs), TAL effector nucleases (TALENs), clustered regularly interspaced short palindromic repeats (CRISPR), base editing, prime editing. Base editing with different Cas variants for expanding the target sites was adopted to create precise mutations to achieve the desired phenotype. Also the direct delivery of the CRISPR-Cas9 system as a ribonucleoprotein (RNP) complex, composed of the Cas9 protein and single guide RNA (sgRNA), has become a widely adopted and potent approach for genome editing, where delivering the Cas9 protein and sgRNA as an RNP complex directly into the target cells achieves precise and efficient genome modifications without the need for long-term expression of the Cas9 gene. Additionally, the transient nature of the

RNP complex minimizes the potential off-target effects, thereby enhancing the specificity and safety of the genome editing process. Overall, direct delivery of the CRISPR-Cas9 system as an RNP complex represents a powerful and versatile tool for precise genome editing in various organisms. These tools play pivotal roles in plant research by enabling precise modifications in the plant genome, offering opportunities to engineer traits for improved yield, resilience and nutritional value in future crops. Meganucleases (MegNs) are endodeoxyribonucleases that exist naturally in various microbial life forms and the mitochondria and chloroplasts of eukaryotic cells. These endonucleases are the first ones identified as possessing the capability to cause DSBs (double-stranded breaks) and increase homology-directed repair. Self-splicing elements contain meganuclease genes. These endonucleases are very specific for the areas they want to cut because they have a long recognition site ranging from 4 to 40 base pairs. One advantage of meganucleases is their high specificity for target DNA, which is facilitated by their lengthy recognition sites. Additionally, their relatively small size facilitates ease of delivery. They also help make more recombinant DNA, which improves homology-directed repair (HDR) because they leave a 3′ overhang after DNA cleavage (Mandip and Steer 2019). Developing sequence-specific enzymes for all possible target sequences is laborious due to its time-consuming and expensive nature. Every novel genome engineering target necessitates an initial protein engineering step to create a customized MegN.

Zinc finger nucleases and TALENS are created to generate mutations at the targeted site, which leads to the editing of genes for novel traits. Zinc fingers are structural motifs in proteins that can bind to specific DNA sequences. Each zinc finger typically recognizes a short stretch of DNA, 3–4 bases. These zinc finger motifs are often part of transcription factors, which are proteins that regulate the transcription of genes by binding to specific DNA sequences. Engineered hybrid proteins and zinc finger nucleases combine the DNA-binding capability of zinc finger domains with the ability to induce double-strand breaks (DSBs) in DNA. Each ZFN subunit has several zinc finger domains that bind to DNA and a Fok1 endonuclease domain that cuts DNA (Bibikova et al. 2001). The specificity of zinc fingers depends on the target sequence and the adjacent sequence, which causes genome fragmentation and instability due to non-specific cleavages. This led to the development of alternate editing technology, TALENs, with a conserved repeat-based DNA-binding domain originating from transcription activator-like effectors and a FokI restriction endonuclease domain. In this method, sequence-specific targeting was achieved by 12–16 fused TAL domains, each targeting a single nucleotide in the DNA. As in the ZFNs, two complementary TALEN monomers, one binding sense and the other binding to the antisense strand, fused with FOK-I endonuclease, target the region of interest and create DSBs after homodimerization. The TALENs are simple to design and they produce a smaller number of off-target breaks. Indeed, one of the challenges associated with TALENs is their relatively large size compared to other genome editing tools, including ZFNs and CRISPR-Cas9 (Huo et al. 2019). The above three methods involve targeting a specific region with proteins having DNA binding domains; thus, it involves protein engineering that takes more time, is resource intensive, and requires a greater understanding of the DNA binding activity of these nucleases.

9.2.1 CRISPR

CRISPR stands for clustered, regularly interspaced short palindromic DNA sequences, and it is the most important platform for genome editing. The non-repetitive sequences and 32-bp repeat elements present downstream of the *iap gene were* identified by Ishino and team in *Escherichia coli* and later it was named CRISPR in 2002 (Ishino et al. 1987). The guide RNA sequence identifies the target loci through Watson-Crick base pairing and the endonuclease activity of Cas9 can induce gene modification by nicking the target DNA, causing the formation of double-strand breaks. These breaks set off non-homologous end joining (NHEJ) DNA repair mechanisms in living cells, which cause gene mutations like adding, removing and replacing genes.

Researchers mostly use the Type II CRISPR-Cas9 system compared to the other two. To edit target genes, the CRISPR-Cas9 system needs the CAS9 protein, crRNA (CRISPR RNA), tracrRNA (transactivating crRNA RNA) and RNase III (Ribonuclease III). The single guide RNA (sgRNA) was created by combining crRNA with tracrRNA, which serves the same function as a hybrid of crRNA and tracrRNA (Jinek et al. 2012). After the usage of CRISPR cas9 technology in plants was demonstrated, it was widely used by researchers worldwide to modify metabolic pathways, enhancing yield and crop quality.

9.2.2 Base Editors

Researchers combined a rat cytidine deaminase (rAPOBEC1) with an impaired dead Cas9 to develop the first form of Cytosine Base Editors (CBE), known as CBE1. The dead Cas9 refers to Cas9 with D10A and H840A mutations. The cytosine in the nontarget strand was replaced by uracil, and the cytosine was converted to thymine using the DNA repair mechanism. Then, the cytosine base editor is modified to rAPOBEC1-dCas9-UGI, which was constructed by adding a uracil DNA glycosylase inhibitor to the C-terminal end of CBE1 to prevent UNG activity (Komor et al. 2017). To further improve the efficiency of CBE, the dcas9 was replaced by cas9 nickase (ncas9 D10A) and two molecules of UGI were fused to the C-terminal end of cas9 nickase. The developed constructs were also validated in rice and generated mutants with a base editing efficiency of up to 13.3% (Lu and Zhu 2017).

Adenosine base editors were developed with nCas9 (D10A) fused with an artificially synthesized adenosine deaminase to edit bases A to G. Researchers initially developed ABE7.10 by fusing nCas9 (D10A) with a dimer consisting of wild-type adenine deaminase TadA and an evolved adenine deaminase called TadA7.10. This engineering resulted in the creation of ABE7.10, with specific attention given to defining the editing window of 4–8 nucleotides from the PAM region for precise adenine base editing. The subsequent advancement of ABE7.1 was ABE8e, which was created by incorporating a more efficient adenine deaminase variant, TadA8e (Gaudelli et al. 2020). PhieABE, an improved version of the ABE8e adenine base editor, was developed by fusing TadA8e and a single-stranded DNA binding domain. The PhieABE exhibits higher base editing efficiency and a broader base editing window than other ABE base editing systems (Tan et al. 2022).

9.2.3 SWISS: Base Editing at Multiple Sites

A multiplex base editing system called simultaneous and wide editing induced by a single system was developed for wide-scale editing of plant genomes using nCas9 and both ABE and CBE base editors. This system comprises sgRNA scaffolds possessing distinct RNA aptamers that recruit cognate-binding proteins. Fusing these binding proteins with cytidine or adenosine deaminases enables concurrent cytidine base editing and adenine base editing at different target sites. Additionally, by employing another set of sgRNAs within this dual-function system, SWISS can introduce a double-strand break at a third target site, thereby achieving tri-functional genome editing at multiple sites (Li et al. 2020a).

Using hygromycin resistance genes, two universal discriminated sgRNA-based Surrogate systems (DisSUGs) were created to decrease off-target mutations and improve base editing efficiency. Two systems are identified: one has a hygromycin resistance gene (Hyg−ATG) that does not contain the initiation codon ATG and the other has a coding sequence mutation (Hyg−TP). When the DisSUGs targeted endogenous genomic sites with enhanced sgRNA (esgRNA) and Hyg with wild-type sgRNA were employed, they were the most successful at enriching base-edited plant cells. Without increasing the frequency of off-target mutations across the rice genome, the frequency of homologous mutations induced by both cytosine and adenine base editors was markedly increased (Xu et al. 2020a).

9.2.4 Prime Editing

Prime editing is one of the genome editing techniques designed to introduce precise small insertions or deletions (indels) and various single or multiple base substitutions, including transitions and transversions, without the need for double-strand breaks (Anzalone et al. 2019). The reverse transcriptase (M-MLV-RT, Moloney murine leukemia virus reverse transcriptase) fused with a catalytically impaired nCas9 (H840A) at the C-terminus makes up the prime editor. As part of the prime editing process, the reverse transcriptase copies genetic information from the prime editing guide RNA (pegRNA) to the target genomic locus. These unique changes were created without the necessity of double-strand breaks or donor DNA templates. A reverse transcriptase fused to an RNA-programmable nickase and a prime editing guide RNA (pegRNA) are used in the first version of Prime Editors (PE1). This mechanism directly copies genetic data into the target genomic locus from an extension on the pegRNA. PE2, an advancement of the PE1 prime editor, incorporates an engineered RT to enhance editing efficiencies. Additionally, PE3 introduces nicks in the non-edited strand to induce its replacement, further amplifying editing efficiency. Typically, these prime editing approaches achieve editing efficiencies ranging from 20% to 50%, with 1–10% of indel formation.

Ni and co-workers (2023) developed a collection of new prime editors for common wheat use. This development involved engineering the prime editing guide RNA (pegRNA) and the protein components of the prime editor. The efficiency of prime editing in wheat was greatly increased by the introduction of a V223A mutation into the M-MLV RT (Moloney murine leukaemia virus reverse transcriptase) and the incorporation of an engineered pegRNA (epegRNA). These modifications altered the architecture of the PE protein by altering SpCas9 activity and nuclear localization signals (NLSs). They showed that they could edit 4–10 genes at once with an efficiency of up to 74.5% (Ni et al. 2023).

9.2.5 Multiplex Gene Editing

Multiplexed genome editing is a valuable tool for regulating gene expression, stacking desirable traits and controlling regulatory pathways. Consequently, it has played a pivotal role in advancing crop improvement, breeding and domestication endeavors, as it allows for simultaneous and targeted modifications of multiple genomic loci. Multiplex genome editing has also made it possible to combine traits that are important for farming in better plant germplasm. This has a significant influence on improving the intricate agronomic features.

Expressing Cas9 or Cas9-derived effectors along with multiple guide RNAs (gRNAs) targeting distinct sites accomplishes multiplex genome editing. Microinjection or the expression of gene constructs containing multiple single-guide RNA (sgRNA)-expressing cassettes has been a common way to deliver drugs in the past (Cong et al. 2013).

The prevailing strategy for achieving multiplex genome editing involves stacking multiple single-guide RNA (sgRNA)-expressing cassettes in a single plasmid construct, conventional delivery of guide RNAs and cas9 protein using multiple individual constructs and Csy4 endoribonuclease (endo-RNase) with Cas9 or a tRNA processing system. In the construct with multiple guide RNA expression system, each sgRNA-expressing cassette typically has a size of about 400–500 base pairs, comprising the RNA polymerase and other necessary elements for effective expression and targeting of specific genomic loci. This approach facilitates the simultaneous editing of multiple target sites within the genome.

Haurwitz and team identified Csy4 endoribonuclease as responsible for pre-crRNA processing in *Pseudomonas aeruginosa* (Haurwitz et al. 2010). Csy4 creates interactions specific to specific sequences inside the CRISPR repeat stem loop's major groove through this mechanism. Csy4 can precisely bind and cut pre-crRNAs. Xie and co-workers developed a novel tRNA processing system where multiple guide RNAs can be produced using a tRNA-gRNA construct with a specific architecture. The guide RNA expression units can be produced in vivo using an inbuilt RNA processing system. This approach for expressing gRNAs not only facilitates multiplex targeting but also enhances the editing efficiency of the CRISPR-Cas9 system in plants (Xie et al. 2015).

9.2.6 Ribonucleoproteins

The problems of delivering pDNA or mRNA can be avoided by directly delivering the RNP complex (Zuris et al. 2015). RNP delivery stands out for its rapid genome editing since it eliminates the necessity for intracellular transcription and translation. Simultaneously, transient genome editing not only allows for high editing efficiency but also mitigates the risks of off-target effects, insertional mutagenesis and immune responses. Physical approaches such as PEG-mediated transformation, microinjection, biolistics electroporation, iTOP, microfluidics and nanotube spearing have been employed for the cytosolic delivery of macromolecules. Even Microinjection, electroporation and microfluidics achieve high efficiency in RNP delivery and are applicable to almost all cell types. Still, they may induce cell damage and necessitate skilled manipulation and expensive equipment. Biolistic and PEG-mediated transformations are well-known technologies primarily used for plant cells (Zhang et al. 2021a). The materials used for delivering ribonucleoproteins include chitosan (CS) nanoparticles, PBAEs, nanogels and inorganic particles like gold nanoparticles, calcium phosphate nanoparticles and graphene oxides.

9.3 Target Traits for Rice Improvement

Several studies have proved the effective use of genome editing in rice improvement (Table 9.1). Typically, researchers express target-specificity-enhancing sgRNA and Cas9 for genome editing from transgenes incorporated into the host genome. To sustain the stable phenotypes of edited plants, researchers must remove the CRISPR-Cas9 construct inserted into the genome, as its presence makes it difficult to distinguish between newly generated and previously generated mutations. The presence of CRISPR-Cas9 in the genome also raises the possibility of off-target effects. Furthermore, crops that have undergone genome editing should not contain transgenes if regulatory approval is needed for commercial applications. Researchers use a range of methods to eliminate CRISPR-Cas9 components after altering the genome. The societal acceptance of genome editing in agriculture also depends on completing this step. Figure 9.1 shows the overall strategy for improving rice via genome editing.

9.3.1 Biotic Stress Resistance in Rice

Rice, a vital staple food for over 50% of the global population, is threatened by pathogens, including *Magnaporthe oryzae* (*M. oryzae*) causing rice blast and *Xanthomonas oryzae pv. oryzae* (Xoo) causing bacterial blight (Yin et al. 2018a).

Table 9.1 Details of genes used in genome editing approaches with their function in improving functional traits in rice

Trait	Target gene(s)	Editing tool	Function	Reference
1. Disease resistance				
Blast	*Pi21* and *OsSULTR3*	CRISPR-Cas9	Increased resistance to blast and bacterial leaf streak	(Yang et al. 2023)
	OsWRKY7	CRISPR-Cas9	Sensitivity to blast	Tun et al. (2023)
	RBL1	CRISPR-Cas9	Multi-pathogen resistance	Sridhar (2023)
	OsS5H1, *OsS5H2* and *OsS5H3*	CRISPR-Cas9	Broad spectrum disease resistance	Liu et al. (2023)
	OsUMP1	CRISPR-Cas9	Resistance to blast, BLB and sheath blight	Hu et al. (2023)
	OsHRC	CRISPR-Cas9	Resistance	Ding et al. (2023)
	OsPib	CRISPR-Cas9	More symptoms after pathogen inoculation	Xie et al. (2022)
	OsDjA2 and *OsERF104*	CRISPR-Cas9	Resistance	Távora et al. (2022)
	LLM9428/OsCATC	CRISPR-Cas9	Enhanced resistance to blast	Liao et al. (2022)
	OsMORE1a	CRISPR-Cas9	Blast and BLB resistance and susceptibility to *Cochliobolus miyabeanus*	Kim et al. (2022b)
	OsPita-Fuhui2663	CRISPR-Cas9	Resistance	He et al. (2022)
	OsGRF18	CRISPR-Cas9	Improved resistance	Yudong et al. (2021)
	OsUGT74J1	CRISPR-Cas9	Salicylic acid accumulation & resistance	Tezuka et al. (2021)
	Pi21, *Bsr-d1* and *OsERF922*	CRISPR-Cas9	Enhanced resistance to blast	Tao et al. (2021)
	OsUBC26	CRISPR-Cas9	Compromised resistance	Liu et al. (2021a)
	OsWRKY93	CRISPR-Cas9	Sensitivity to pathogen	Li et al. (2021a, b)
	OsECBS	CRISPR-Cas9	Impaired resistance	Liang et al. (2021)
	OsXa7	CRISPR-Cas9	Sensitivity to pathogen	Chen et al. (2021b)
	OsBsr-d1	CRISPR-Cas9	Enhanced blast resistance	Zhu et al. (2020)
	Ospi-ta	CRISPR-Cas9 (intron targeted insertion)	Resistance	Xu et al. (2020b)
	OsPi21	CRISPR-Cas9	Resistance	Nawaz et al. (2020)
	OsPiPR1	CRISPR-Cas9	Partially reduced blast resistance	Liu et al. (2020b, c, d)
	OsFLR1 and *OsFLR13*	CRISPR-Cas9	Increased susceptibility	Huang et al. (2020a, b, c)
	OsFLR2 and *OsFLR11*	CRISPR-Cas9	Enhanced resistance	Yang et al. (2020), Huang et al. (2020a, b, c)
	OsFLR1	CRISPR-Cas9	Enhanced pathogen attack	Huang et al. (2020a, b, c)
	OsFLR2	CRISPR-Cas9	Delayed pathogen attack	Huang et al. (2020a, b, c)
	Pi21, *Xa13* and TMS5	CRISPR-Cas9	Enhanced resistance to blast, BLB disease	Li et al. (2019)
	CSD1, *CSD2* and *Os11g09780*	CRISPR-Cas9	Enhanced resistance to blast	Li et al. (2019)
	OsSEC3A	CRISPR-Cas9	Enhanced resistance to blast	Ma et al. (2018)
	OsERF922	CRISPR-Cas9	Enhanced resistance to blast	Wang et al. (2016a)
	Pi21	CRISPR-Cas9	Enhanced resistance to blast	Wang et al. (2016b)

(continued)

Table 9.1 (continued)

Trait	Target gene(s)	Editing tool	Function	Reference
BLB	*OsSWEET11*	CRISPR-Cas9	Disease resistance	Sree et al. (2023), Kim et al. (2019)
	OsCPK24	CRISPR-Cas9	More susceptibility to BLB	Lu et al. (2023)
	OsXa13	CRISPR-Cas9	Disease resistance	Li et al. (2022a, b, c)
	OsSWEET13	CRISPR-Cas9	Resistance	Diana et al. (2022)
	OsTFIIAγ1 or *OsTFIIAγ5 (TF)*	CRISPR-Cas9	Resistance	Xu et al. (2021a, b, c, d)
	OsUPT box	CRISPR-Cas12a	Improved resistance	Yu et al. (2021)
	OsEBEAvrXa23	CRISPR-Cas9 knock-in	Resistance	Wei et al. (2021a, b)
	OsSWEET14	CRISPR-Cas9	Less susceptibility	Arulganesh et al. (2021), Duy et al. (2021), Zeng et al. (2020a, b), Kim et al. (2021), Zafar et al. (2020)
	OsSWEET11, OsSWEET14 and *OsSULTR3;6*	CRISPR-Cas9	BLB and leaf streak resistance	Ni et al. (2021)
	OsPrx30	CRISPR-Cas9	Increased resistance	Liu et al. (2021a, b, c)
	OsXa13 promoter	CRISPR-Cas9	Disease resistance	Li et al. (2020a, b)
	OsCUL3a	CRISPR-Cas9	Enhanced resistance to BLB	Gao et al. (2020b)
	OsSWEET11, OsSWEET13 and *OsSWEET14*	CRISPR-Cas9	Enhanced resistance to BLB	Oliva et al. (2019)
	Bsrk-1	CRISPR-Cas9	Enhanced resistance to BLB and blast	Zhou et al. (2018)
	Os09g29100 (Tal7)	TALENs	Enhanced resistance to BLB	Cai et al. (2017)
	OsSWEET14	TALENs	Enhanced resistance to BLB	Blanvillain-Baufumé et al. (2017)
	OsSWEET13	TALENs	Enhanced resistance to BLB	Zhou et al. (2015)
	SWEET14, SWEET11	NHEJ	Rice bacterial blight resistance	Jiang et al. (2013)
	OsSWEET14	TALENs	Enhanced resistance to BLB	Li et al. (2012)
Sheath blight	*OsPP2A-1*	CRISPR-Cas9	Susceptible	Lin et al. (2021)
	OsMESL	CRISPR-Cas9	Enhanced resistance to BLB, blast and sheath blight	Hu et al. (2021a)
RTV	*OseIF4G*	CRISPR-Cas9	Enhanced tolerance to RTV and black-streaked dwarf virus	Wang et al. (2021a, b, c, d)
	OseIF4G	CRISPR-Cas9	Enhanced tolerance	Nithya et al. (2020)
Rice black-streaked dwarf virus	*OsAGO2*	CRISPR-Cas9	Over-expression enhances susceptibility & mutants showed resistance	Wang et al. (2021a, b, c, d)

(continued)

Table 9.1 (continued)

Trait	Target gene(s)	Editing tool	Function	Reference
Southern rice black-streaked dwarf virus	*OsAP47*	CRISPR-Cas9	Enhances resistance against Rice black-streaked dwarf virus and southern rice black-streaked dwarf virus	Wang et al. (2022a)
	OsARF17 (TF)	CRISPR-Cas9	Increased susceptibility	Zhang et al. (2020a)
Bacterial leaf streak	*OsSULRT3; 6*	CRISPR-Cas9	Resistance	Xu et al. (2021a, b, c, d)
Meloidogyne incognita	*OsBet v1*	CRISPR-Cas9	Increased susceptibility	Li et al. (2022a, b, c)
Yellow mottle virus	*OsCPR5.1*	CRISPR-Cas9	Disruption resulted in RYMV resistance	Arra et al. (2023)
	OsCPR5.2	CRISPR-Cas9	Susceptibility after disruption	Arra et al. (2023)
BPH	*OsCKXs*	CRISPR-Cas9	Enhanced resistance	Zhang et al. (2022a, b)
	OsAOC and *OsMYC2*	CRISPR-Cas9	Better performance of pathogen in knock out lines	Xu et al. (2021a, b, c, d)
2. Abiotic stress tolerance				
Drought	*OsWRKY76*	CRISPR-Cas9	Weak drought tolerance	Zhang et al. (2023)
	OsLKP2	CRISPR-Cas9	Enhanced drought tolerance	Shim et al. (2023)
	OsSAPK3	CRISPR-Cas9	Reduced drought sensitivity	Lou et al. (2023)
	OsIPA1	CRISPR-Cas9	Regulates *SNAC1* to improve drought tolerance	Chen et al. (2023)
	OsMADS26	CRISPR-Cas9	Proof of concept study	Anjala and Augustine (2022)
	OsBC1L1 or *OsBC1L8*	CRISPR-Cas9	Stomatal clustering & stomatal production	Li et al. (2021b)
	OsEPFL10 and *STOMAGEN*	CRISPR-Cas9	Increased tolerance	Karavolias et al. (2021)
	OsbZIP72	CRISPR-Cas9	Sensitivity to salinity & drought	Baoxiang et al. (2021)
	OsABA8ox2	CRISPR-Cas9	Enhanced drought tolerance	Zhang et al. (2020a, b)
	OsmiR535	CRISPR-Cas9	Enhanced tolerance against drought, salinity and cold stresses	Yue et al. (2020)
	OsSDD1 and *OsRSD1*	CRISPR-Cas9	Reduced stomatal density	Yu et al. (2020)
	OsNAC006	CRISPR-Cas9	Drought & heat sensitivity	Wang et al. (2020a, b, c, d)
	OsPYL9	CRISPR-Cas9	Enhanced ABA accumulation & reduced stomatal conductance	Usman et al. (2020a)
	OsDST	CRISPR-Cas9	Drought and salt tolerance	Kumar et al. (2020)
	OsAFP1	CRISPR-Cas9	Decreased water loss & ABA sensitivity	Tianshun et al. (2021)
	OsERA1	CRISPR-Cas9	Enhanced drought tolerance	Ogata et al. (2020)
	OsAAA-1 and *OsAAA-2*	CRISPR-Cas9	Enhanced drought tolerance	Lu et al. (2020)
	OsmiR818b	CRIPSR-Cas9	Lower resistance	Chung et al. (2020)
	SRL 1,2	CRISPR-Cas9	Drought tolerance	Liao et al. (2019)
	OsSAPK2	CRISPR-Cas9	Drought stress response	Lou et al. (2017)

(continued)

Table 9.1 (continued)

Trait	Target gene(s)	Editing tool	Function	Reference
Heat	*OsLRK1*	CRISPR-Cas9	Compromised growth at 35 °C	Qu et al. (2020)
	OsNTL3	CRISPR-Cas9	Heat sensitivity	Liu et al. (2020b, c, d)
Salinity	*OsNPF8.1*	CRISPR-Cas9	Less salinity and drought tolerance in knock-out lines	Diyang et al. (2023)
	OsGL1–11	CRISPR-Cas9	Salinity sensitivity	Zhao et al. (2022)
	OsVDE	CRISPR-Cas9	Higher survival rate & stomatal conductance	Wang et al. (2022b)
	OsBEAR1	CRISPR-Cas9 & artificial miRNA	Salt sensitivity	Teng et al. (2022)
	OsBadh$_2$	CRISPR-Cas9	Improved tolerance in non-aromatic variety	Prodhan et al. (2022)
	OsGLYI3	CRISPR-Cas9	Salinity sensitivity	Liu et al. (2022)
	OsRR22	CRISPR-Cas9	Improved salt tolerance	Zhang et al. (2019), Han et al. (2022), Tripathy et al. (2021)
	OsXLG2 and *OsXLG4*	CRISPR-Cas9	Salinity tolerance	Biswal et al. (2022)
	OsbHLH024 (TF)	CRISPR-Cas9	Improved salt tolerance	Alam et al. (2022a, b)
	OsbHLH044	CRISPR-Cas9	Salinity sensitivity	Alam et al. (2022a, b)
	OsHKT2;1	CRISPR-Cas9	Enhanced salt tolerance by reduction of sodium ion & ROS accumulation	Wei et al. (2021a, b)
	OsMKK1, OsMKK6 and *OsMKK1/6*	CRISPR-Cas9	Increment in auxin contents with enhanced lateral roots under salinity	Yang et al. (2021)
	OsqSOR1	CRISPR-Cas9	Reduced stress in saline fields	Kitomi et al. (2020)
	OsGTγ-2	CRISPR-Cas9	Salinity hypersensitive plants	Liu et al. (2020b, c, d)
	OsSST	CRISPR-Cas9	Better plant adaptation in saline conditions	Lian et al. (2020)
	OsMPT3;1 and *OsMPT3;2*	CRISPR-Cas9	Increased Na+/K + ratio	Huang et al. (2020a, b, c)
	OsSERK2	CRISPR-Cas9	Salinity sensitivity	Dong et al. (2020)
	Osgs3 and *Osdep1*	CRISPR-Cas9	Improved salinity tolerance	Cui et al. (2020)
	OsPQT3	CRISPR-Cas9	Improved grain yield and enhanced resistance	Alfatih et al. (2020)
	OsRAV2	CRISPR-Cas9	Salinity stress response	Duan et al. (2016)
Cold	*OsCOLD11*	CRISPR-Cas9	Reduce chilling tolerance	Li et al. (2023)
	OsWRKY76	CRISPR-Cas9	Decreased cold tolerance	Zhang et al. (2022a)
	OsPIN5b, OsGS3, OsMYB30	CRISPR-Cas9	Improved cold tolerance with enhanced yield	Zeng et al. (2020a, b)
	OsAnn5	CRISPR-Cas9	Cold sensitivity	Que et al. (2023)
	OsTCD3	CRISPR-Cas9	Cold sensitivity	Lin et al. (2020)
	OsANN3	CRISPR-Cas9	Cold stress response	Shen et al. (2017)
	TIFY1a, TIFY1b	CRISPR-Cas9	Cold tolerance	Huang et al. (2017)

Table 9.1 (continued)

Trait	Target gene(s)	Editing tool	Function	Reference
Herbicide	*OsHPPD3*	CRISPR-Cas12a	Herbicide tolerance	Wu et al. (2023)
	OsALS	CRISPR-Cas9	Herbicide tolerance	Zafar et al. (2023)
	OsCYP76C6	CRISPR-Cas9	Increased conjugates & isoproturon metabolites	Zhai et al. (2022)
	OsPUT1/2/3	CRISPR-Cas9	Paraquat resistance	Lyu et al. (2022)
	OsALS	CRISPR-Cas9	Strong resistance	Liying et al. (2022)
	OsALS	Base editing	Improve herbicide tolerance	Zhang et al. (2021b)
	OsACC1	Prime-editing-library-mediated saturation mutagenesis	Herbicide resistance	Xu et al. (2021a, b, c, d)
	OsALS	CRISPR-Cas9	Herbicide tolerance	Wang et al. (2021a, b, c, d)
	OsAFB4	CRISPR-Cas9	Improved picloram resistance	Guo et al. (2021)
	OsHIS1	CRISPR-Cas9	Herbicide sensitive	Komatsu et al. (2020)
	OsALS1	Base editing	Herbicide tolerant	Kuang et al. (2020)
	OsALS	CRISPR-Cas12a	Efficient biallelic gene targeting	Li et al. (2020a)
	OsALS	CRISPR-Cas9	Herbicide tolerance	Butt et al. (2020)
	OsEPSPS	Prime editing	Herbicide resistance	Butt et al. (2020)
	OsALS	Base editing	Herbicide resistance	Shimatani et al. (2017)
	OsALS	CRISPR-Cas9	Herbicide resistance	Sun et al. (2016)
	OsEPSPS	CRISPR-Cas9	Herbicide resistance	Li et al. (2016b)
Heavy metals	*OsmiR535*	CRISPR-Cas9	Tolerance to Cd toxification	Yue et al. (2023)
	OsACE2	CRISPR-Cas9	Detoxification to oxyfluorfen	Xu et al. (2023)
	OsNramp5	CRISPR-Cas9	Mn, Cu and Se reduced in knock-out lines	Feng et al. (2023)
	OsLCD	CRISPR-Cas9	Enhanced Cd tolerance	Elkonin et al. (2023)
	OsCERK1	CRISPR-Cas9	Cupric oxide nanoparticles tolerance	Chen et al. (2022b)
	OsPMEI12	CRISPR-Cas9	Enhanced fresh weight and dry weight in Cd stress	Li et al. (2022a, b, c)
	OsNRAMP5	CRISPR-Cas9	Decreases in root uptake of Pb	Chang et al. (2022)
Lodging	*SD1*	CRISPR-Cas9	Enhanced resistance to blast and lodging resistance	Gang et al. (2023)
	OsSd-1	CRISPR-Cas9	Lodging resistance	Beyene et al. (2022)
	OsGW2	CRISPR-Cas9	Lodging resistance	Yamaguchi et al. (2020)
	OsRhoGDI2	CRISPR-Cas9	Semi dwarfism	Wang et al. (2020a, b, c, d)
3. Quality improvement and nutritional enhancement				

(continued)

Table 9.1 (continued)

Trait	Target gene(s)	Editing tool	Function	Reference
	LyCE	CRISPR-Cas9	β-Carotene accumulation	Kim et al. (2022a)
	OsBADH2	CRISPR-Cas9	Aroma	Shanthinie et al. (2019)
	Or	CRISPR-Cas9	β-Carotene accumulation	Endo et al. (2019)
	WAXY	CRISPR-Cas9	Starch metabolism	Zhang et al. (2018a)
	OsSBEIIb and *OsPDS*	CRISPR-Cas12a	Encodes starch branching enzyme and phytoene desaturase	Li et al. (2018a)
	FAD2–1	CRISPR-Cas9	Fatty acid metabolism	Abe et al. (2018)
	SBEI, SBEIIb	CRISPR-Cas9	Starch metabolism	Sun et al. (2017)
	Badh2	TALEN	Aroma	Shan et al. (2015)
4. Yield optimization				
	OsMADS17	CRISPR-Cas9	Increases grain number and altered grain weight	Li et al. (2023)
	OsKRN2	CRISPR-Cas9	Increased grain number	Huang et al. (2021)
	OsHXK1	CRISPR-Cas9	Increased photosynthetic efficiency, yield	Zheng et al. (2021)
	OsSPL16/OsGW8	CRISPR-Cas9	Altered grain length	Usman et al. (2020b)
	LARGER2	CRISPR-Cas9	Altered panicle size and increased grain number	Chen et al. (2022a)
	OsSPL4	CRISPR-Cas9	Increased grain number and altered grain size	Hu et al. (2021b)
	OsPDCD5	CRISPR-Cas9	Altered plant architecture and enhanced grain yield	Dong et al. (2021)
	FZP	CRISPR-Cas9	Altered panicle branching and increased grain number	Chen et al. (2021a)
	OsER1	CRISPR-Cas9	Increased grain number	Guo et al. (2020)
	FWL4	CRISPR-Cas9	Enhanced tiller number and yield	Gao et al. (2020a)
	GS3, GW2, Gn1a	CRISPR-Cas9	Yield improvement	Zhou et al. (2019)
	OsGS9	CRISPR-Cas9	Altered grain length	Zhao et al. (2018)
	GS3, Gn1a	CRISPR-Cas9	Enhanced yield	Shen et al. (2018), Ashikari et al. (2005)
	PYL 1, PYL 4, PYL 6	CRISPR-Cas9	Plant growth & stress response	Miao et al. (2018)
	OsAAP3	CRISPR-Cas9	Enhanced grain yield	Lu et al. (2018)
	Gn1a, DEP1	CRISPR-Cas9	Enhanced yield	Huang et al. (2018)
	OsSPL14	Base editing	Ideal plant architecture	Hua et al. (2019)
	OsCCD7	CRISPR-Cas9	Enhanced tillering	Butt et al. (2018)
	OsCDC48	Base editing	Regulates senescence & cell death	Zong et al. (2017)
	GW2, GW5, TGW6	CRISPR-Cas9	Yield improvement	Xu et al. (2016)
	Gn1a, DEP1, GS3, IPA1	CRISPR-Cas9	Yield improvement	Li et al. (2016b)
5. Other key traits				

Trait	Target gene(s)	Editing tool	Function	Reference
	OsLPR3	CRISPR-Cas9	Improved pi tolerance	Lin et al. (2023)
	EPF1	CRISPR-Cas9	Altered stomatal density	Rathnasamy et al. (2023)
	OsHAK8	CRISPR-Cas9	Reduced K^+ uptake	Wang et al. (2021a, b, c, d)
	OsIRO3	CRISPR-Cas9	Accumulation of magnesium, iron and ROS in leaves	Wang et al. (2020a, b, c, d)
	OsSAW1	CRISPR-Cas9	Male sterility	Wang et al. (2020a, b, c, d)
	OsDOFs (TFs)	CRISPR-Cas9	11 DOFs regulate heading date in long day conditions & 9 DOFs control heading date in short-day conditions	Huang et al. (2020a, b, c)
	REC8, PAIR, MTL and *OSD1*	CRISPR-Cas9	Asexual reproduction	Wang et al. (2019)
	BBM	CRISPR-Cas9	Formation of an embryo from a fertilized egg	Khanday et al. (2019)
	OsVP1	CRISPR-Cas	Dormancy	Jung et al. (2019)
	NRT1.1B	CRISPR-Cas9	Nitrogen efficiency use	Li et al. (2018b)
	OsCDC48	Base editing (C to T)	Regulates senescence & cell death	Zong et al. (2017)
	EPFL9	CRISPR-Cas9	Stomatal development	Yin et al. (2017)
	Hd2, Hd4, Hd5	CRISPR-Cas9	Flowering time	Li et al. (2017)
	LOX3	TALEN	Improves storability	Ma et al. (2015)
	ROC5, SPP, YSA	CRISPR-Cas9	Albino phenotype	Feng et al. (2013)

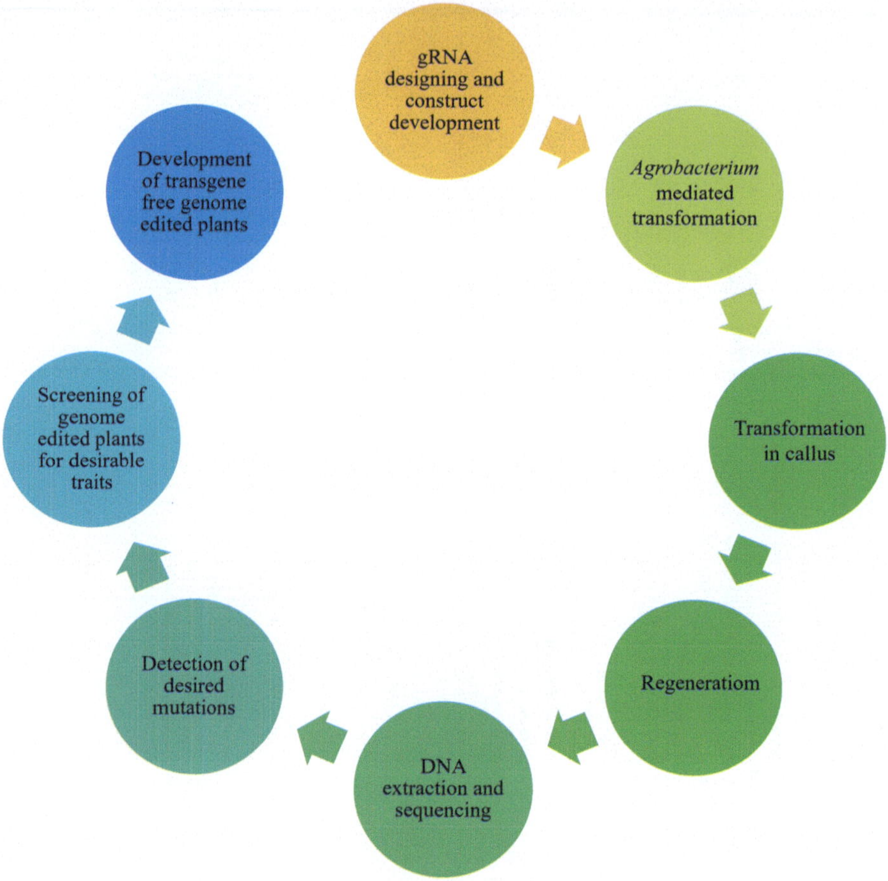

Fig. 9.1 The overall strategy for rice improvement via genome editing

The deadly impact of these diseases has prompted many scientists to do much research on this aspect. To economically and effectively control these diseases, creating rice cultivars with broad-spectrum resistance to both pathogens is a key strategy (Deng et al. 2017). Rice susceptibility genes modified by CRISPR-Cas9 act as a possible alternative breeding method to develop disease resistance (Shaheen et al. 2023).

9.3.1.1 Bacterial Leaf Blight (BLB)

Targeted mutations in the UPT (upregulated by transcription activator-like 1 effector box) box of the Xa13 promoter with CRISPR-Cas12a improved resistance to BLB. The pollen-specific expression of *Zmaa1* and the *α-amylase* gene eliminated transgenic pollen contamination and DsRed allowed for the visual identification of transgenic seeds and their subsequent disposal (Yu et al. 2021). *OsSWEET14* promoter disruption in EBE leads to the development of BLB-resistant cultivars like CO 51 (*indica* rice cultivar) and TBR225 (Vietnamese rice variety) (Arulganesh et al. 2021; Duy et al. 2021). Similarly, the EBE promoters of *OsSWEET11* and

OsSWEET13 were edited and resistant to moderately resistant CO 51 cultivars were developed through *Agrobacterium*-mediated transformation (Sree et al. 2023; Diana et al. 2022). By means of multiplex CRISPR-Cas9 genome editing, Kitaake rice is engineered to be resistant to all known strains of *Xoo* by the targeted disruption of the *SWEET* gene, TALe EBEs (*SWEET11, SWEET13* and *SWEET14*). Furthermore, the modified mega variety lines Ciherang-Sub1 and IR64 developed normally without any reduction in yield (Oliva et al. 2019).

9.3.1.2 Blast

Tavora and team reported *OsDjA2* and *OsERF104* gene deletion using CRISPR-Cas9, significantly leading to blast resistance (Távora et al. 2022). *OsERF922* targeted mutagenesis resulted in a lesser number of blast lesions in edited lines compared to non-edited lines (Wang et al. 2016a). Enhanced resistance to blast was observed with the mutants of *FLR2* and *FLR11* (FERONIA-like receptor) genes, which could be due to changed defense-related gene expression and ROS production (Yang et al. 2020). In addition to gene disruption, rice lines have also been designed to develop blast resistance through the CRISPR-Cas gene knock-in strategy. By inserting exon #2 of the *Pi-ta* gene (blast-resistant gene) into the *pi-ta* gene (susceptible gene), Xu and team successfully achieved blast resistance (Xu et al. 2020b). An *indica* TGMS (thermosensitive genic male sterile) line, Longke638S, was simultaneously edited with *Bsr-d1, Pi21* and *ERF922* genes for the development of multiple disease-tolerant rice. Single mutants and triple mutants exhibited enhanced resistance to blast and BLB, with no negotiations between resistance and agronomic traits. Unexpectedly, *Pi21/erf922* single mutants displayed increased resistance to bacterial blight. All these were attributed to the increased expression of genes related to the SA and JA pathways (Zhou et al. 2022).

9.3.1.3 Sheath Blight

According to Jung and co-workers, *Osamt1* lines (an ammonium transporter) exhibited hypersensitivity to isolates of *R. solani*, demonstrating a positive correlation between NH_4+ uptake and sheath blight disease (Jung et al. 2023). In addition, the *Protein Phosphate (OsPP2A1)* gene has been subjected to the CRISPR-Cas9 system, which produced five mutants: pp2a-1-1 with 1 and 2 bp insertions in the very first exon, while pp2a-2 exhibited a one bp loss in the 11th exon. Furthermore, pp2a-5 had a one bp loss in the first exon in addition to the one bp insertion in pp2a-3 and pp2a-4's first and second exons, respectively. Interestingly, in contrast to the control, all these mutants showed hypersensitivity to the sheath blight illness (Lin et al. 2021). A novel gene named methyl esterase-like (*OsMESL*) affecting reactive oxygen species (ROS) accumulation confers resistance to bacterial blight, rice blast and sheath blight (Hu et al. 2021a).

9.3.1.4 Rice Tungro Virus

A novel in-frame mutation in the *eIF4G* (translation initiation factor 4 gamma) gene in the SVLFPNLAGKSYVV motif successfully imparts resistance to rice tungro disease caused by rice tungro spherical virus (RTSV) (Kumam et al. 2022).

9.3.2 Abiotic Stress Tolerance in Rice

Abiotic stresses like salt, drought and cold significantly reduce yields and make it difficult to generate sufficient food to sustain the world's expanding population. Breeding cultivars resistant to these stresses seems like a long-term fix for this issue. Rapid, reliable trait modification for crop improvement has recently become feasible due to recent advancements in genome editing technology.

9.3.2.1 Salinity Tolerance

Salinity is one of the primary concerns influencing rice production globally. The most effective way to manage salinity is to grow salt-tolerant rice cultivars (Saminadane et al. 2023; Changappa et al. 2023). Takagi and team observed a drastic increase in salt tolerance when the *OsRR22* gene failed to produce a 696 amino acid type-B response regulator transcription factor involved in metabolism and cytokinin signal transduction (Takagi et al. 2015). With this evidence, CRISPR-Cas knockout was attempted by Zhang and co-workers, resulting in the development of saline-tolerant homozygous mutant lines with no significant differences in agronomic traits in comparison with wild-type plants (Zhang et al. 2019). A single base pair alteration in the *OsqSOR1* gene's third exon resulted in premature truncation and led to the production of soil surface roots with increased salinity tolerance (Kitomi et al. 2020). The extra-large GTP-binding protein (*osxlg1/osxlg4*) double mutants probably grew longer roots, which made them better able to handle salt (Biswal et al. 2022). On the other hand, it has also been documented that positive regulators for salinity tolerance are disrupted, resulting in the rice lines becoming sensitive to salt. According to Yang and team, the disruption of the Ca^{2+} sensor calmodulin (*OsCaM1*) gene resulted in a weakened lateral root length and lower primary root, indicating its sensitivity to salt stress (Yang et al. 2021). Similarly, saline hypersensitivity was produced by a CRISPR-Cas9-based mutation in the first exon of the glyoxalase (*OsGLYi3*) gene, which was followed by increased methylglyoxal and glyoxalase I activity (Liu et al. 2022).

9.3.2.2 Drought Tolerance

Drought and salt tolerance (*DST*) gene-edited homozygous mutant plants phenocopied the mutation N69D, reported in a Japanese cultivar with a 366 bp deletion, leading to moderate-level osmotic stress tolerance and high-level salt stress tolerance at the seedling stage. The mutants exhibited reduced stomatal density, broader leaves and increased leaf water retention in dehydration (Kumar et al. 2020). Ogata and team created *ENHANCED RESPONSE TO ABA1* (*ERA1*) mutants that were more sensitive to ABA and had a stronger response to drought stress through stomatal regulation. The mutants exhibited enhanced root growth with comparable leaf growth to wild-type plants (Ogata et al. 2020). Homozygous mutants obtained from semi-rolled leaves 1 and 2 (*SRL1* and *SRL2*) revealed reduced chlorophyll content, stomatal number, stomatal conductance and increased panicle number. Moreover, there is a reduction in malondialdehyde content accompanied by an increased survival rate, grain filling percentage, ABA content, SOD and CAT activities related to wild types, giving rise to enhanced yield under drought stress with curled leaves (Liao et al. 2019). A vital regulator of ABA signaling and osmotic stress responses, SNF-1-RELATED PROTEIN KINASE 2 (SnRK2) is a member of the plant-specific protein kinase family and an essential regulator of ABA signaling and osmotic stress responses. Lou and co-workers used CRISPR-Cas9 to conduct a thorough analysis of *SAPK2*'s activities. Given that the recently generated mutants were ABA-insensitive and lacking functional *SAPK2*, it is clear that *SAPK2* is required for ABA-mediated seed dormancy (Lou et al. 2017). The fact that the *SAPK2* mutants were more vulnerable to oxidative stress and drought than wild-type plants underscored the significance of *SAPK2* for rice drought tolerance. Moreover, it has been discovered that a large number of stress-related factors (TFs) positively correlate with drought tolerance. For example, under drought stress, CRISPR-Cas9-generated knock-out lines of *NAM*, *ATAF1/2* and *CUC2* (OsNAC006) showed decreased growth of antioxidants (catalase, peroxidase and peroxidase dismutase) and morpho-physiological features (Wang et al. 2020a, b, c, d). Similar to this, in cases of 20% PEG treatment, *OsWRKY76* (w76-1 and w76-2) TF disrupted by CRISPR-Cas9 resulted in symptoms like wilting, chlorosis, electrolyte leakage and water loss (Zhang et al. 2023).

9.3.2.3 Heat Tolerance

Scientists have attempted to create heat-resistant rice varieties using the CRISPR-Cas9 system to mitigate the negative effects of extended heat stress. Null mutants of *OsNAC006* induced by CRISPR-Cas9 have shown thermotolerance in rice, which was accompanied by increased levels of H_2O_2 and O_2 (Wang et al. 2020a, b, c, d). Additional genes that have been positively associated with heat tolerance must also be expressed to ensure heat tolerance. For example, there is a positive correlation between leaf dark respiration and *LRK1* (leucine-rich repeat receptor kinase). CRISPR-Cas9 produced a premature protein termination by mutating a single base pair at nucleotide 508, resulting in the production of a stop codon at amino acid 180. When *lrk1* mutants were exposed to 35 °C for 10 days, their morphological growth was slowed and their dark respiration was reduced (Qu et al. 2020).

9.3.2.4 Cold Tolerance

Since rice is highly susceptible to cold, particularly when it's still in the seedling stage, low temperatures are thought to be the primary factor influencing rice growth and productivity. It is imperative to engineer cold tolerance in the plant in order to improve rice production. Zeng and colleagues successfully knocked down the native cold tolerance gene *MYB30* along with two yield-related genes at the same time. The triple mutant outperformed wild-type plants in terms of yield and cold tolerance (Zeng et al. 2020a). Similarly, *osatp4* and *ostcd3* mutants exposed to lower temperatures showed lowered chlorophyll levels and enhanced cold resistance (Lin et al. 2020). Furthermore, cold tolerance is closely associated with WRKY TFs. For example, cold tolerance was introduced in rice using CRISPR-Cas9-directed disruption of OSWRKY63 TF (Li et al. 2021a). Additionally, researchers have also identified positive regulators in this context. For instance, the fifth exon of the annexin (*OsAnn5*) gene in edited rice seedlings was sensitive to cold (Qu et al. 2020). Zhand and team used CRISPR-Cas9 to alter the OsWRKY76 TF and create cold-sensitive lines (Zhang et al. 2022a, b).

9.3.2.5 Herbicide Tolerance

The development of herbicide-tolerant genotypes through the use of CRISPR-Cas systems may prove to be an effective approach to combating the challenge of weeds and boosting rice production worldwide. The Acetolactase Synthase (*ALS*) gene initiates the first step of branched amino acid biosynthesis, which is the primary target of bispyribac sodium (BS) along with other herbicides. Basmati rice's ability to withstand BS has recently been achieved by tryptophan to leucine conversion at the 548th position of the *ALS* gene (Zafar et al. 2023). To enhance the substitution rate at the target site, Xu and co-workers created saturation mutagenesis mediated by prime-editing libraries. As a result, they were able to identify 16 different mutations in *OsACC1* that correspond to herbicide resistance (Xu et al. 2021a, b, c, d; Liu et al. 2020a). Moreover, *OsPUT* knock-out lines yielded triple mutants showing resistance to the paraquat (Alfatih et al. 2020). Glyphosate is a commonly used chemical that harms the main crop, in addition to BS. A single base pair substitution in the glyphosate resistance gene (*EPSPS*) from 96G to A generated herbicide tolerance (Jiang et al. 2022).

9.3.2.6 Heavy Metal Tolerance

Heavy metal contamination in food could harm human health. Removing heavy metals from the soil would reduce the amount of metal that accumulates in rice grains. Furthermore, genome editing has the potential to decrease the concentrations of harmful metals in food crops. By knocking out *OsNramp5*, a metal transporter gene, low-Cd rice lines were successfully created using CRISPR-Cas9, which showed an enormous reduction in Cd concentration compared to the wild-type. Field studies showed that the mutants had <0.05 mg/kg of Cd, while the control plants had 0.33–2.90 mg/kg without affecting the yield (Tang et al. 2017). Yu and co-workers found that when Cd levels were controlled at 2 μmol/L, the mir535 CRISPR-Cas9 mutants of brown rice kept 35% less Cd than the control (Yue et al. 2023). Elkonin and team found that shoots of plants lacking the low cadmium (*oslcd*) gene took in less Cd when grown in soil high in Cd (Elkonin et al. 2023).

9.3.3 Grain and Nutritional Quality Improvement in Rice

CRISPR technology has revolutionized rice breeding, contributing significantly to the enhancement of grain and nutritional quality. By precisely editing the rice genome, CRISPR allows researchers to target specific genes associated with grain characteristics and nutritional content. This targeted approach facilitates the development of rice varieties with improved traits such as larger grain size, enhanced nutritional value and increased resistance to pests and diseases. Through CRISPR-mediated genome editing, scientists can fine-tune the expression of key genes involved in nutrient metabolism, ultimately leading to rice varieties that offer improved nutritional benefits and contribute to global food security.

9.3.3.1 High Oleic/Low Linoleic Rice

A vital constituent in rice bran oil is oleic acid. Raising the amount of oleic acid in rice bran oil might benefit people's health and aid in illness prevention. Plants require the enzyme fatty acid desaturase 2 (FAD2) to transform oleic acid into linoleic acid. *OsFAD2-1* predominantly expresses in seeds among the three functional FAD2 genes identified in rice. Rice bran oil's fatty acid composition was improved by targeted mutagenesis of *OsFAD2-1* using CRISPR-Cas9, which produced *OsFAD2-1* knockout plants with twice as much oleic acid and no detectable linoleic acid (Abe et al. 2018).

9.3.3.2 β-Carotene Accumulation

It is important to increase the nutritional content of rice through carotene biofortification. To target the *OsOr* gene, which is an ortholog of the *Orange* (*Or*) gene found in cauliflower, scientists engineered rice to express CRISPR-Cas9. Furthermore, it results in a build-up of beta-carotene in cauliflower curd. The intentional alteration of *OsOr* through CRISPR-Cas9 genome editing resulted in an increased accumulation of β-carotene within the rice callus (Endo et al. 2019). A critical enzyme in higher plants' carotenoid biosynthesis pathway is lycopene epsilon-cyclase (*LcyE*). Kim and team established an approach for targeted mutagenesis in rice using CRISPR-Cas9 and geminiviral replicon encompassing golden SNP substitution of the *LcyE* gene. In contrast to wild-type (WT), the total carotenoid content of the LcyEsg2-HDR1 and LcyEsg2-HDR2 lines was 6.8–9.6 times greater and there was a decline in the level of reactive oxygen species (Kim et al. 2022a).

9.3.3.3 Aromatic Rice

A key attribute that impacts rice grains' acceptability and marketability is their fragrance. Mutations of betaine aldehyde dehydrogenase (*OsBADH$_2$*) cause rice to produce aroma. Fragrant rice was also developed using CRISPR-Cas9-based genome editing of *Badh$_2$* on the ASD 16 (*indica* rice variety). On editing the seventh exon of *Badh$_2$*, novel aromatic chemical compounds, such as pyrrolidine, pyrazine, pyridine, pyrozole and pyradazine, were found in the grains of T_1 progenies by adopting comparative volatile profiling. These research efforts may facilitate the breeding of aromatic rice more swiftly (Shanthinie et al. 2019). *Badh$_2$* gene editing by TALEN resulted in 2AP levels rising from 0.35 to 0.75 mg/kg. It is nearly identical to the aromatic rice variety, which served as a positive control (Shan et al. 2015).

9.3.3.4 Red Rice

The RcRd genotype in wild rice (*O. rufipogon*) determines the richness of health-promoting proanthocyanidins and anthocyanins in red rice. Many cultivated rice varieties have white grains due to a 14-bp frame-shift deletion in the seventh exon of the *Rc* gene. Using CRISPR-Cas9, researchers reversed this deletion, converting white pericarp rice into red pericarp. Mutants showed no significant differences in agronomic traits compared to the wild type, suggesting that modifying *Rc* doesn't negatively impact major traits (Zhu et al. 2019).

9.3.3.5 Increased Amylose Content

High-amylose cereals serve as an excellent reservoir of resistant starch (RS) (Raja et al. 2017; Gurunathan et al. 2019). CRISPR-Cas9, employed to mutate the *SBEI* and *SBEIIb* genes in rice, regulates amylose content (AC). While *SBEI* edits showed no distinct difference from the wild type, *SBEIIb* edits led to significant increases in AC (up to 25%) and RS (up to 9.8%) (Sun et al. 2017).

9.3.4 Yield Optimization in Rice

Enhancing grain yield in rice is a vital trait that involves manipulating numerous quantitative trait loci (QTLs) governed by multiple genes. Researchers have applied CRISPR-Cas9 technology in diverse strategies to enhance rice grain yield.

9.3.4.1 Altering Panicle Architecture and Grain-Related Attributes

CRISPR-Cas9 was employed to successfully mutate four key yield-related genes in rice: *Gn1a* (grain number), *GS3* (grain size), *DEP1* (panicle architecture) and *IPA1* (plant architecture) (Li et al. 2016a). *ERECTA1 (OsER1)*, a negative regulator of spikelet number per panicle, influences the OsMKKK10-OsMKK4-OsMPK6 cascade, which is crucial for maintaining cytokinin homeostasis. The OsER1-OsMKKK10-OsMKK4-OsMPK6 pathway regulates panicle morphology by impacting cytokinin metabolism. OsMPK6 phosphorylates DST, a zinc finger transcription factor, enhancing its activation of CYTOKININ OXIDASE2 (OsCKX2) (Guo et al. 2020; Gouda et al. 2020). Li and team identified a gene *COG1*, encoding the TF OsMADS17, with a 65-bp deletion in the 5′ UTR in cultivated rice, which increased grain number and grain weight. Hence, downregulating OsMADS17 or expression of OsAP2–39 further enhances grain yield by increasing both grain number and weight (Li et al. 2023). The CRISPR-Cas9 gene editing system generated two elite alleles of *OsSPL4*, which showed an increase in grain number per panicle and grain size, resulting in increased grain yield. A study of cells showed that *OsSPL4* could control the growth of spikelets by speeding up cell division (Hu et al. 2021b). *FRIZZY PANICLE (FZP)* regulates panicle branching in rice. The 5′ UTR in Nipponbare targeted by the CRISPR-Cas9 multiplex editing system created novel beneficial FZP alleles that increased grain counts. While deletion of 38-bp (−44 to −7 bp) in the upstream part of FZP has no effect on modifying grain counts, deletions of 149-bp (−157 to −9 bp) or 85-bp (−156 to −72 bp) in the upstream portion of *FZP* can dramatically increase grain numbers without reduction in grain size. These results concluded that the 113-bp segment (−157 to −45 bp) upstream of *FZP* plays an important role in its activity, and grain numbers can be increased by deleting this region (Chen et al. 2021a). The *LARGE2* gene encodes an E3 ubiquitin ligase, OsUPL2, with a HECT domain. Mutants with a large loss of function have significant panicles with a higher number of grains, wide leaves and grains and thick culms. *LARGE2* suppresses meristematic activity to control grain number and panicle size. These results suggest that *LARGE2-APO1/APO2* is a suitable target for enhancing panicle size and grain number in crops (Chen et al. 2022a). Chen and co-workers discovered a QTL in maize that enhanced the number of kernel rows. Based on fine mapping, researchers have identified *KRN2*, an intriguing gene, at this locus. Similarly, *KRN2* and its homolog's gene editing in rice revealed increased grains per plant (Huang et al. 2021). *OsPDCD5* encodes the rice protein called programmed cell death. OsPDCD5-targeted mutagenesis improved plant architecture and grain yield. When compared to the wild types, statistical analysis showed that the yield of knockout lines rose from 6.25% to 20.13%. These lines displayed a decline in amylose content in addition to an increase in the percentage of milled rice and gel consistency (Dong et al. 2021).

9.3.4.2 Altering Tiller Number

Agrobacterium-mediated rice transformation targeted eight rice *FWL* genes, resulting in transgenic T_0 lines for 15 of the 16 targets with an average mutation rate of 81.6%. Scientists looked at different versions of the *OsFWL4* gene and found that some had a lot more tillers and plants than the wild type. Additionally, flag leaves of *OsFWL4* mutants showed increased width due to a rise in cell number. Furthermore, mutants of the *OsFWL1* gene exhibited greater grain length than the wild type, and it was concluded that *OsFWL4* is a negative regulator of tiller number and plant yield (Gao et al. 2020a). According to Butt and co-workers, modifying the strigolactone biosynthesis pathway by mutating *OsCCD7* using CRISPR-Cas9 resulted in a decrease in height and an increase in tillering (Butt et al. 2018). The hexokinase gene *OsHXK1* edited in the *indica* rice varieties Huanghuazhan, Meixiangzhan and Wushansimiao resulted in the production of high-yielding rice varieties with enhanced photosynthetic efficiency using the CRISPR-Cas9 knockout technique (Zheng et al. 2021).

9.3.5 Other Key Traits

9.3.5.1 Hybrid Breeding

Hybrid breeding, an important strategy to enhance rice production involving male-sterile lines, is a key to successful cross-breeding. Han and team developed high-yielding glutinous CMS (Cytoplasmic Male Sterile) lines in rice by causing targeted disruption of the Wx and TGW6 genes, which confer 1000 grain weight (GW) and amylose content (AC). The pollen sterility in CMS lines was caused by the absence of mitochondrial NADP+-dependent malic enzymes (insufficient energy) (Han et al. 2018). On the other hand, the upregulation of acetyl-CoA synthetase and isoamylase in both maintainer and CMS lines has a strong link with CMS and amylose content. The mutants showed increased GW and gel consistency (GC), decreased AC and gelatinization temperature (GT), opaque grain appearance and no change in starch content (SC). To evade the potential risks of maintaining three lines in three-line breeding, there arises a two-line breeding system for hybrid seed production in rice, i.e., "turning three lines into two lines." Commercial TGMS (thermo-sensitive male sterility) rice production by inducing specific mutations in the *TMS5* gene (sterility-fertility transition gene) has been reported by Zhou and co-workers (2016). Pak and team edited the second and fourth exons of *OsOPR7* through the CRISPR-Cas9 genome editing technique, which resulted in the production of male-sterile plants that were restored to fertility by exogenous application of methyl jasmonate and paved the way for the establishment of two-line hybrid rice production (Pak et al. 2021). Gu and co-workers created a rPGMS (reverse photoperiod-sensitive genic male sterility) line by targeting the *CSA* (carbon starved anther) gene in *japonica* rice (Gu et al. 2019).

9.3.5.2 Reducing Seed Dormancy

Researchers used CRISPR-Cas9 to decrease seed dormancy in rice by knocking out the viviparous-1 (*OsVP1*) gene, a transcription factor linked to seed development and ABA signaling. Edited lines had higher levels of ABA signaling genes, which suggests that genome editing might be able to help rice seeds come out of dormancy (Jung et al. 2019).

9.3.5.3 Improving Storability

Grain deterioration during storage reduces the lifespan and quality of seeds, which results in significant losses for farmers. Lipoxygenases (LOXs) catalyze the dioxygenation of PUFA (polyunsaturated fatty acids) to hydroperoxide, which has a detrimental effect on regulating the storability of seeds. Rice's *LOX3* gene was successfully mutated via TALEN-based genome editing, leading to a *LOX3* deficit and better seed storability (Ma et al. 2015).

9.3.5.4 Asexual Reproduction in Rice

Using genome editing, researchers disrupted meiosis in plants, prompting the production of mother-derived egg cells via mitosis. By expressing *BBM1*, these cells developed into embryos, leading to the production of clonal seeds through seed propagation (Khanday et al. 2019). The same strategy, involving CRISPR-Cas9 editing of *REC8, PAIR, MTL* and *OSD1* genes, enabled clonal reproduction of F_1 rice hybrids. Plants made this way could easily grow more plants by using seeds, which could be a good way for improved rice varieties that were created through breeding to spread themselves (Wang et al. 2019).

9.4 Molecular Mechanisms and Challenges

9.4.1 Molecular Mechanism Behind Genome Editing for Yield Improvement in Rice

The influence of multiple genetic loci is involved in the regulation of quantitative variables, such as yield and quality. Three critical elements that primarily influence rice yield are the number of panicles per plant, the number of grains per panicle and the weight of individual grains (Xing and Zhang 2010). Genetic engineering has made significant strides toward increasing rice productivity. Zhang and co-workers identified genes such as *GS3, DEP1, GS5, GW2, Gn1a* and *TGW6* as negative regulators of grain size, quantity and weight (Zhang et al. 2017). Li and team said that the CRISPR-Cas9 system was used to change four genes individually: *Gn1a, DEP1, GS3* and *IPA1*. This led to the expected changes in phenotype, such as more grains, panicles that stand upright more densely and bigger grains (Li et al. 2016a). Additionally, it was reported that a triple null mutant containing simultaneous mutations in three-grain weight-related genes (*GW2, GW5* and *TGW6*) showed a 29.3% increase in 1000-grain weight (Xu et al. 2016). This shows how important it is to use multiplex gene editing to add null mutants of key genes that affect yield to a single cultivar. This is an important way to control different aspects of rice productivity. Additionally, CRISPR-Cas9-mediated multiplex genome editing has been successfully used to control the heading date, a crucial agronomic feature affecting rice production, by targeting three important genes (*Hd2, Hd4* and *Hd5*) that are known to influence the heading date in rice negatively (Li et al. 2017). The results showed that this high-efficiency breeding strategy could work by showing that the heading date dropped significantly in mutant derivatives from nine rice lines.

The development of hybrid rice, which is acknowledged for its yield superiority of 10–15% above inbred lines, significantly increases rice yield. Zhou and team showed how to quickly make 11 new TGMS *indica* rice lines in just 1 year using the CRISPR-Cas9 system to make specific changes to the temperature-sensitive gene TMS5 (Zhou et al. 2016). Li and co-workers employed CRISPR-Cas9 technology to specifically target the Carbon Starved Anther (CSA) gene in *japonica* rice. This gene is a crucial locus exhibiting male sterility in short days and male fertility in long days. This approach resulted in the development of two reverse PGMS lines (9522csa and JY5Bcsa) and one rP(T)GMS145 line (KY131csa-4). These results highlight the efficacy of CRISPR-Cas technology in accelerating the breeding of TGMS lines, which is an important first step towards breeding two-line hybrid rice on a large scale (Li et al. 2016c).

Grain deterioration can significantly affect the quality and shelf life of stored rice seeds, leading to substantial economic losses. The 14 lipoxygenases (LOXs) enzymes in rice, especially *LOX1*, *LOX2* and *LOX3*, are very important for seed storage because they act as negative regulators. Ma and team successfully mutated the *LOX3* gene by using TALEN technology and a modified ligation-independent cloning technique. This method resulted in the generation of LOX3 knockout lines with enhanced seed storability (Ma et al. 2015). The production of 2-acetyl-1-pyrroline content (2AP), the main aroma component that regulates aromatic rice, is attributed to the protein betaine aldehyde dehydrogenase ($BADH_2$). According to Ma and co-workers, a 35–75% increase in 2AP content was seen due to a TALEN-induced mutation in the $OsBADH_2$ gene. Furthermore, Shao and team used CRISPR-Cas9 to edit the fragrant gene $OsBADH_2$ in the *indica* rice line Zhonghua 11, leading to an increase in fragrance due to the presence of an additional T base in the first exon of $OsBADH_2$ (Shao et al. 2017).

Different phytohormones and their interconnected signaling networks control rice development and yield in a complex way, with abscisic acid (ABA) being a key player. To improve rice growth and productivity, Miao and co-workers successfully modified the group I (*PYL1–PYL6* and *PYL12*) and group II (*PYL7–PYL11* and *PYL13*) *PYL* genes. Significantly, group I gene mutations, specifically *pyl1/4/6*, exhibited robust growth and improved grain productivity without compromising seed dormancy or other agronomic traits (Miao et al. 2018).

9.4.2 Molecular Mechanism Behind Genome Editing for Nutritional Improvement

Rice serves as a primary source of nutrients, supplying over half of the global population with up to 70% of their daily caloric intake. Abundant Amylose content (AC) and Resistance starch (RS) in rice enhance human health by reducing the risk of major diseases such as hypertension, diabetes and colon cancers (Chen et al. 2012). Researchers have successfully shown that CRISPR-Cas9 technology can be used to make high-amylose rice, which meets the growing demand for rice varieties that improve nutrition and address public health issues. Sun and team accomplished this by concentrating on two rice branching enzymes (SBE), *SBEI* and *SBEIIb*. The wild-type and *sbeI* mutants showed no discernible differences; however, the *sbeII* mutants had significantly higher amounts of AC and RS content, up to 25 and 9.8, respectively. This highlights the critical function of *SBEII* in determining the specific structure and nutritional properties of starch. Therefore, using CRISPR-Cas9 to change *SBEIIb* stands out as an important way to make rice varieties with higher levels of RS and AC (Sun et al. 2017).

Cadmium (Cd), a highly toxic heavy metal, poses significant health hazards for those who rely on rice as a staple food. *Indica* cultivars have a higher Cd content compared to *japonica* rice cultivars, necessitating measures to ensure food safety (Arao and Ishikawa 2006). Even if techniques like soil treatment, phytoremediation, field flooding and the application of charcoal can reduce Cd levels to some extent, it is still highly challenging to develop low-Cd rice cultivars using conventional methods. Therefore, there is a pressing need for innovative strategies to create Cd-free rice lines to safeguard public health. Researchers recently used the CRISPR-Cas9 editing method to produce *indica* rice lines with reduced Cd accumulation by mutating the metal transporter gene *OsNramp5* (Tang et al. 2017). Cd levels in wild-type *indica* rice ranged from 0.33 mg/kg to 2.90 mg/kg. However, tests done in the field showed that Cd levels in *OsNramp5* mutant grains were always <0.05 mg/kg, which did not affect plant production. In the near future, the use of genome editing tools may be able to mitigate a number of concerns related to heavy metal contamination in rice grains.

9.4.3 Molecular Mechanism Behind Genome Editing for Biotic Stress in Rice

The productivity and quality of rice production are significantly affected by diseases caused by biotic agents such as bacteria, fungi, viruses and insects (Wang et al. 2022). Scientists have applied advances in genome editing to mutate disease-related genes, enhancing resistance in rice. A global threat to food security is *Xanthomonas oryzae* pv. *oryzae* (*Xoo*), which causes the vascular disease known as bacterial leaf blight (BLB) in rice. While traditional breeding has been effective, the emergence of new virulent pathotypes calls for advanced strategies. Transcription activator-like effectors (TALEs) from *Xanthomonas* go after the *SWEET* gene family and change the way sugar moves through rice cells (Gupta et al. 2021). TALEN technology disrupted the promoter of *OsSWEET14*, providing resistance against bacterial blight (Li et al. 2012). CRISPR-Cas9 added a null mutation to *OsSWEET13*, which made *indica* rice IR24 more resistant (Zhou et al. 2015). Further TALEN modifications targeting the EBEtal7 binding site show promise in reducing BLB severity (Cai et al. 2017). These strategies offer the potential for engineering rice cultivars with reduced susceptibility to *Xoo*.

The worldwide rice blast is a destructive disease caused by the fungus *Magnaporthe oryzae* (Zhang et al. 2014). Efforts have been made to use molecular tools to produce blast-resistant cultivars, though it is labor-intensive. Using CRISPR-Cas9 to delete the ERF transcription factor gene *OsERF922* made rice plants more resistant to rice blast (Wang et al. 2016a). With a mutation rate of up to 42%, T_2 homozygous mutant lines had a lot fewer blast lesions than wild-type plants. This shows that CRISPR-Cas9 might be able to make rice plants more resistant to blasts.

The interaction between rice tungro bacilliform virus (RTBV) and rice tungro spherical virus (RTSV) results in rice tungro disease (RTD), which poses a serious threat to rice farming in Asian countries. The translation initiation factor 4 gamma (*eIF4G*) gene largely confers RTSV resistance, as indicated by breeding attempts for resistance. The *eIF4G* gene in the RTSV-susceptible rice variety IR64 is being changed with CRISPR-Cas9 to make a new source of resistance to RTD (Macovei et al. 2018). As a useful tool for diversifying the development of RTSV-resistant types, the subsequently transmitted unique *eIF4G* alleles to the following generations remain unmodified at off-target loci.

9.4.4 Molecular Mechanism Behind Genome Editing for Abiotic Stress in Rice

The rice bentazon-sensitive lethal (BEL) gene makes the plant resistant to bentazon and sulfonylurea herbicides. On the other hand, the loss-of-function mutant bel is sensitive to these herbicides (Pan et al. 2016). Farmers commonly utilize the bel mutant to identify seed contamination in a two-line hybrid rice production system. Though natural genetic resources are scarce, BEL plays a crucial role in guaranteeing the safety of hybrid rice farming. The BEL gene was, therefore, modified using CRISPR-Cas9 in rice by *Agrobacterium*-mediated gene transfer (Xu et al. 2014). The mutagenic efficiency of stable CRISPR-Cas9 transformants ranged from 2% to 16% and phenotypic analysis showed that the biallelic mutated transgenic plant became sensitive to bentazon.

Researchers have introduced genome editing-based mutations into the *ALS* gene to develop herbicide-tolerant rice varieties. According to Li and team, they used TALEN to replace genes in rice using homologous recombination (HR), which led to 6.3% efficient double point mutations in the *OsALS* gene. The produced stable mutant exhibited strong resistance to herbicides. Similarly, Suna and co-workers introduced several unique point mutations in the rice *ALS* gene using CRISPR-Cas9-mediated HR. Phenotypic screening showed that while the wild-type plant succumbed after 36 days of bispyribac sodium (BS) spraying, the edited lines exhibited tolerance to BS and grew normally. Genome editing technologies can efficiently produce rice plants homozygous for herbicide tolerance in just one generation (Sun et al. 2016; Li et al. 2016b).

Rice seedlings exhibit heightened sensitivity to low temperatures, particularly during the seedling stage, accentuating the significance of enhancing cold tolerance for augmenting rice productivity. One of the genes linked to rice's ability to withstand low temperatures is the transcription factor TIFY1b. CRISPR-Cas9 technology was utilized to modify TIFY1b and its homologous gene, TIFY1a (Huang et al. 2017). Site-specific mutations found in T_0 rice plants enable the use of TIFY1 mutant lines to investigate the function of *TIFY1* genes in promoting rice adaptation to low temperatures.

9.5 Strategies for Reducing Off-Target Mutations

In recent times, the optimization of CRISPR technology has emerged as a pivotal aspect for scientists aiming to mitigate its inherent off-target effects and enhance editing efficiency. Enhancing sgRNA's specificity, especially in the PAM seed region, is one of the main strategies. Ren and co-workers demonstrated that a higher GC content in the sgRNA seed region is linked to improved gene editing efficacy and reduced off-target effects, highlighting its importance. Base preferences that affect Cas9 configuration are taken into account during design. Adding two guanines to the 5′ end of sgRNAs, known as the "GG20" alteration, has been shown to significantly reduce off-target effects, highlighting ongoing efforts to enhance the precision of sgRNA design. ZiFiT, Cas9 and the CRISPR design tool are a few examples of online resources useful for creating sgRNAs and determining possible off-target sites (Ren et al. 2016). The list of online tools to design and evaluate sgRNA efficiency for CRISPR-Cas9 mutagenesis is listed in Table 9.2. Another vital avenue for optimization involves the regulation of the dosage of the Cas9-sgRNA complex. Pattanayak and team emphasize the importance of managing the expression quantity of Cas9 protein or sgRNA to minimize off-target effects. The gRNA: Cas9 complex ratio holds significance, as studies suggest that maintaining a ratio of 2:1 or 3:1 enhances the efficiency of target gene knockout while effectively reducing off-target effects. Continuously expressing Cas9 in cells has been linked to a higher risk of side effects. This has led to suggestions for controlling Cas9 activity by using blocking antibodies or proteins that stop the Cas9 protein from working. Striking a balance between gene editing efficiency and off-target effects remains a key consideration in optimizing CRISPR technology (Pattanayak et al. 2013).

Table 9.2 List of available tools to evaluate and design the single guide RNA (sgRNA) efficiency for CRISPR

Tool	Method	Web source	Reference
CasOT	Alignment	http://eendb.zfgenetics.org/casot/	Xiao et al. (2014)
Cas-OFFinder	Alignment	http://www.rgenome.net/cas-offinder	Bae et al. (2014)
FlashFry	Alignment	http://aaronmck.github.io/FlashFry/	McKenna and Shendure (2018)
CrisFlash	Alignment	https://github.com/crisflash	Jacquin et al. (2019)
MIT	Scoring	http://www.genome-engineering.org/	Hsu et al. (2013)
CFD	Scoring	http://www.broadinstitute.org/rnai/ public/software/index	Doench et al. (2016)
CRISTA	Scoring	http://crista.tau.aC.il/	Abadi et al. (2017)
Elevation	Scoring	https://crispr.mL	Listgarten et al. (2018)
DeepCRISPR	Scoring	http://www.deepcrispr.net/	Chuai et al. (2018)

9.5.1 Length, Mismatches and GC Contents of Guide RNAs

For editing plant genomes, the CRISPR-Cas9 system has proven useful in various applications, including knock-out, knock-in and gene expression modulation. Effective applications have been observed in a wide range of plants, including rice, *Petunia*, tomato, *A. thaliana*, maize, poplar, grape, cucumber and poplar (Zhang et al. 2018b; Malzahn et al. 2017). It is important for functional genome analysis to be able to predict and reduce off-target mutations. This is especially true in plant breeding, where whole-genome sequences are still not complete (Endo et al. 2015). To maximize on-target mutagenesis and minimize off-target alterations, research on CRISPR-Cas9 in plants highlights the critical necessity of creating highly precise single-guide RNA (sgRNA) (Wang et al. 2014). To do this, it is important to stay away from sgRNAs whose seeds are homologous to many genome loci. This is because carefully designed sgRNAs can still target specific genes even when there are many homologous loci in the genome being studied (Xie et al. 2015; Zhang et al. 2018b). Things like GC content above 70% in the target sequence may make off-target effects stronger, and the best ranges have been found to be between 65% and 80% (Li et al. 2016b). Conversely, GC-rich sgRNAs (>50%) are efficient in enhancing on-target mutagenesis due to their strong binding to target sites (Hsu et al. 2013). Additionally, sgRNA length influences genome editing efficiency and off-target mutations, with 18-bp sgRNAs recommended to reduce the frequency of off-target effects (Fu et al. 2014; Sugano et al. 2018). Researchers have explored strategies such as utilizing chimeric guide RNA with partial DNA replacement to decrease off-target effects in human cells, which may also be relevant in plant applications (Yin et al. 2018b). Using more than one sgRNA for a target gene can also make mutations more common in plants like rice and tomato. This shows how important it is to have complete sgRNA design strategies (Xu et al. 2014). Lin and team proposed guidelines to reduce potential off-target effects, applicable to both human cells and plants, such as avoiding target sequences with more than three mismatches within 7–10 bp of the PAM and steering clear of sgRNA bulges within 12 bp of the PAM (Lin et al. 2014). Despite the potential for off-target mutations, the frequency in plants is generally lower than somatic mutations during tissue culture, highlighting the CRISPR system's capability to minimize unwanted changes (Woo et al. 2015). Li and co-workers observed that the lowest off-target rate (2.5%) was found for sgRNA with six mismatches, underscoring the significance of carefully designing sgRNAs to prevent unwanted changes. The highest off-target changes (67.5%) were found in rice plants when there were two mismatched bases between sgRNA and its off-target site (Li et al. 2016b).

9.5.2 Concentration of sgRNA/Cas9

A key strategy for mitigating off-target effects in the CRISPR-Cas9 system involves controlling the concentration of the sgRNA/Cas9 complex (Xie et al. 2015). To achieve this, the selection of promoters is crucial for regulating Cas9 expression. Notably, in *A. thaliana*, Cas9, under the control of the egg-cell-specific promoter (ECS), exhibited a higher editing frequency compared to the constitutive CaMV35S promoter (Woo et al. 2015). The stronger expression of Cas9 facilitated by the ECS promoter enhances the efficiency of the CRISPR-Cas9 genome editing system in tested plants (Wang et al. 2014). By modifying sgRNA structure and using UBIQUITIN1 (UBI1) and ACTIN1 (ACT1) endogenous promoters, there is an enhanced editing efficiency in the VQR (D1135V/R1335Q/T1337R) Cas9 variant, which cleaves NGA PAM in rice. Hence, the improved CRISPR-Cas9-VQR system offers a robust tool for multiplex genome editing at sites with noncanonical NGA PAM (Hu et al. 2018). High levels of sgRNA have been correlated with reduced editing potential in the CRISPR-Cas9 system in tomato and *A. thaliana* (Hsu et al. 2013; Woo et al. 2015). In monocots, the use of plant endogenous promoters for Cas9 expression led to higher on-target mutations than the CaMV35S promoter (Xie et al. 2015). Moreover, the selection of suitable T_1 plants where CRISPR components have been segregated can potentially avoid off-target mutations (Woo et al. 2015). Additionally, optimizing the ratio of sgRNA to cas9 is crucial, with studies indicating that a 1:1 ratio is optimal for targeted mutagenesis in *A. thaliana* (Li et al. 2013). The ratio of Cas9/sgRNA varies for different genes in different plant species, recommending specific ratios for maximum mutation efficiency depending on the targeted gene (Malnoy et al. 2016).

9.5.3 Transformation Methods

The selection of the transformation method significantly influences on-target accuracy in genome editing. The transformation method followed in recent CRISPR-Cas9 mediated genome editing is listed in Table 9.3. An optimized, efficient rapid rice

Table 9.3 List of recent CRISPR studies with transformation method in *Arabidopsis thaliana* and *Oryza sativa*

Plant species	sgRNA promoter	Cas9 or Cpf1 promoter	On-target effects (%)	Off-target mutation (%)	Transformation method/off-target decreasing strategy	Reference
Arabidopsis thaliana	AtU6	35DPPDK	1.1–5.6	Not detected	Protoplast transfection/highly specific sgRNA	Li et al. (2013)
	AtU6	AtRPS5a	85	<0.1	*Agrobacterium*-mediated transformation/ appropriate promoter	Kang et al. (2018)
	AtU6	CaMV35S	6.5	N/A	Protoplast transfection/–	Lin et al. (2018)
	AtU6	AtYAO	86	Not detected	*Agrobacterium*-mediated transformation/ heat stress	LeBlanc et al. (2018)
	AtU626	2X CaMV35S	24	Not detected	*Agrobacterium*-mediated transformation/ dual sgRNAs-Cas9 vector	Zhao et al. (2016)
Oryza sativa	OsU3	CaMV35S	7.3	N/A	Protoplast transfection/–	Lin et al. (2018)
	OsU6	2X CaMV35S	8	Not detected	Particle bombardment transformation/Cpf1	Begemann et al. (2017)
	OsU3	CaMV35S	5	Not detected	Particle bombardment transformation/ highly specific sgRNA	Baysal et al. (2016)
	OsU3	ZmUbi	85	0–2.2	*Agrobacterium*-mediated transformation/ careful sgRNA design	Wang et al. (2015)
	OsU3	ZmUbi	90.6	N/A	Particle bombardment transformation/dual sgRNAs-Cas9 vector	Sun et al. (2016)
	OsU3	ZmUbi	75	N/A	*Agrobacterium*-mediated transformation/ dual sgRNAs-Cas9 vector	Sun et al. (2016)

transformation protocol has to be employed to rapidly regenerate transgenic rice plants (Thakur et al. 2022). Using electroporation to deliver sgRNA and nuclease as ribonucleoprotein (RNP) complexes into plant protoplasts has also shown a low rate of mutations that are not intended. The Cas9-sgRNA complexes quickly and effectively cut the target site, reducing the number of cuts that happen at unwanted locations (Zhang et al. 2017; Woo et al. 2015). While the RNP strategy has been successful in several plants, some limitations, such as challenging regeneration and the need for expensive equipment, are associated with protoplast transformation. Alternative methods, like nanocarriers, are suggested to overcome these limitations. Proposing a nanoparticle-mediated RNP delivery system into intact plants appears to be a species-independent, economical, time-efficient and equipment-independent solution, given that RNP has the ability to reduce undesirable changes (Zhang et al. 2018b; Woo et al. 2015).

9.5.4 Different *Cas* Variants

The CRISPR-Cas9 and Cas12a (Cpf1) systems, two well-known tools for editing plant genomes, have differences that affect how well they work and what they can edit (Zhang et al. 2018b). Lee and co-workers looked at these nucleases side by side in maize and focused on the glossary2 (*gl2*) gene, which has parts that recognize both Cas9 and Cpf1. The researchers found that the CRISPR-Cas9 system works better and more precisely than Cpf1. For example, Cas9 was able to make 90–100% of the targeted mutations it was supposed to, while Cpf1 only managed 0–60% in T_0 plants (Lee et al. 2018). The effectiveness of Cpf1 depends a lot on the sequence of the crRNA; crRNAs with stable secondary structures in the targeting region work less well (Zetsche et al. 2017). Researchers have explored natural variants of Cas9, such as SaCas9 and StCas9, to enhance specific mutagenesis and reduce off-target changes in plants by recognizing longer PAM sequences (Wolter et al. 2018). Yamato and team demonstrated the potential of modified versions of SpCas9, namely SpCas9-VQR and SpCas9-EQR, in recognizing atypical targets in *A. thaliana*, thereby showcasing increased sgRNA design flexibility and efficient genome editing (Yamato et al. 2019). Furthermore, researchers have utilized Cas9-paired nickases to reduce off-target effects, with studies showing that the 3′-overhang structure leads to lower on-target mutation frequencies compared to the 5′-overhang structure (Woo et al. 2015). Nuclease-deactivated variants of Cas9 (dCas9) have gained popularity for their use in the CRISPR system to avoid off-target mutations. Combining dCas9 with the cytidine deaminase enzyme makes it possible to change specific nucleotides in the target plant gene. This makes it a useful tool for controlling genes specifically (Piatek et al. 2015). When used with transcription effectors like VP64 and SRDX, the CRISPR-dCas9 system gives scientists a lot of freedom to change how genes are expressed in plants (Zhang et al. 2018b; Wyvekens et al. 2015). Researchers have employed strategies such as truncated guide RNAs (tru-gRNAs) and the dimerization-dependent RNA guide FokI-dCas9 nuclease (RFN) to reduce unintended mutations, demonstrating the potential to enhance the specificity of CRISPR in plants (Zhang et al. 2018b; Wyvekens et al. 2015).

9.5.5 Temperature Effects on On- and Off-Targets

Recent research has emphasized the influence of temperature on enhancing target mutagenesis in plants utilizing the CRISPR-Cas9 system. LeBlanc and team demonstrated that exposing *A. thaliana* plants to heat stress at 37 °C significantly increased on-target mutagenesis while reducing off-target effects. Plants subjected to heat stress exhibited a higher frequency of on-target mutations compared to those maintained at the standard temperature of 22 °C, with only 12% of GFP-positive plants under heat stress versus 89% at 22 °C. A heat-shock (HS)-inducible CRISPR/Cas9 system was developed by inducing targeted mutations at two loci to assess the targeted mutagenesis efficiency in rice. The mutagenesis rate was very low (~16%), but it significantly increased to 50–63% post-HS (Thakur et al. 2022). Given the common use of low temperatures (20–28 °C) in *Agrobacterium*-mediated transformation, alternative temperature-independent transformation methods are preferable to maintain the efficiency of Cas9 and Cpf1 (LeBlanc et al. 2018).

9.5.6 Aptazyme Overcomes CRISPR-Cas9 Limitations

There are problems with the CRISPR-Cas9 system that make it less effective. Two major ones are the lack of promoters that can be used to make sgRNA in living cells and the inability to find Cas9-free stable plant mutants (He et al. 2017). In vivo, sgRNA production usually depends on RNA polymerase III (Pol III) because RNA polymerase II (Pol II) isn't very good at what it does and changes in ways that make sgRNA and genome editing less effective (Piatek et al. 2015; He et al. 2017). He and team identified drawbacks associated with the utilization of U3 and U6 promoters for sgRNA generation in various plants, including constitutive expression, sequence constraints for transcription initiation and the presence of a poly-U tail in primary transcripts, which may potentially reduce editing efficiency. He and co-workers proposed a ribozyme-based sgRNA strategy to address these limitations, introducing a ribozyme-flanked artificial sgRNA (RGR). Conventional promoters can control the RGR, which features ribozymes at both ends, ensuring nuclease activity that cleaves modifications like polyadenylation. Important considerations for ribozyme-flanked artificial sgRNA design involve positioning cleavage sites before and after the first and last nucleotides of the sgRNA and selecting ribozymes with small size and flank sequence-independent cleavage activity, such as a hammerhead ribozyme <50 bp in length with 6 bp reverse complementary to the sgRNA's initial 6 bp (He et al. 2017). Ribozymes, small RNA molecules with potential in genome editing, were discovered and recognized over 30 years ago (Tang and Breaker 1997; Saragliadis et al. 2012; Chen et al. 2018). With conserved single-stranded sections essential for catalytic activity, the hammerhead ribozyme, which is characterized by intra- and intermolecular helices, acts as a molecular scaffold, enclosing the ribosome binding site (RBS) (Tang and Breaker 1997; Chen et al. 2018). Target gene expression can be controlled by certain ribozymes that display ligand-dependent behavior (Saragliadis et al. 2012). Aptazymes are allosteric ribozymes that have been joined with aptamers. They allow gene expression to be controlled by ligands (Tang and Breaker 1997; Saragliadis et al. 2012; Chen et al. 2018). Chen and team described a useful way to stop mutations that aren't supposed to happen by adding a theophylline-dependent aptazyme (AZ) to the Cas9/sgRNA structure. This aptazyme, inserted in both the tetraloop and stem-loop II of sgRNA, efficiently regulates sgRNA activity without impacting on-target mutation, providing a reliable approach to minimize off-target effects in plants (Chen et al. 2018; Konermann et al. 2015). Future research is needed to design aptazymes with ligands that are not naturally occurring in plants, although theophylline is natural in plants. This method, which uses aptazymes to control sgRNA after transcription, could help make the CRISPR-Cas9 system more accurate (Chen et al. 2018). It is suggested that alternative temperature-independent methods, such as mesoporous silica nanoparticle (MSN)-mediated gene transformation, offer advantages. It is recommended to employ MSN: Cas9/sgRNA-ribozyme at a transformation temperature ranging from 34 to 37 °C to significantly reduce undesired mutations. These results reveal a new, simple and quick way to change genes and edit the genome. It has many benefits, including using less time and energy, avoiding unwanted mutations, increasing the number of on-target changes and not needing outside forces or expensive equipment (Hajiahmadi et al. 2019).

9.6 Ensuring Stable Inheritance of Edited Traits Across Generations in Rice

Ensuring stable inheritance of edited traits across generations in rice is a crucial aspect of genome editing. Studies have shown that genome editing in plants can result in stable, heritable and non-mosaic modifications in a genome, allowing the resulting trait or phenotype to be reliably maintained and propagated into subsequent generations. Researchers have used CRISPR-Cas9-mediated genome editing in rice to generate inheritable and "transgene clean" targeted gene edits in later generations.

Key findings related to the stable inheritance of edited traits in rice include:

1. Transmission of mutations: Regardless of the presence of the CRISPR-Cas9 transgene, the mutations in T_1 lines were stably transmitted to later generations, indicating stable inheritance of edited traits.
2. Mendalian inheritance: The study demonstrated that mutations in T_0 transgenic rice are likely to be mostly somatic mutations, following the basic genetic rules of Mendelian inheritance. This means that the mutations in T_0 transgenic rice are probably mostly somatic mutations.
3. Elimination of transgenes: The CRISPR-Cpf1 system and base editing have made it possible to remove transgenes from the genome through genetic segregation. This means that there are no differences between plants that have been gene-edited and plants that have been bred in the usual way.
4. Stress tolerance: Genome editing technologies improve stress tolerance in rice, enabling the development of crop varieties with stable or higher yields under stress conditions.

In conclusion, genome editing in rice has allowed for the stable inheritance of edited traits across generations. Researchers have used the CRISPR-Cas9 system to generate inheritable and "transgene clean" targeted gene edits in later generations, and these mutations largely conform to classical genetic laws. The development of the CRISPR-Cpf1 system and base editing have enabled the elimination of transgenes from the genome, resulting in no differences between gene-edited plants and those developed through conventional breeding.

Ensuring stable inheritance of edited traits across generations in plants is a crucial aspect of genome editing. Various mechanisms, including DNA methylation and histone modifications, achieve the stable inheritance of epigenetic modifications necessary for the stable inheritance of edited traits. Edits in plants are predominantly inherited in a Mendelian fashion over multiple generations, ensuring a stable inheritance of modifications. The elimination of transgenes from genome-edited plants can contribute to the achievement of precise genome editing, in which unnecessary changes do not exist in the genome. Although crossing in successive generations in plants can also remove unintended mutations at off-target sites, several tools and approaches have been developed to reduce off-target effects. In conclusion, genome editing in plants can result in stable, heritable and non-mosaic modifications in a genome, allowing the resulting trait or phenotype to be reliably maintained and propagated into subsequent generations.

9.7 Future Directions

Advances in many contributing areas are fueling explosive progress in plant genome editing. New CRISPR-Cas nucleases, such as Cas12-based proteins with different PAM solutions (Fonfara et al. 2014), greatly increase the potential flexibility of the system. Cas12-based type V and Cas13-based type VI systems (Murugan et al. 2017) could also increase this adaptability. The prime editing system, recently developed, seems to be a potentially flexible and effective genome modification method. It is based on a Cas9 fusion with reverse transcriptase and a priming gRNA (Anzalone et al. 2019). Researchers are still refining type II and type I systems to function as transcriptional activators or repressors, in addition to modifying the genome (Gilbert et al. 2013; Farzadfard et al. 2013; Rath et al. 2015; Pickar-Oliver et al. 2019). Depending on the cargo and desired result, the current mainstays for crop genome editing across species are *Agrobacterium* and physical delivery via particle bombardment. *Agrobacterium* is widely employed for crops where stably integrated T-DNA can subsequently be segregated away. However, diverse species where transgene introduction is not desired, such as asexually or vegetatively propagated species, have successfully used particle gun delivery of RNPs (Metje-Sprink et al. 2019). Genome modification is made possible by providing morphogenic gene activity through Agrobacterium-mediated co-transformation without integrating genes like *Wus2* (Hoerster et al. 2020).

The main challenge in CRISPR-based gene editing is the specificity of PAM (Hsu et al. 2014). The alternate PAM sequences (3′ NAG and NGA) and cas9 variants

(StCas9 and SaCas9) along with VQR and VRER variants have broadened the application of genome editing in crop plants (Zhang et al. 2014b; Kaya et al. 2016). A recent study found that wild-type SpCas9 effectively recognizes both 5′-NAG-3′ and 5′-NGG-3′ PAMs in rice. The study also reported that using 5′-NAG-3′ alone or with 5′-NGG-3′ leads to efficient genome editing in rice with low off-target effects (Meng et al. 2018). Researchers focused on enhancing the editing efficiency of VQR variants in rice. A new SpCas9 variant (xCas9) has been reported to recognize a broad range of PAM sequences, including NG, GAA and GAT (Hu et al. 2016). These findings will expand the scope of genome editing by using various engineered Cas9 variants with different PAM specificities. However, not all Cas9 variants are effective in plants, highlighting the need to develop more variants that can recognize a wider range of PAM sites to enhance genome editing in cereal crops, especially rice. Another challenge is to increase the efficiency of gene replacement in genome editing via gain-of-function mutations. Precision editing through HR is vital but demanding due to low HR efficiency in plants. Enhancing HR-based genome editing and optimizing donor template DNA delivery methods addresses this challenge. Finally, yet importantly the gene-edited rice must reach farmers' fields at the earliest. Most genome editing work has been proof of concept in controlled environments, raising uncertainties about performance in natural conditions. More field trials are needed to address this. Additionally, biosafety regulations limit adoption. While the USDA has exempted many CRISPR-edited crops from strict GMO regulations, the European Union treats genome-edited organisms as GMOs, influencing rice-growing countries' regulatory decisions (Mishra et al. 2018). Fortunately, India's landmark decision in March 2022 with the revision of "LIGHTLY REGULATED" crops, the Union Ministry of Environment and Forest granted access to SDN-1 and SDN-2 crops for open field trials leading to commercial release. Although no gene-edited crops have been commercially released under newly relaxed regulations, many are in development and could soon reach the market. Indian public sector labs are working on drought-tolerant rice and maize, salt-tolerant rice, blast-resistant rice, nitrogen and water-use efficient high-yield rice and so on. Nonetheless, genome editing can create crops indistinguishable from those bred conventionally, and CRISPR/Cas9 has revolutionized rice improvement, essential for future food demands. In future, CRISPR/Cas9 technology in agriculture is poised to revolutionize crop productivity, potentially sparking a second green revolution to meet the dietary and nutritional needs of the world's growing population.

9.8 Conclusion

CRISPR-Cas9-mediated genome editing in plants has become a widely publicized field with far-ranging potential for the future of agriculture. However, genome editing per se represents only the tip of the technological iceberg. Below the surface, gene editing relies on the fundamentals and techniques of plant transformation, which are true for all plants. The combination of component methods to achieve successful rice transformation represents the often-underappreciated foundation of current successes. These component areas include establishing reliable culture systems, optimizing DNA delivery, selecting transgenic events and, finally, regenerating plants that contain the desired edits. Although the accuracy of CRISPR-Cas technology is very high, it is not immune to unwanted changes, or "off-targets," from occurring. When the edit is well-characterized molecularly and the selection of plants with the desired edit is made, it reduces the possible hazards associated with introducing plants with undesirable off-targets into the market. There are still certain issues, even though new genomic techniques (NGTs) can provide faster and more accurate edits. The dangers of genome editing are considered in relation to off-target effects. According to recent research, their frequency is either lower or comparable to conventional breeding methods. The traceability of crops and products that have undergone genome editing may face difficulties due to the detectability of alterations induced by CRISPR Cas. These challenges impact the legality of laws and necessitate the creation of reliable detection techniques.

Author Contributions Ambothi Rathnasamy Sakthi and Shobica Priya Ramasamy contributed to drafting specific sections related to the chapter and provided critical revisions for those sections. Shubham Rajaram Salunkhe and Ameena Premnath: Assisted with collecting relevant literature and contributed to writing specific subsections. Also, responsible for managing references, creating tables and figures, and assisting with the final formatting of the chapter. Bharathi Raja Ramadoss and Anandhan Tamilselvan: Conceptualized and structured the book chapter, oversaw the entire chapter development process, coordinated the contributions of all authors, and was responsible for the final approval and submission of the chapter.

References

Abadi S, Yan WX, Amar D, Mayrose I (2017) A machine learning approach for predicting CRISPR-Cas9 cleavage efficiencies and patterns underlying its mechanism of action. PLoS Comput Biol 13(10):e1005807. https://doi.org/10.1371/journal.pcbi.1005807

Abe K, Araki E, Suzuki Y et al (2018) Production of high oleic/low linoleic rice by genome editing. Plant Physiol Biochem 131:58–62. https://doi.org/10.1016/j.plaphy.2018.04.033

Alam MS, Kong J, Tao R et al (2022a) CRISPR/Cas9 mediated knockout of the OsbHLH024 transcription factor improves salt stress resistance in rice (*Oryza sativa* L.). Plan Theory 11(9):1184. https://doi.org/10.3390/plants11091184

Alam MS, Yang ZK, Li C et al (2022b) Loss-of-function mutations of OsbHLH044 transcription factor lead to salinity sensitivity and a greater chalkiness in rice (*Oryza sativa* L.). Plant Physiol Biochem 193:110–123. https://doi.org/10.1016/j.plaphy.2022.10.033

Alfatih A, Wu J, Jan SU et al (2020) Loss of rice *PARAQUAT TOLERANCE 3* confers enhanced resistance to abiotic stresses and increases grain yield in field. Plant Cell Environ 43(11):2743–2754. https://doi.org/10.1111/pce.13856

Anjala K, Augustine R (2022) Designing of guide RNA constructs for CRISPR/Cas9-mediated editing of rice transcription factor OsMADS26 for enhancing drought tolerance. J Appl Biol Biotechnol 11(1):176–182. https://doi.org/10.7324/JABB.2023.110124

Anzalone AV, Randolph PB, Davis JR et al (2019) Search-and-replace genome editing without double-strand breaks or donor DNA. Nature 576(7785):149–157. https://doi.org/10.1038/s41586-019-1711-4

Arao T, Ishikawa S (2006) Genotypic differences in cadmium concentration and distribution of soybean and rice. Japan Agric Res Quart 40(1):21–30. https://doi.org/10.6090/jarq.40.21

Arra Y, Auguy F, Stiebner M et al (2023) Rice yellow mottle virus resistance by genome editing of the *Oryza sativa* L. ssp. *japonica* nucleoporin gene *OsCPR5.1* but not *OsCPR5.2*. Plant Biotechnol J 22(5):1299–1311. https://doi.org/10.1111/pbi.14266

Arulganesh T, Kumam Y, Kumar KK et al (2021) Genome editing of elite rice cultivar CO51 for bacterial leaf blight resistance. Electron J Plant Breed 12(4):1060–1068. https://doi.org/10.37992/2021.1204.147

Ashikari M, Sakakibara H, Lin S et al (2005) Cytokinin oxidase regulates rice grain production. Science 309(5735):741–745. https://doi.org/10.1126/science.1113373

Bae S, Park J, Kim JS (2014) Cas-OFFinder: a fast and versatile algorithm that searches for potential off-target sites of Cas9 RNA-guided endonucleases. Bioinformatics 30(10):1473–1475. https://doi.org/10.1093/bioinformatics/btu048

Baoxiang W, Yan L, Yifeng W et al (2021) *OsbZIP72* is involved in transcriptional gene-regulation pathway of abscisic acid signal transduction by activating rice high-affinity potassium transporter *OsHKT1;1*. Rice Sci 28(3):257–267. https://doi.org/10.1016/j.rsci.2021.04.005

Baysal C, Bortesi L, Zhu C et al (2016) CRISPR/Cas9 activity in the rice *OsBEIIb* gene does not induce off-target effects in the closely related paralog *OsBEIIa*. Mol Breed 36:1–11. https://doi.org/10.1007/s11032-016-0533-4

Begemann MB, Gray BN, January E et al (2017) Precise insertion and guided editing of higher plant genomes using Cpf1 CRISPR nucleases. Sci Rep 7(1):11606. https://doi.org/10.1038/s41598-017-11760-6

Beyene G, Chauhan RD, Villmer J et al (2022) CRISPR/Cas9-mediated tetra-allelic mutation of the 'green revolution' SEMIDWARF-1 (*SD-1*) gene confers lodging resistance in tef (*Eragrostis tef*). Plant Biotechnol J 20(9):1716–1729. https://doi.org/10.1111/pbi.13842

Bibikova E, Skiöld T, Bogdanova S et al (2001) Titanite-rutile thermochronometry across the boundary between the Archaean craton in Karelia and the Belomorian Mobile Belt, eastern Baltic shield. Precambrian Res 105(2–4):315–330. https://doi.org/10.1016/S0301-9268(00)00117-0

Biswal AK, Wu TY, Urano D et al (2022) Novel mutant alleles reveal a role of the extra-large G protein in rice grain filling, panicle architecture, plant growth, and disease resistance. Front Plant Sci 12:2821. https://doi.org/10.3389/fpls.2021.782960

Blanvillain-Baufumé S, Reschke M, Solé M et al (2017) Targeted promoter editing for rice resistance to *Xanthomonas oryzae* pv. *Oryzae* reveals differential activities for *SWEET 14*-inducing TAL effectors. Plant Biotechnol J 15(3):306–317. https://doi.org/10.1111/pbi.12613

Burlando B, Cornara L (2014) Therapeutic properties of rice constituents and derivatives (*Oryza sativa* L.): a review update. Trends Food Sci Technol 40(1):82–98. https://doi.org/10.1016/j.tifs.2014.08.002

Butt H, Jamil M, Wang JY et al (2018) Engineering plant architecture *via* CRISPR/Cas9-mediated alteration of strigolactone biosynthesis. BMC Plant Biol 18(1):1–9. https://doi.org/10.1186/s12870-018-1387-1

Butt H, Rao GS, Sedeek K et al (2020) Engineering herbicide resistance via prime editing in rice. Plant Biotechnol J 18(12):2370. https://doi.org/10.1111/pbi.13399

Cai L, Cao Y, Xu Z et al (2017) A transcription activator-like effector Tal7 of *Xanthomonas oryzae* pv. *Oryzicola* activates rice gene *Os09g29100* to suppress rice immunity. Sci Rep 7(1):5089. https://doi.org/10.1038/s41598-017-04800-8

Capecchi MR (1989) Altering the genome by homologous recombination. Science 244(4910):1288–1292. https://doi.org/10.1126/science.2660260

Chang JD, Gao W, Wang P, Zhao FJ (2022) *OsNRAMP5* is a major transporter for lead uptake in rice. Environ Sci Technol 56(23):17481–17490. https://doi.org/10.1021/acs.est.2c06384

Changappa V, Tamilselvan A, Ramadoss BR et al (2023) Identification of novel allelic combination for salt tolerance through targeted association mapping in rice. Genet Resour Crop Evol 71:129–143. https://doi.org/10.1007/s10722-023-01611-x

Chen L, Magliano DJ, Zimmet PZ (2012) The worldwide epidemiology of type 2 *diabetes mellitus*—present and future perspectives. Nat Rev Endocrinol 8(4):228–236. https://doi.org/10.1038/nrendo.2011.183

Chen H, Li Y, Du C et al (2018) Aptazyme-mediated direct modulation of post-transcriptional sgRNA level for conditional genome editing and gene expression. J Biotechnol 288:23–29. https://doi.org/10.1016/j.jbiotec.2018.10.011

Chen H, Cai Y, Zhang S et al (2021a) Identification of a novel mutant allele, *fzp-15*, involved in panicle branch pattern of rice (*Oryza sativa*). Plant Breed 140(4):595–602. https://doi.org/10.1111/pbr.12928

Chen X, Liu P, Mei L et al (2021b) *Xa7*, a new executor R gene that confers durable and broad-spectrum resistance to bacterial blight disease in rice. Plant Commun 2(3):100143. https://doi.org/10.1016/j.xplc.2021.100143

Chen W, Chen L, Zhang X et al (2022a) Convergent selection of a WD40 protein that enhances grain yield in maize and rice. Science 375(6587):eabg7985. https://doi.org/10.1126/science.abg7985

Chen Y, Liu Z, Meng S et al (2022b) *OsCERK1* contributes to cupric oxide nanoparticles induced phytotoxicity and basal resistance against blast by regulating the anti-oxidant system in rice. J Fungi 9(1):36. https://doi.org/10.3390/jof9010036

Chen F, Zhang H, Li H et al (2023) *IPA1* improves drought tolerance by activating *SNAC1* in rice. BMC Plant Biol 23(1):1–12. https://doi.org/10.1186/s12870-023-04062-9

Chuai G, Ma H, Yan J et al (2018) DeepCRISPR: optimized CRISPR guide RNA design by deep learning. Genome Biol 19:1–18. https://doi.org/10.1186/s13059-018-1459-4

Chung PJ, Chung H, Oh N et al (2020) Efficiency of recombinant CRISPR/rCas9-mediated miRNA gene editing in rice. Int J Mol Sci 21(24):9606. https://doi.org/10.3390/ijms21249606

Cong L, Ran FA, Cox D, Lin S et al (2013) Multiplex genome engineering using CRISPR/Cas systems. Science 339(6121):819–823. https://doi.org/10.1126/science.123114

Cui Y, Jiang N, Xu Z, Xu Q (2020) Heterotrimeric G protein are involved in the regulation of multiple agronomic traits and stress tolerance in rice. BMC Plant Biol 20:1–13. https://doi.org/10.1186/s12870-020-2289-6

Deng Y, Zhai K, Xie Z et al (2017) Epigenetic regulation of antagonistic receptors confers rice blast resistance with yield balance. Science 355(6328):962–965. https://doi.org/10.1126/science.aai8898

Dhaliwal HS, Uchimiya H (1999) Genetic engineering for disease and pest resistance in plants. Plant Biotechnol 16(4):255–261. https://doi.org/10.5511/plantbiotechnology.16.255

Diana PA, Shanthinie A, Arulganesh T et al (2022) Targeted editing of *Ossweet13*, a bacterial leaf blight susceptible gene in rice using crispr tool. Electron J Plant Breed 13(3):772–779. https://doi.org/10.37992/2022.1303.139

Ding Y, Zhang F, Sun F et al (2023) Loss of *OsHRC* function confers blast resistance without yield penalty in rice. Plant Biotechnol J 21(8):1516. https://doi.org/10.1111/pbi.14061

Diyang Q, Rui H, Ji L et al (2023) Peptide transporter *OsNPF8.1* contributes to sustainable growth under salt and drought stresses, and grain yield under nitrogen deficiency in rice. Rice Sci 30(2):113–126. https://doi.org/10.1016/j.rsci.2023.01.004

Doench JG, Fusi N, Sullender M et al (2016) Optimized sgRNA design to maximize activity and minimize off-target effects of CRISPR-Cas9. Nat Biotechnol 34(2):184–191. https://doi.org/10.1038/nbt.3437

Dong N, Yin W, Liu D et al (2020) Regulation of brassinosteroid signaling and salt resistance by *SERK2* and potential utilization for crop improvement in rice. Front Plant Sci 11:621859. https://doi.org/10.3389/fpls.2020.621859

Dong S, Dong X, Han X et al (2021) *OsPDCD5* negatively regulates plant architecture and grain yield in rice. Proc Natl Acad Sci 118(29):e2018799118. https://doi.org/10.1073/pnas.2018799118

Duan YB, Li J, Qin RY et al (2016) Identification of a regulatory element responsible for salt induction of rice OsRAV2 through *ex situ* and *in situ* promoter analysis. Plant Mol Biol 90:49–62. https://doi.org/10.1007/s11103-015-0393-z

Duy PN, Lan DT, Thu P et al (2021) Improved bacterial leaf blight disease resistance in the major elite Vietnamese rice cultivar TBR225 via editing of the OsSWEET14 promoter. PLoS One 16(9):e0255470. https://doi.org/10.1371/journal.pone.0255470

Elkonin LA, Gerashchenkov GA, Borisenko NV et al (2023) Development of sorghum mutants with improved *in vitro* protein digestibility by CRISPR/Cas9 editing of kafirin genes. Crop J 11(5):1411–1418. https://doi.org/10.1016/j.cj.2023.02.005

Endo M, Mikami M, Toki S (2015) Multigene knockout utilizing off-target mutations of the CRISPR/Cas9 system in rice. Plant Cell Physiol 56(1):41–47. https://doi.org/10.1093/pcp/pcu154

Endo A, Saika H, Takemura M et al (2019) A novel approach to carotenoid accumulation in rice callus by mimicking the cauliflower Orange mutation *via* genome editing. Rice 12(1):1–5. https://doi.org/10.1186/s12284-019-0345-3

Farooq MS, Fatima H, Rehman OU et al (2023) Major challenges in widespread adaptation of aerobic rice system and potential opportunities for future sustainability. S Afr J Bot 159:231–251. https://doi.org/10.1016/j.sajb.2023.06.017

Farzadfard F, Perli SD, Lu TK (2013) Tunable and multifunctional eukaryotic transcription factors based on CRISPR/Cas. ACS Synth Biol 2(10):604–613. https://doi.org/10.1021/sb400081r

Feng Z, Zhang B, Ding W et al (2013) Efficient genome editing in plants using a CRISPR/Cas system. Cell Res 23(10):1229–1232. https://doi.org/10.1038/cr.2013.114

Feng PEI, Guangda WANG, Peng GAO et al (2023) Evaluation of new *japonica* rice lines with low cadmium accumulation and good quality generated by knocking out *OsNramp5*. Chin J Rice Sci 37(1):16. https://doi.org/10.16819/j.1001-7216.2023.220503

Fonfara I, Le Rhun A, Chylinski K et al (2014) Phylogeny of Cas9 determines functional exchangeability of dual-RNA and Cas9 among orthologous type II CRISPR-Cas systems. Nucleic Acids Res 42(4):2577–2590. https://doi.org/10.1093/nar/gkt1074

Fu Y, Sander JD, Reyon D et al (2014) Improving CRISPR-Cas nuclease specificity using truncated guide RNAs. Nat Biotechnol 32(3):279–284. https://doi.org/10.1038/nbt.2808

Gang LI, Qingsong GAO, Wei LI et al (2023) Directed knockout of *SD1* gene improves lodging resistance and blast resistance of rice. Chin J Rice Sci 37(4):359. https://doi.org/10.16819/j.1001-7216.2023.221113

Gao Q, Li G, Sun H et al (2020a) Targeted mutagenesis of the rice FW 2.2-like gene family using the CRISPR/Cas9 system reveals *OsFWL4* as a regulator of tiller number and plant yield in rice. Int J Mol Sci 21(3):809. https://doi.org/10.3390/ijms21030809

Gao Z, Liu Q, Zhang Y et al (2020b) *OsCUL3a*-associated molecular switches have functions in cell metabolism, cell death, and disease resistance. J Agric Food Chem 68(19):5471–5482. https://doi.org/10.1021/acs.jafc.9b07426

Gaudelli NM, Lam DK, Rees HA et al (2020) Directed evolution of adenine base editors with increased activity and therapeutic application. Nat Biotechnol 38(7):892–900. https://doi.org/10.1101/2020.03.13.990630

Gilbert LA, Larson MH, Morsut L et al (2013) CRISPR-mediated modular RNA-guided regulation of transcription in eukaryotes. Cell 154(2):442–451. https://doi.org/10.1016/j.cell.2013.06.044

Gouda G, Gupta MK, Donde R et al (2020) Computational approach towards understanding structural and functional role of cytokinin oxidase/dehydrogenase 2 (*CKX2*) in enhancing grain yield in rice plant. J Biomol Struct Dyn 38(4):1158–1167. https://doi.org/10.1080/07391102.2019.1597771

Goufo P, Trindade H (2014) Rice antioxidants: phenolic acids, flavonoids, anthocyanins, proanthocyanidins, tocopherols, tocotrienols, γ-oryzanol, and phytic acid. Food Sci Nutr 2(2):75–104. https://doi.org/10.1002/fsn3.86

Gu W, Zhang D, Qi Y et al (2019) Generating photoperiod-sensitive genic male sterile rice lines with CRISPR/Cas9. Plant Genome Edit CRISPR Syst: Methods Protoc 1917:97–107. https://doi.org/10.1007/978-1-4939-8991-1_8

Guo N, Gu M, Hu J et al (2020) Rice *OsLHT1* functions in leaf-to-panicle nitrogen allocation for grain yield and quality. Front Plant Sci 11:1150. https://doi.org/10.3389/fpls.2020.01150

Guo F, Huang Y, Qi P et al (2021) Functional analysis of auxin receptor OsTIR1/OsAFB family members in rice grain yield, tillering, plant height, root system, germination, and auxinic herbicide resistance. New Phytol 229(5):2676–2692. https://doi.org/10.1111/nph.17061

Gupta PK, Balyan HS, Gautam T (2021) *SWEET* genes and TAL effectors for disease resistance in plants: present status and future prospects. Mol Plant Pathol 22(8):1014–1026. https://doi.org/10.1111/mpp.13075

Gurunathan S, Ramadoss BR, Mudili V et al (2019) Single nucleotide polymorphisms in starch biosynthetic genes associated with increased resistant starch concentration in rice mutant. Front Genet 10:946. https://doi.org/10.3389/fgene.2019.00946

Hajiahmadi Z, Movahedi A, Wei H et al (2019) Strategies to increase on-target and reduce off-target effects of the CRISPR/Cas9 system in plants. Int J Mol Sci 20(15):3719. https://doi.org/10.3390/ijms20153719

Han Y, Luo D, Usman B et al (2018) Development of high yielding glutinous cytoplasmic male sterile rice (*Oryza sativa* L.) lines through CRISPR/Cas9 based mutagenesis of *Wx* and *TGW6* and proteomic analysis of anther. Agronomy 8(12):290. https://doi.org/10.3390/agronomy8120290

Han X, Chen Z, Li P et al (2022) Development of novel rice germplasm for salt-tolerance at seedling stage using CRISPR-Cas9. Sustain For 14(5):2621. https://doi.org/10.3390/su14052621

Haurwitz RE, Jinek M, Wiedenheft B et al (2010) Sequence-and structure-specific RNA processing by a CRISPR endonuclease. Science 329(5997):1355–1358. https://doi.org/10.1126/science.119227

He Y, Zhao Y (2020) Technological breakthroughs in generating transgene-free and genetically stable CRISPR-edited plants. Abiotic 1(1):88–96. https://doi.org/10.1007/s42994-019-00013-x

He Y, Wang R, Dai X, Zhao Y (2017) On improving CRISPR for editing plant genes: ribozyme-mediated guide RNA production and fluorescence-based technology for isolating transgene-free mutants generated by CRISPR. Prog Mol Biol Transl Sci 149:151–166. https://doi.org/10.1016/bs.pmbts.2017.03.012

He N, Huang F, Yu M et al (2022) Analysis of a rice blast resistance gene *Pita-Fuhui2663* and development of selection marker. Sci Rep 12(1):14917. https://doi.org/10.1038/s41598-022-19004-y

Hoerster G, Wang N, Ryan L et al (2020) Use of non-integrating Zm-Wus2 vectors to enhance maize transformation. In Vitro Cell Dev Biol-Plant 56(3):265–279. https://doi.org/10.1007/s11627-019-10042-2

Hsu PD, Scott DA, Weinstein JA et al (2013) DNA targeting specificity of RNA-guided Cas9 nucleases. Nat Biotechnol 31(9):827–832. https://doi.org/10.1038/nbt.2647

Hsu PD, Lander ES, Zhang F (2014) Development and applications of CRISPR-Cas9 for genome engineering. Cell 157:1262–1278. https://doi.org/10.1093/hmg/ddy120

Hu X, Wang C, Fu Y et al (2016) Expanding the range of CRISPR/Cas9 genome editing in rice. Mol Plant 9:943–945. https://doi.org/10.1016/j.molp.2016.03.003

Hu X, Meng X, Liu Q et al (2018) Increasing the efficiency of CRISPR-Cas9-VQR precise genome editing in rice. Plant Biotechnol J 16(1):292–297. https://doi.org/10.1111/pbi.12771

Hu B, Zhou Y, Zhou Z et al (2021a) Repressed *OsMESL* expression triggers reactive oxygen species-mediated broad-spectrum disease resistance in rice. Plant Biotechnol J 19(8):1511–1522. https://doi.org/10.1111/pbi.13566

Hu J, Huang L, Chen G et al (2021b) The elite alleles of *OsSPL4* regulate grain size and increase grain yield in rice. Rice 14:1–18. https://doi.org/10.1186/s12284-021-00531-7

Hu XH, Shen S, Wu JL et al (2023) A natural allele of proteasome maturation factor improves rice resistance to multiple pathogens. Nat Plants 9(2):228–237. https://doi.org/10.1038/s41477-022-01327-3

Hua K, Tao X, Zhu JK (2019) Expanding the base editing scope in rice by using Cas9 variants. Plant Biotechnol J 17(2):499–504. https://doi.org/10.1111/pbi.12993

Huang X, Zeng X, Li J et al (2017) Construction and analysis of tify1a and tify1b mutants in rice (*Oryza sativa*) based on CRISPR/Cas9 technology. J Agric Biotechnol 25(6):1003–1012

Huang L, Zhang R, Huang G et al (2018) Developing superior alleles of yield genes in rice by artificial mutagenesis using the CRISPR/Cas9 system. Crop J 6(5):475–481. https://doi.org/10.1016/j.cj.2018.05.005

Huang S, Xin S, Xie G et al (2020a) Mutagenesis reveals that the rice *OsMPT3* gene is an important osmotic regulatory factor. Crop J 8(3):465–479. https://doi.org/10.1016/j.cj.2020.02.001

Huang Y, Han Z, Cheng N et al (2020b) Minor effects of 11 Dof family genes contribute to the missing heritability of heading date in rice (*Oryza sativa* L.). Front. Plant Sci 10:1739. https://doi.org/10.3389/fpls.2019.01739

Huang YY, Liu XX, Xie Y et al (2020c) Identification of FERONIA-like receptor genes involved in rice-Magnaporthe oryzae interaction. Phytopathol Res 2(1):1–10. https://doi.org/10.1186/s42483-020-00052-z

Huang L, Hua K, Xu R et al (2021) The *LARGE2-APO1/APO2* regulatory module controls panicle size and grain number in rice. Plant Cell 33(4):1212–1228. https://doi.org/10.1093/plcell/koab041

Huo Z, Tu J, Xu A et al (2019) Generation of a heterozygous p53 R249S mutant human embryonic stem cell line by TALEN-mediated genome editing. Stem Cell Res 34:101360. https://doi.org/10.1016/j.scr.2018.101360

Indiastat (2022). https://www.indiastat.com/Home/DataSearchKeyword=rice%20production

Ishino Y, Shinagawa H, Makino K et al (1987) Nucleotide sequence of the *iap* gene, responsible for alkaline phosphatase isozyme conversion in *Escherichia coli*, and identification of the gene product. J Bacteriol 169(12):5429–5433. https://doi.org/10.1128/jb.169.12.5429-5433.1987

Jacquin AL, Odom DT, Lukk M (2019) Crisflash: open-source software to generate CRISPR guide RNAs against genomes annotated with individual variation. Bioinformatics 35(17):3146–3147. https://doi.org/10.1093/bioinformatics/btz019

Jiang W, Bikard D, Cox D et al (2013) RNA-guided editing of bacterial genomes using CRISPR-Cas systems. Nat Biotechnol 31(3):233–239. https://doi.org/10.1038/nbt.2508

Jiang Y, Chai Y, Qiao D et al (2022) Optimized prime editing efficiently generates glyphosate-resistant rice plants carrying homozygous TAP-IVS mutation in EPSPS. Mol Plant 15(11):1646–1649. https://doi.org/10.1016/j.molp.2022.09.006

Jinek M, Chylinski K, Fonfara I et al (2012) A programmable dual-RNA–guided DNA endonuclease in adaptive bacterial immunity. Science 337(6096):816–821. https://doi.org/10.1126/science.122582

Jung YJ, Lee HJ, Bae S et al (2019) Acquisition of seed dormancy breaking in rice (*Oryza sativa* L.) *via* CRISPR/Cas9-targeted mutagenesis of *OsVP1* gene. Plant Biotechnol Rep 13:511–520. https://doi.org/10.1007/s11816-019-00580-x

Jung JH, Li Z, Chen H, Yang S et al (2023) Mutation of phytochrome B promotes resistance to sheath blight and saline–alkaline stress *via* increasing ammonium uptake in rice. Plant J 113(2):277–290. https://doi.org/10.1111/tpj.16046

Kang BC, Yun JY, Kim ST et al (2018) Precision genome engineering through adenine base editing in plants. Nat Plants 4(7):427–431. https://doi.org/10.1038/s41477-018-0115-z

Karavolias NG, Patel D, Seong K et al (2021) CRISPR/Cas9 knockout of *EPFL10* reduces stomatal density while maintaining photosynthesis and enhancing water conservation in rice. Plant Physiol 192(2):1168–1182. https://doi.org/10.1101/2021.12.21.473329

Kaya H, Mikami M, Endo A et al (2016) Highly specifictargeted mutagenesis in plants using *Staphylococcus aureus* Cas9. Sci Rep 6:26871. https://doi.org/10.1038/srep26871

Khanday I, Skinner D, Yang B et al (2019) A male-expressed rice embryogenic trigger redirected for asexual propagation through seeds. Nature 565(7737):91–95. https://doi.org/10.1038/s41586-018-0785-8

Kim YA, Moon H, Park CJ (2019) CRISPR/Cas9-targeted mutagenesis of *Os8N3* in rice to confer resistance to *Xanthomonas oryzae* pv. *Oryzae*. Rice 12(1):1–13. https://doi.org/10.1186/s12284-019-0325-7

Kim P, Xue CY, Song HD et al (2021) Tissue-specific activation of *DOF11* promotes rice resistance to sheath blight disease and increases grain weight *via* activation of *SWEET14*. Plant Biotechnol J 19(3):409. https://doi.org/10.1111/pbi.13489

Kim JH, Yu J, Kim HK et al (2022a) Genome editing of golden SNP-carrying lycopene epsilon-cyclase (*LcyE*) gene using the CRSPR-Cas9/HDR and geminiviral replicon system in rice. Int J Mol Sci 23(18):10383. https://doi.org/10.3390/ijms231810383

Kim CY, Park JY, Choi G et al (2022b) A rice gene encoding glycosyl hydrolase plays contrasting roles in immunity depending on the type of pathogens. Mol Plant Pathol 23(3):400–416. https://doi.org/10.1111/mpp.13167

Kitomi Y, Hanzawa E, Kuya N et al (2020) Root angle modifications by the *DRO1* homolog improve rice yields in saline paddy fields. Proc Natl Acad Sci 117(35):21242–21250. https://doi.org/10.1073/pnas.2005911117

Komatsu A, Ohtake M, Shimatani Z, Nishida K (2020) Production of herbicide-sensitive strain to prevent volunteer rice infestation using a CRISPR-Cas9 cytidine deaminase fusion. Front Plant Sci 11:925. https://doi.org/10.3389/fpls.2020.00925

Komor AC, Badran AH, Liu DR (2017) CRISPR-based technologies for the manipulation of eukaryotic genomes. Cell 168(1):20–36. https://doi.org/10.1016/j.cell.2016.10.044

Konermann S, Brigham MD, Trevino AE et al (2015) Genome-scale transcriptional activation by an engineered CRISPR-Cas9 complex. Nature 517(7536):583–588. https://doi.org/10.1038/nature14136

Kuang Y, Li S, Ren B et al (2020) Base-editing-mediated artificial evolution of *OsALS1* in planta to develop novel herbicide-tolerant rice germplasms. Mol Plant 13(4):565–572. https://doi.org/10.1016/j.molp.2020.01.010

Kumam Y, Rajadurai G, Kumar KK et al (2022) Genome editing of indica rice ASD16 for imparting resistance against rice tungro disease. J Plant Biochem Biotechnol 31:880–893. https://doi.org/10.1007/s13562-021-00765-y

Kumar VV, Verma RK, Yadav SK et al (2020) CRISPR-Cas9 mediated genome editing of drought and salt tolerance (*OsDST*) gene in indica mega rice cultivar MTU1010. Physiol Mol Biol Plants 26:1099–1110. https://doi.org/10.1007/s12298-020-00819-w

LeBlanc C, Zhang F, Mendez J et al (2018) Increased efficiency of targeted mutagenesis by CRISPR/Cas9 in plants using heat stress. Plant J 93(2):377–386. https://doi.org/10.1111/tpj.13782

Lee JK, Jeong E, Lee J et al (2018) Directed evolution of CRISPR-Cas9 to increase its specificity. Nat Commun 9(1):3048. https://doi.org/10.1038/s41467-018-05477-x

Li T, Liu B, Spalding MH et al (2012) High-efficiency TALEN-based gene editing produces disease-resistant rice. Nat Biotechnol 30(5):390–392. https://doi.org/10.1038/nbt.2199

Li JF, Norville JE, Aach J et al (2013) Multiplex and homologous recombination–mediated genome editing in Arabidopsis and *Nicotiana benthamiana* using guide RNA and Cas9. Nat Biotechnol 31(8):688–691. https://doi.org/10.1038/nbt.2654

Li M, Li X, Zhou Z et al (2016a) Reassessment of the four yield-related genes *Gn1a, DEP1, GS3,* and *IPA1* in rice using a CRISPR/Cas9 system. Front Plant Sci 7:377. https://doi.org/10.3389/fpls.2016.00377

Li J, Meng X, Zong Y et al (2016b) Gene replacements and insertions in rice by intron targeting using CRISPR–Cas9. Nat Plants 2(10):1–6. https://doi.org/10.1038/nplants.2016.139

Li Q, Zhang D, Chen M et al (2016c) Development of japonica photo-sensitive genic male sterile rice lines by editing carbon starved anther using CRISPR/Cas9. J Genet Genomics 43(6):415–419. https://doi.org/10.1016/j.jgg.2016.04.011

Li X, Zhou W, Ren Y et al (2017) High-efficiency breeding of early-maturing rice cultivars via CRISPR/Cas9-mediated genome editing. J Genet Genomics 44(3):175–178. https://doi.org/10.1016/j.jgg.2017.02.001

Li C, Zong Y, Wang Y et al (2018a) Expanded base editing in rice and wheat using a Cas9-adenosine deaminase fusion. Genome Biol 19:1–9. https://doi.org/10.1186/s13059-018-1443-z

Li J, Zhang X, Sun Y et al (2018b) Efficient allelic replacement in rice by gene editing: a case study of the *NRT1.1B* gene. J Integr Plant Biol 60(7):536–540. https://doi.org/10.1111/jipb.12650

Li Y, Cao XL, Zhu Y et al (2019) Osa-*miR398b* boosts H_2O_2 production and rice blast disease-resistance *via* multiple superoxide dismutases. New Phytol 222(3):1507–1522. https://doi.org/10.1111/nph.15678

Li S, Zhang Y, Xia L, Qi Y (2020a) CRISPR-Cas12a enables efficient biallelic gene targeting in rice. Plant Biotechnol J 18(6):1351. https://doi.org/10.1016/j.molcel.2021.12.032

Li C, Li W, Zhou Z et al (2020b) A new rice breeding method: CRISPR/Cas9 system editing of the *Xa13* promoter to cultivate transgene-free bacterial blight-resistant rice. Plant Biotechnol J 18(2):313. https://doi.org/10.1111/pbi.13217

Li Y, Liao S, Mei P, Pan Y et al (2021a) *OsWRKY93* dually functions between leaf senescence and in response to biotic stress in rice. Front Plant Sci 12:643011. https://doi.org/10.3389/fpls.2021.643011

Li Z, Sun P, Sun P et al (2021b) *OsBC1L1* and *OsBC1L8* function in stomatal development in rice. Biochem Biophys Res Commun 576:40–47. https://doi.org/10.1016/j.bbrc.2021.08.074

Li C, Zhou L, Wu B et al (2022a) Improvement of bacterial blight resistance in two conventionally cultivated rice varieties by editing the noncoding region. Cells 11(16):2535. https://doi.org/10.3390/cells11162535

Li Z, Huang Q, Lin B et al (2022b) CRISPR/Cas9-targeted mutagenesis of a representative member of a novel PR10/bet v1-like protein subfamily significantly reduces rice plant height and defense against *Meloidogyne graminicola*. Phytopathol Res 4(1):38. https://doi.org/10.1186/s42483-022-00143-z

Li Z, Rao MJ, Li J et al (2022c) CRISPR/Cas9 mutant rice *Ospmei12* involved in growth, cell wall development, and response to phytohormone and heavy metal stress. Int J Mol Sci 23(24):16082. https://doi.org/10.3390/ijms232416082

Li Y, Wu S, Huang Y et al (2023) *OsMADS17* simultaneously increases grain number and grain weight in rice. Nat Commun 14:3098. https://doi.org/10.1038/s41467-023-38726-9

Lian T, Huang Y, Xie X et al (2020) Rice SST variation shapes the rhizosphere bacterial community, conferring tolerance to salt stress through regulating soil metabolites. MSystems 5(6):10–1128. https://doi.org/10.1128/mSystems.00721-20

Liang J, Shen Q, Wang L et al (2021) Rice contains a biosynthetic gene cluster associated with production of the casbane-type diterpenoid phytoalexin ent-10-oxodepressin. New Phytol 231(1):85–93. https://doi.org/10.1111/nph.17406

Liao S, Qin X, Luo L et al (2019) CRISPR/Cas9-induced mutagenesis of semi-rolled leaf1, 2 confers curled leaf phenotype and drought tolerance by influencing protein expression patterns and ROS scavenging in rice (*Oryza sativa* L.). Agronomy 9(11):728. https://doi.org/10.3390/agronomy9110728

Liao Y, Ali A, Xue Z et al (2022) Disruption of LLM9428/OsCATC represses starch metabolism and confers enhanced blast resistance in rice. Int J Mol Sci 23(7):3827. https://doi.org/10.3390/ijms23073827

Lin Y, Cradick TJ, Brown MT et al (2014) CRISPR/Cas9 systems have off-target activity with insertions or deletions between target DNA and guide RNA sequences. Nucleic Acids Res 42(11):7473–7485. https://doi.org/10.1093/nar/gku402

Lin CS, Hsu CT, Yang LH et al (2018) Application of protoplast technology to CRISPR/Cas9 mutagenesis: from single-cell mutation detection to mutant plant regeneration. Plant Biotechnol J 16(7):1295–1310. https://doi.org/10.1111/pbi.12870

Lin D, Kong R, Chen L et al (2020) Chloroplast development at low temperature requires the pseudouridine synthase gene *TCD3* in rice. Sci Rep 10(1):8518. https://doi.org/10.1038/s41598-020-65467-2

Lin QJ, Chu J, Kumar V et al (2021) Protein phosphatase 2A catalytic subunit PP2A-1 enhances rice resistance to sheath blight disease. Front Genome Edit 3:632136. https://doi.org/10.3389/fgeed.2021.632136

Lin W, Kuang H, Bai M et al (2023) Multiplex genome editing targeting soybean with ultra-low anti-nutritive oligosaccharides. Crop J 11(3):825–831. https://doi.org/10.1016/j.cj.2023.01.001

Listgarten J, Weinstein M, Kleinstiver BP et al (2018) Prediction of off-target activities for the end-to-end design of CRISPR guide RNAs. Nat Biomed Eng 2(1):38–47. https://doi.org/10.1038/s41551-017-0178-6

Liu X, Qin R, Li J et al (2020a) A CRISPR-Cas9-mediated domain-specific base-editing screen enables functional assessment of ACCase variants in rice. Plant Biotechnol J 18(9):1845. https://doi.org/10.1111/pbi.13348

Liu MH, Kang H, Xu Y et al (2020b) Genome-wide association study identifies an *NLR* gene that confers partial resistance to *Magnaporthe oryzae* in rice. Plant Biotechnol J 18(6):1376–1383. https://doi.org/10.1111/pbi.13300

Liu X, Wu D, Shan T et al (2020c) The trihelix transcription factor OsGTγ-2 is involved adaption to salt stress in rice. Plant Mol Biol 103:545–560. https://doi.org/10.1007/s11103-020-01010-1

Liu XH, Lyu YS, Yang W et al (2020d) A membrane-associated NAC transcription factor OsNTL3 is involved in thermotolerance in rice. Plant Biotechnol J 18(5):1317–1329. https://doi.org/10.1111/pbi.13297

Liu H, Dong S, Li M et al (2021a) The class III peroxidase gene *OsPrx30*, transcriptionally modulated by the AT-hook protein OsATH1, mediates rice bacterial blight-induced ROS accumulation. J Integr Plant Biol 63(2):393–408. https://doi.org/10.1111/jipb.13040

Liu X, Song L, Zhang H et al (2021b) Rice ubiquitin-conjugating enzyme OsUBC26 is essential for immunity to the blast fungus *Magnaporthe oryzae*. Mol Plant Pathol 22(12):1613–1623. https://doi.org/10.1111/mpp.13132

Liu Y, Luo W, Linghu Q et al (2021c) *In planta* genome editing in commercial wheat varieties. Front Plant Sci 12:648841. https://doi.org/10.3389/fpls.2021.648841

Liu S, Liu W, Lai J et al (2022) *OsGLYI3*, a glyoxalase gene expressed in rice seed, contributes to seed longevity and salt stress tolerance. Plant Physiol Biochem 183:85–95. https://doi.org/10.1016/j.plaphy.2022.04.028

Liu X, Yu Y, Yao W et al (2023) CRISPR/Cas9-mediated simultaneous mutation of three salicylic acid 5-hydroxylase (*OsS5H*) genes confers broad-spectrum disease resistance in rice. Plant Biotechnol J 21:1873. https://doi.org/10.1111/pbi.14099

Liying YIN, Yuanye ZHANG, Rongtian LI et al (2022) Improvement of herbicide resistance in rice by using CRISPR/Cas9 system. Chin J Rice Sci 36(5):459. https://doi.org/10.16819/j.1001-7216.2022.211004

Lou D, Wang H, Liang G et al (2017) *OsSAPK2* confers abscisic acid sensitivity and tolerance to drought stress in rice. Front Plant Sci 8:993. https://doi.org/10.3389/fpls.2017.00993

Lou D, Lu S, Chen Z et al (2023) Molecular characterization reveals that *OsSAPK3* improves drought tolerance and grain yield in rice. BMC Plant Biol 23(1):53. https://doi.org/10.1186/s12870-023-04071-8

Lu Y, Zhu JK (2017) Precise editing of a target base in the rice genome using a modified CRISPR/Cas9 system. Mol Plant 10(3):523–525. https://doi.org/10.1016/j.cell.2016.10.044

Lu K, Wu B, Wang J et al (2018) Blocking amino acid transporter OsAAP3 improves grain yield by promoting outgrowth buds and increasing tiller number in rice. Plant Biotechnol J 16(10):1710–1722. https://doi.org/10.1111/pbi.12907

Lu G, Wang C, Wang G et al (2020) Knockouts of drought sensitive genes improve rice grain yield under both drought and well-watered field conditions. Adv Crop Sci Tech 8(3):1000444

Lu H, Shen Z, Xu Y et al (2023) Immune mechanism of ethylicin-induced resistance to *Xanthomonas oryzae* pv. *Oryzae* in rice. J Agric Food Chem 71(1):288–299. https://doi.org/10.1021/acs.jafc.2c07385

Lyu YS, Cao LM, Huang WQ et al (2022) Disruption of three polyamine uptake transporter genes in rice by CRISPR/Cas9 gene editing confers tolerance to herbicide paraquat. Abiotic 3(2):140–145. https://doi.org/10.1007/s42994-022-00075-4

Ma L, Zhu F, Li Z et al (2015) TALEN-based mutagenesis of lipoxygenase *LOX3* enhances the storage tolerance of rice (*Oryza sativa*) seeds. PLoS One 10(12):e0143877. https://doi.org/10.1371/journal.pone.0143877

Ma J, Chen J, Wang M et al (2018) Disruption of *OsSEC3A* increases the content of salicylic acid and induces plant defense responses in rice. J Exp Bot 69(5):1051–1064. https://doi.org/10.1093/jxb/erx458

Macovei A, Sevilla NR, Cantos C et al (2018) Novel alleles of rice eIF4G generated by CRISPR/Cas9-targeted mutagenesis confer resistance to rice tungro spherical virus. Plant Biotechnol J 16(11):1918–1927. https://doi.org/10.1111/pbi.12927

Majhi PK, Singh SK, Anandan A et al (2022) Parental polymorphism survey for evaluation and selection of contrasting parents for drought tolerance in rice (*Oryza sativa* L.) by using

SSR markers. Int J Environ Clim 12(11):2507–2519. https://doi.org/10.9734/IJECC/2022/v12i1131244

Malnoy M, Viola R, Jung MH et al (2016) DNA-free genetically edited grapevine and apple protoplast using CRISPR/Cas9 ribonucleoproteins. Front Plant Sci 7:1904. https://doi.org/10.3389/fpls.2016.01904

Malzahn A, Lowder L, Qi Y (2017) Plant genome editing with TALEN and CRISPR. Cell Biosci 7(1):1–18. https://doi.org/10.1186/s13578-017-0148-4

Mandip KC, Steer CJ (2019) A new era of gene editing for the treatment of human diseases. Swiss Med Wkly 149(0304):w20021. https://doi.org/10.1080/07352689.2017.1402989

McKenna A, Shendure J (2018) FlashFry: a fast and flexible tool for large-scale CRISPR target design. BMC Biol 16(1):1–6. https://doi.org/10.1186/s12915-018-0545-0

Meng X, Hu X, Liu Q et al (2018) Robust genome editing of CRISPR-Cas9 at NAG PAMs in rice. Sci China Life Sci 61:122–125. https://doi.org/10.1007/s11427-017-9247-9

Metje-Sprink J, Menz J, Modrzejewski D, Sprink T (2019) DNA-free genome editing: past, present and future. Front Plant Sci 9:1957. https://doi.org/10.3389/fpls.2018.01957

Miao J, Guo D, Zhang J et al (2013) Targeted mutagenesis in rice using CRISPR-Cas system. Cell Res 23(10):1233–1236. https://doi.org/10.1038/cr.2013.123

Miao C, Xiao L, Hua K et al (2018) Mutations in a subfamily of abscisic acid receptor genes promote rice growth and productivity. Proc Natl Acad Sci 115(23):6058–6063. https://doi.org/10.1073/pnas.1804774115

Mishra R, Joshi RK, Zhao K (2018) Genome editing in rice: recent advances, challenges, and future implications. Front Plant Sci 9:409924. https://doi.org/10.3389/fpls.2018.01361

Murugan K, Babu K, Sundaresan R et al (2017) The revolution continues: newly discovered systems expand the CRISPR-Cas toolkit. Mol Cell 68(1):15–25. https://doi.org/10.1016/j.molcel.2017.09.007

Nawaz G, Usman B, Peng H et al (2020) Knockout of *Pi21* by CRISPR/Cas9 and iTRAQ-based proteomic analysis of mutants revealed new insights into *M. Oryzae* resistance in elite rice line. Genes 11(7):735. https://doi.org/10.3390/genes11070735

Ni Z, Cao Y, Jin X et al (2021) Engineering resistance to bacterial blight and bacterial leaf streak in rice. Rice 14(1):38. https://doi.org/10.1186/s12284-021-00482-z

Ni P, Zhao Y, Zhou X et al (2023) Efficient and versatile multiplex prime editing in hexaploid wheat. Genome Biol 24(1):156. https://doi.org/10.1186/s13059-023-02990-1

Nithya S, Kumam Y, Varanavasiappan S et al (2020) Targeted mutation in *eIF4G* gene in rice. Electron J Plant Breed 11(04):1194–1199. https://www.ejplantbreeding.org/index.php/EJPB/article/view/3699

Ogata T, Ishizaki T, Fujita M et al (2020) CRISPR/Cas9-targeted mutagenesis of *OsERA1* confers enhanced responses to abscisic acid and drought stress and increased primary root growth under nonstressed conditions in rice. PLoS One 15(12):e0243376. https://doi.org/10.1371/journal.pone.0243376

Oliva R, Ji C, Atienza-Grande G, Huguet-Tapia JC et al (2019) Broad-spectrum resistance to bacterial blight in rice using genome editing. Nat Biotechnol 37(11):1344–1350. https://doi.org/10.1038/s41587-019-0267-z

Pak H, Wang H, Kim Y et al (2021) Creation of male-sterile lines that can be restored to fertility by exogenous methyl jasmonate for the establishment of a two-line system for the hybrid production of rice (*Oryza sativa* L.). Plant Biotechnol J 19(2):365–374. https://doi.org/10.1111/pbi.13471

Pan C, Ye L, Qin L et al (2016) CRISPR/Cas9-mediated efficient and heritable targeted mutagenesis in tomato plants in the first and later generations. Sci Rep 6(1):24765. https://doi.org/10.1038/srep46916

Pattanayak V, Lin S, Guilinger JP et al (2013) High-throughput profiling of off-target DNA cleavage reveals RNA-programmed Cas9 nuclease specificity. Nat Biotechnol 31(9):839–843. https://doi.org/10.1038/nbt.2673

Piatek A, Ali Z, Baazim H et al (2015) RNA-guided transcriptional regulation in planta via synthetic dC as9-based transcription factors. Plant Biotechnol J 13(4):578–589. https://doi.org/10.1111/pbi.12284

Pickar-Oliver A, Black JB, Lewis MM et al (2019) Targeted transcriptional modulation with type I CRISPR–Cas systems in human cells. Nat Biotechnol 37(12):1493–1501. https://doi.org/10.1038/s41587-019-0235-7

Premnath A, Ramalingam AP, Ramasamy SP et al (2022) Genome editing and miRNA-based approaches in cereals under abiotic stress. In: Sustainable remedies for abiotic stress in cereals. Springer Nature Singapore, pp 647–673. https://doi.org/10.1007/978-981-19-5121-3_25

Prodhan ZH, Islam SA, Alam MS et al (2022) Impact of *OsBadh2* mutations on salt stress response in rice. Plan Theory 11(21):2829. https://doi.org/10.3390/plants11212829

Qu M, Essemine J, Li M et al (2020) Genome-wide association study unravels *LRK1* as a dark respiration regulator in rice (*Oryza sativa* L.). Int J Mol Sci 21(14):4930. https://doi.org/10.3390/ijms21144930

Que Z, Lu Q, Li Q, Shen C (2023) The rice annexin gene *OsAnn5* is involved in cold stress tolerance at the seedling stage. Plant Direct 7(11):e539. https://doi.org/10.1002/pld3.539

Raja RB, Agasimani S, Jaiswal S et al (2017) EcoTILLING by sequencing reveals polymorphisms in genes encoding starch synthases that are associated with low glycemic response in rice. BMC Plant Biol 17:1–13. https://doi.org/10.1186/s12870-016-0968-0

Ramadoss BR, Gangola MP, Agasimani S et al (2019) Starch granule size and amylopectin chain length influence starch *in vitro* enzymatic digestibility in selected rice mutants with similar amylose concentration. J Food Sci Technol 56:391–400. https://doi.org/10.1007/s13197-018-3500-8

Rath D, Amlinger L, Hoekzema M et al (2015) Efficient programmable gene silencing by cascade. Nucleic Acids Res 43(1):237–246. https://doi.org/10.1093/nar/gku1257

Rathnasamy SA, Kambale R, Elangovan A et al (2023) Altering stomatal density for manipulating transpiration and photosynthetic traits in rice through CRISPR/Cas9 mutagenesis. Curr Issues Mol Biol 45(5):3801–3814. https://doi.org/10.3390/cimb45050245

Ren C, Liu X, Zhang Z et al (2016) CRISPR/Cas9-mediated efficient targeted mutagenesis in chardonnay (*Vitis vinifera* L.). Sci Rep 6(1):32289. https://doi.org/10.1038/srep32289

Saminadane T, Geddan S, Krishnaswamy P et al (2023) Development of early maturing salt tolerant rice variety KKL (R) 3 using combination of conventional and molecular breeding approaches. Front Genet 14:1332691. https://doi.org/10.3389/fgene.2023.1332691

Saragliadis A, Klauser B, Hartig JS (2012) *In vivo* screening of ligand-dependent hammerhead ribozymes. Ribozymes: Methods Protoc 848:455–463. https://doi.org/10.1007/978-1-61779-545-9_28

Shaheen N, Ahmad S, Alghamdi SS et al (2023) CRISPR-Cas system, a possible "savior" of rice threatened by climate change: an updated review. Rice 16(1):39. https://doi.org/10.1186/s12284-023-00652-1

Shan Q, Zhang Y, Chen K et al (2015) Creation of fragrant rice by targeted knockout of the *OsBADH2* gene using TALEN technology. Plant Biotechnol J 13(6):791–800. https://doi.org/10.1111/pbi.12312

Shanthinie A, Dilip KR, Raveendran M (2019) Comparative profiling of volatile compounds in the grains of rice varieties differing in their aroma. Electron J Plant Breed 10(2):614–619. https://doi.org/10.5958/0975-928X.2019.00077.2

Shao G, Xie L, Jiao G et al (2017) CRISPR/CAS9-mediated editing of the fragrant gene *Badh2* in rice. Chin J Rice Sci 31(2):216–222. https://doi.org/10.1111/ppl.13871

Shen C, Que Z, Xia Y et al (2017) Knock out of the annexin gene *OsAnn3 via* CRISPR/Cas9-mediated genome editing decreased cold tolerance in rice. J Plant Biol 60:539–547. https://doi.org/10.1007/s12374-016-0400-1

Shen L, Wang C, Fu Y et al (2018) QTL editing confers opposing yield performance in different rice varieties. J Integr Plant Biol 60(2):89–93. https://doi.org/10.1111/jipb.12501

Shim Y, Seong G, Choi Y et al (2023) Suppression of cuticular wax biosynthesis mediated by rice LOV KELCH REPEAT PROTEIN 2 supports a negative role in drought stress tolerance. Plant Cell Environ 46(5):1504–1520. https://doi.org/10.1111/pce.14549

Shimatani Z, Kashojiya S, Takayama M et al (2017) Targeted base editing in rice and tomato using a CRISPR-Cas9 cytidine deaminase fusion. Nat Biotechnol 35(5):441–443. https://doi.org/10.1038/nbt.3833

Singh SK, Majhi PK, Anandan A et al (2022) Microsatellites based parental polymorphism survey for moisture stress in rice (*Oryza sativa* L.) between the parental genotypes HUR-1309 and CR Dhan 801. Emergent Life Sci Res 8:248–258. https://doi.org/10.31783/elsr.2022.82248258

Singh A, Rami E, Upadhyay P, Gangawane AK (2023) The impact of climate change on crop production and combat strategies in climate change and sustainable development. CRC Press, pp 103–115

Sree DB, Shanthinie A, Vignesh P et al (2023) Targeted editing of *OsSWEET11* promoter for imparting bacterial leaf blight resistance in rice. Electron J Plant Breed 14(3):938–947. https://doi.org/10.37992/2023.1403.106

Sridhar PR (2023) First CRISPR'd rice with multi-pathogen resistance. https://doi.org/10.13140/RG.2.2.30469.12001

Sugano SS, Nishihama R, Shirakawa M et al (2018) Efficient CRISPR/Cas9-based genome editing and its application to conditional genetic analysis in *Marchantia polymorpha*. PLoS One 13(10):e0205117. https://doi.org/10.1371/journal.pone.0205117

Sun Y, Zhang X, Wu C et al (2016) Engineering herbicide-resistant rice plants through CRISPR/Cas9-mediated homologous recombination of acetolactate synthase. Mol Plant 9(4):628–631. https://doi.org/10.1016/j.molp.2016.01.001

Sun Y, Jiao G, Liu Z et al (2017) Generation of high-amylose rice through CRISPR/Cas9-mediated targeted mutagenesis of starch branching enzymes. Front Plant Sci 8:298. https://doi.org/10.3389/fpls.2017.00298

Takagi H, Tamiru M, Abe A et al (2015) MutMap accelerates breeding of a salt-tolerant rice cultivar. Nat Biotechnol 33(5):445–449. https://doi.org/10.1038/nbt.3188

Tan R, Krueger RK, Gramelspacher MJ et al (2022) Cas11 enables genome engineering in human cells with compact CRISPR-Cas3 systems. Mol Cell 82(4):852–867. https://doi.org/10.1016/j.molcel.2021.12.032

Tang J, Breaker RR (1997) Rational design of allosteric ribozymes. Chem Biol 4(6):453–459. https://doi.org/10.1016/s1074-5521(97)90197-6

Tang L, Mao B, Li Y et al (2017) Knockout of *OsNramp5* using the CRISPR/Cas9 system produces low cd-accumulating indica rice without compromising yield. Sci Rep 7(1):14438. https://doi.org/10.1038/s41598-017-14832-9

Tao H, Shi X, He F et al (2021) Engineering broad-spectrum disease-resistant rice by editing multiple susceptibility genes. J Integr Plant Biol 63(9):1639–1648. https://doi.org/10.1111/jipb.13145

Távora FT, Meunier AC, Vernet A et al (2022) CRISPR/Cas9-targeted knockout of rice susceptibility genes *OsDjA2* and *OsERF104* reveals alternative sources of resistance to *Pyricularia oryzae*. Rice Sci 29(6):535–544. https://doi.org/10.1016/j.rsci.2022.04.001

Teng Y, Lv M, Zhang X et al (2022) *BEAR1*, a bHLH transcription factor, controls salt response genes to regulate rice salt response. J Plant Biol 65(3):217–230. https://doi.org/10.1007/s12374-022-09347-4

Tezuka D, Matsuura H, Saburi W et al (2021) A ubiquitously expressed UDP-glucosyltransferase, *UGT74J1*, controls basal salicylic acid levels in rice. Plan Theory 10(9):1875. https://doi.org/10.3390/plants10091875

Thakur T, Sinha K, Kaur T et al (2022) Efficient genetic transformation of rice for CRISPR/Cas9 mediated genome-editing and stable overexpression studies: a case study on rice lipase 1 and galactinol synthase encoding genes. Agronomy 12:179. https://doi.org/10.3390/agronomy12010179

Tianshun ZHOU, Dong YU, Ling LIU et al (2021) CRISPR/Cas9-mediated editing of Afp1 improves rice stress tolerance. Chin J Rice Sci 35(1):11. https://doi.org/10.16819/j.1001-7216.2021.0503

Tripathy SP, Majhi PK, Patra B et al (2021) Editing the genome for salt tolerance in rice. Biot Res Today 312:1143–1145

Tun W, Yoon J, Vo KTX et al (2023) Sucrose preferentially promotes expression of *OsWRKY7* and *OsPR10a* to enhance defense response to blast fungus in rice. Front Plant Sci 14:1117023. https://doi.org/10.3389/fpls.2023.1117023

Usman B, Nawaz G, Zhao N, Liao S, Liu Y, Li R (2020a) Precise editing of the *OsPYL9* gene by RNA-guided Cas9 nuclease confers enhanced drought tolerance and grain yield in rice (*Oryza sativa* L.) by regulating circadian rhythm and abiotic stress responsive proteins. Int J Mol Sci 21(21):7854. https://doi.org/10.3390/ijms21217854

Usman B, Nawaz G, Zhao N et al (2020b) Programmed editing of rice (*Oryza sativa* L.) *OsSPL16* gene using CRISPR/Cas9 improves grain yield by modulating the expression of pyruvate enzymes and cell cycle proteins. Int J Mol Sci 22(1):249. https://doi.org/10.3390/ijms22010249

Wang T, Wei JJ, Sabatini DM, Lander ES (2014) Genetic screens in human cells using the CRISPR-Cas9 system. Science 343(6166):80–84. https://doi.org/10.1126/science.1246981

Wang ZP, Xing HL, Dong L et al (2015) Egg cell-specific promoter-controlled CRISPR/Cas9 efficiently generates homozygous mutants for multiple target genes in Arabidopsis in a single generation. Genome Biol 16:1–12. https://doi.org/10.1186/s13059-015-0715-0

Wang F, Wang C, Liu P et al (2016a) Enhanced rice blast resistance by CRISPR/Cas9-targeted mutagenesis of the ERF transcription factor gene *OsERF922*. PLoS One 11(4):e0154027. https://doi.org/10.1371/journal.pone.0154027

Wang F, Fan F, Li W et al (2016b) Knock-out efficiency analysis of *Pi21* gene using CRISPR/Cas9 in rice. Chin J Rice Sci 30(5):469–478. https://doi.org/10.16819/j.1001-7216.2016.6009

Wang C, Liu Q, Shen Y et al (2019) Clonal seeds from hybrid rice by simultaneous genome engineering of meiosis and fertilization genes. Nat Biotechnol 37(3):283–286. https://doi.org/10.1038/s41587-018-0003-0

Wang B, Fang R, Chen F et al (2020a) A novel CCCH-type zinc finger protein *SAW1* activates *OsGA20ox3* to regulate gibberellin homeostasis and anther development in rice. J Integr Plant Biol 62(10):1594–1606. https://doi.org/10.1111/jipb.12924

Wang B, Zhong Z, Wang X et al (2020b) Knockout of the *OsNAC006* transcription factor causes drought and heat sensitivity in rice. Int J Mol Sci 21(7):2288. https://doi.org/10.3390/ijms21072288

Wang F, Itai RN, Nozoye T et al (2020c) The *bHLH* protein *OsIRO3* is critical for plant survival and iron (Fe) homeostasis in rice (*Oryza sativa* L.) under Fe-deficient conditions. Soil Sci Plant Nutr 66(4):579–592. https://doi.org/10.1080/00380768.2020.1783966

Wang K, An W, Liu Y et al (2020d) Disruption of *OsRhoGDI2* by CRISPR/Cas9 technology leads to semi-dwarf in rice. Chin J Biotechnol 36(4):707–715. https://doi.org/10.13345/j.cjb.190333

Wang F, Xu Y, Li W et al (2021a) Creating a novel herbicide-tolerance *OsALS* allele using CRISPR/Cas9-mediated gene editing. Crop J 9(2):305–312. https://doi.org/10.1016/j.cj.2020.06.001

Wang W, Ma S, Hu P et al (2021b) Genome editing of rice *eIF4G* loci confers partial resistance to rice black-streaked dwarf virus. Viruses 13(10):2100. https://doi.org/10.3390/v13102100

Wang X, Li J, Li F et al (2021c) Rice potassium transporter *OsHAK8* mediates K^+ uptake and translocation in response to low K^+ stress. Front Plant Sci 12:730002. https://doi.org/10.3389/fpls.2021.730002

Wang Z, Chen D, Sun F et al (2021d) *ARGONAUTE 2* increases rice susceptibility to rice black-streaked dwarf virus infection by epigenetically regulating *HEXOKINASE 1* expression. Mol Plant Pathol 22(9):1029–1040. https://doi.org/10.1111/mpp.13091

Wang P, Liu J, Lyu Y et al (2022) A review of vector-borne rice viruses. Viruses 14:2258. https://doi.org/10.3390/v14102258

Wang Z, Zhou L, Lan Y et al (2022a) An aspartic protease 47 causes quantitative recessive resistance to rice black-streaked dwarf virus disease and southern rice black-streaked dwarf virus disease. New Phytol 233(6):2520–2533. https://doi.org/10.1111/nph.17961

Wang X, Ren P, Ji L et al (2022b) *OsVDE*, a xanthophyll cycle key enzyme, mediates abscisic acid biosynthesis and negatively regulates salinity tolerance in rice. Planta 255:1–15. https://doi.org/10.1007/s00425-021-03802-1

Wei H, Wang X, He Y et al (2021a) Clock component *OsPRR73* positively regulates rice salt tolerance by modulating *OsHKT2*; 1-mediated sodium homeostasis. EMBO J 40(3):e105086. https://doi.org/10.15252/embj.2020105086

Wei Z, Abdelrahman M, Gao Y et al (2021b) Engineering broad-spectrum resistance to bacterial blight by CRISPR-Cas9-mediated precise homology directed repair in rice. Mol Plant 14(8):1215–1218. https://doi.org/10.1016/j.molp.2021.05.012

Wolter F, Klemm J, Puchta H (2018) Efficient in planta gene targeting in *Arabidopsis* using egg cell-specific expression of the Cas9 nuclease of *Staphylococcus aureus*. Plant J 94(4):735–746. https://doi.org/10.1111/tpj.13893

Woo JW, Kim J, Kwon SI et al (2015) DNA-free genome editing in plants with preassembled CRISPR-Cas9 ribonucleoproteins. Nat Biotechnol 33(11):1162–1164. https://doi.org/10.1038/nbt.3389

Wu Y, Xiao N, Cai Y et al (2023) CRISPR-Cas9-mediated editing of the *OsHPPD* 3′ UTR confers enhanced resistance to HPPD-inhibiting herbicides in rice. Plant Commun 4:100605. https://doi.org/10.1016/j.xplc.2023.100605

Wyvekens N, Topkar VV, Khayter C et al (2015) Dimeric CRISPR RNA-guided FokI-dCas9 nucleases directed by truncated gRNAs for highly specific genome editing. Hum Gene Ther 26(7):425–431. https://doi.org/10.1089/hum.2015.084

Xiao A, Cheng Z, Kong L et al (2014) CasOT: a genome-wide Cas9/gRNA off-target searching tool. Bioinformatics 30(8):1180–1182. https://doi.org/10.1093/bioinformatics/btt764

Xie K, Minkenberg B, Yang Y (2015) Boosting CRISPR/Cas9 multiplex editing capability with the endogenous tRNA-processing system. Proc Natl Acad Sci 112(11):3570–3575. https://doi.org/10.1073/pnas.1420294112

Xie Y, Wang Y, Yu X et al (2022) *SH3P2*, an SH3 domain-containing protein that interacts with both *Pib* and *AvrPib*, suppresses effector-triggered, Pib-mediated immunity in rice. Mol Plant 15(12):1931–1946. https://doi.org/10.1016/j.molp.2022.10.022

Xing Y, Zhang Q (2010) Genetic and molecular bases of rice yield. Annu Rev Plant Biol 61:421–442. https://doi.org/10.1146/annurev-arplant-042809-112209

Xu R, Li H, Qin R et al (2014) Gene targeting using the *agrobacterium tumefaciens*-mediated CRISPR-Cas system in rice. Rice 7:1–4. https://doi.org/10.1186/s12284-014-0005-6

Xu R, Yang Y, Qin R et al (2016) Rapid improvement of grain weight *via* highly efficient CRISPR/Cas9-mediated multiplex genome editing in rice. J Genet Genomics 43(8):529–532. https://doi.org/10.1016/j.jgg.2016.07.003

Xu W, Yang Y, Liu Y, Kang G et al (2020a) Discriminated sgRNAs-based SurroGate system greatly enhances the screening efficiency of plant base-edited cells. Mol Plant 13(1):169–180. https://doi.org/10.1016/j.molp.2019.10.007

Xu Y, Wang F, Chen Z et al (2020b) Intron-targeted gene insertion in rice using CRISPR/Cas9: a case study of the *pi-ta* gene. Crop J 8(3):424–431. https://doi.org/10.1016/j.cj.2019.03.006

Xu J, Wang X, Zu H et al (2021a) Molecular dissection of rice phytohormone signaling involved in resistance to a piercing-sucking herbivore. New Phytol 230(4):1639–1652. https://doi.org/10.1111/nph.17251

Xu R, Liu X, Li J et al (2021b) Identification of herbicide resistance *OsACC1* mutations *via in planta* prime-editing-library screening in rice. Nat Plants 7(7):888–892. https://www.nature.com/articles/s41477-021-00942-w

Xu X, Xu Z, Li Z et al (2021c) Increasing resistance to bacterial leaf streak in rice by editing the promoter of susceptibility gene *OsSULRT3*. Plant Biotechnol J 19(6):1101. https://doi.org/10.1111/pbi.13602

Xu X, Xu Z, Ma W et al (2021d) TALE-triggered and iTALE-suppressed *Xa1*-mediated resistance to bacterial blight is independent of rice transcription factor subunits *OsTFIIAγ1* or *OsTFIIAγ5*. J Exp Bot 72(8):3249–3262. https://doi.org/10.1093/jxb/erab054

Xu Y, Yan S, Jiang S et al (2023) Identification of a rice leaf width gene narrow leaf 22 (*NAL22*) through genome-wide association study and gene editing technology. Int J Mol Sci 24(4):4073. https://doi.org/10.3390/ijms24044073

Yamaguchi K, Yamamoto T, Segami S et al (2020) *gw2* mutation increases grain width and culm thickness in rice (*Oryza sativa* L.). Breed Sci 70(4):456–461. https://doi.org/10.1270/jsbbs.20018

Yamato T, Handa A, Arazoe T et al (2019) Single crossover-mediated targeted nucleotide substitution and knock-in strategies with CRISPR/Cas9 system in the rice blast fungus. Sci Rep 9(1):7427. https://doi.org/10.1038/s41598-019-43913-0

Yang Z, Xing J, Wang L et al (2020) Mutations of two FERONIA-like receptor genes enhance rice blast resistance without growth penalty. J Exp Bot 71(6):2112–2126. https://doi.org/10.1093/jxb/erz541

Yang J, Ji L, Liu S et al (2021) The CaM1-associated CCaMK–MKK1/6 cascade positively affects lateral root growth *via* auxin signaling under salt stress in rice. J Exp Bot 72(18):6611–6627. https://doi.org/10.1093/jxb/erab287

Yang J, Fang Y, Wu H et al (2023) Improvement of resistance to rice blast and bacterial leaf streak by CRISPR/Cas9-mediated mutagenesis of *Pi21* and *OsSULTR3*;6 in rice (*Oryza sativa* L.). Front Plant Sci 14:1209384. https://doi.org/10.3389/fpls.2023.1209384

Yin X, Biswal AK, Dionora J et al (2017) CRISPR-Cas9 and CRISPR-Cpf1 mediated targeting of a stomatal developmental gene *EPFL9* in rice. Plant Cell Rep 36:745–757. https://doi.org/10.1007/s00299-017-2118-z

Yin X, Zou B, Hong X et al (2018a) Rice copine genes *OsBON1* and *OsBON3* function as suppressors of broad-spectrum disease resistance. Plant Biotechnol J 16(8):1476–1487. https://doi.org/10.1111/pbi.12890

Yin H, Song CQ, Suresh S et al (2018b) Partial DNA-guided Cas9 enables genome editing with reduced off-target activity. Nat Chem Biol 14(3):311–316. https://doi.org/10.1038/nchembio.2559

Yu Q, Chen L, Zhou W, An Y, Luo T, Wu Z et al (2020) RSD1 is essential for stomatal patterning and files in rice. Front Plant Sci 11:600021. https://doi.org/10.3389/fpls.2020.600021

Yu K, Liu Z, Gui H et al (2021) Highly efficient generation of bacterial leaf blight-resistant and transgene-free rice using a genome editing and multiplexed selection system. BMC Plant Biol 21(1):1–10. https://doi.org/10.1186/s12870-021-02979-7

Yudong CAO, Xiangyi XIAO, Naizhong YE et al (2021) Auxin regulator *OsGRF4* simultaneously regulates rice grain shape and blast resistance. Chin J Rice Sci 35(6):629. https://doi.org/10.16819/j.1001-7216.2021.210206

Yue E, Cao H, Liu B (2020) OsmiR535, a potential genetic editing target for drought and salinity stress tolerance in *Oryza sativa*. Plan Theory 9(10):1337. https://doi.org/10.3390/plants9101337

Yue E, Rong F, Liu Z et al (2023) Cadmium induced a non-coding RNA microRNA535 mediates cd accumulation in rice. J Environ Sci 130:149–162. https://doi.org/10.1016/j.jes.2022.10.005

Zafar K, Khan MZ, Amin I et al (2020) Precise CRISPR-Cas9 mediated genome editing in super basmati rice for resistance against bacterial blight by targeting the major susceptibility gene. Front Plant Sci 11:575. https://doi.org/10.3389/fpls.2020.00575

Zafar K, Khan MZ, Amin I et al (2023) Employing template-directed CRISPR-based editing of the *OsALS* gene to create herbicide tolerance in basmati rice. AoB Plants 15(2):plac059. https://doi.org/10.1093/aobpla/plac059

Zeng Y, Wen J, Zhao W et al (2020a) Rational improvement of rice yield and cold tolerance by editing the three genes *OsPIN5b, GS3,* and *OsMYB30* with the CRISPR-Cas9 system. Front Plant Sci 10:1663. https://doi.org/10.3389/fpls.2019.01663

Zeng X, Luo Y, Vu NTQ et al (2020b) CRISPR/Cas9-mediated mutation of *OsSWEET14* in rice cv. Zhonghua11 confers resistance to *Xanthomonas oryzae* pv. *Oryzae* without yield penalty. BMC Plant Biol 20(1):1–11. https://doi.org/10.1186/s12870-020-02524-y

Zetsche B, Heidenreich M, Mohanraju P et al (2017) Multiplex gene editing by CRISPR–Cpf1 using a single crRNA array. Nat Biotechnol 35(1):31–34. https://doi.org/10.1038/nbt.3737

Zhai XY, Chen ZJ, Liu J et al (2022) Expression of CYP76C6 facilitates isoproturon metabolism and detoxification in rice. J Agric Food Chem 70(15):4599–4610. https://doi.org/10.1021/acs.jafc.1c08137

Zhang H, Wu Z, Wang C et al (2014) Germination and infectivity of microconidia in the rice blast fungus *Magnaporthe oryzae*. Nat Commun 5(1):4518. https://doi.org/10.1038/ncomms5518

Zhang H, Zhang J, Wei P et al (2014b) The CRISPR/Cas9 system produces specific and homozygous targeted gene editing in rice in one generation. Plant Biotechnol J 12:797–807. https://doi.org/10.1111/pbi.12200

Zhang H, Zhang J, Lang Z et al (2017) Genome editing—principles and applications for functional genomics research and crop improvement. Crit Rev Plant Sci 36(4):291–309. https://doi.org/10.1080/07352689.2017.1402989

Zhang J, Zhang H, Botella JR, Zhu JK (2018a) Generation of new glutinous rice by CRISPR/Cas9-targeted mutagenesis of the *waxy* gene in elite rice varieties. J Integr Plant Biol 60(5):369–375. https://doi.org/10.1111/jipb.12620

Zhang Q, Xing HL, Wang ZP et al (2018b) Potential high-frequency off-target mutagenesis induced by CRISPR/Cas9 in Arabidopsis and its prevention. Plant Mol Biol 96:445–456. https://doi.org/10.1007/s11103-018-0709-x

Zhang A, Liu Y, Wang F et al (2019) Enhanced rice salinity tolerance via CRISPR/Cas9-targeted mutagenesis of the *OsRR22* gene. Mol Breed 39:1–10. https://doi.org/10.1007/s11032-019-0954-y

Zhang H, Li L, He Y et al (2020a) Distinct modes of manipulation of rice auxin response factor *OsARF17* by different plant RNA viruses for infection. Proc Natl Acad Sci 117(16):9112–9121. https://doi.org/10.1073/pnas.1918254117

Zhang Y, Wang X, Luo Y et al (2020b) *OsABA8ox2*, an ABA catabolic gene, suppresses root elongation of rice seedlings and contributes to drought response. Crop J 8(3):480–491. https://doi.org/10.1016/j.cj.2019.08.006

Zhang S, Shen J, Li D, Cheng Y (2021a) Strategies in the delivery of Cas9 ribonucleoprotein for CRISPR/Cas9 genome editing. Theranostics 11(2):614. https://doi.org/10.7150/thno.47007

Zhang R, Chen S, Meng X et al (2021b) Generating broad-spectrum tolerance to ALS-inhibiting herbicides in rice by base editing. Sci China Life Sci 64:1624–1633. https://doi.org/10.1007/s11427-020-1800-5

Zhang M, Zhao R, Huang K et al (2022a) The *OsWRKY63–OsWRKY76–OsDREB1B* module regulates chilling tolerance in rice. Plant J 112(2):383–398. https://doi.org/10.1111/tpj.15950

Zhang X, Liu D, Gao D et al (2022b) Cytokinin confers brown planthopper resistance by elevating jasmonic acid pathway in rice. Int J Mol Sci 23(11):5946. https://doi.org/10.3390/ijms23115946

Zhang M, Zhao R, Huang K et al (2023) *OsWRKY76* positively regulates drought stress *via* OsbHLH148-mediated jasmonate signaling in rice. Front Plant Sci 14:1168723. https://doi.org/10.3389/fpls.2023.1168723

Zhao Y, Zhang C, Liu W et al (2016) An alternative strategy for targeted gene replacement in plants using a dual-sgRNA/Cas9 design. Sci Rep 6(1):23890. https://doi.org/10.1038/srep23890

Zhao DS, Li QF, Zhang CQ et al (2018) *GS9* acts as a transcriptional activator to regulate rice grain shape and appearance quality. Nat Commun 9(1):1240. https://doi.org/10.1038/s41467-018-03616-y

Zhao W, Xiao W, Sun J et al (2022) An integration of microRNA and transcriptome sequencing analysis reveal regulatory roles of miRNAs in response to chilling stress in wild rice. Plan Theory 11(7):977. https://doi.org/10.3390/plants11070977

Zheng S, Ye C, Lu J et al (2021) Improving the rice photosynthetic efficiency and yield by editing *OsHXK1* via CRISPR/Cas9 system. Int J Mol Sci 22(17):9554. https://doi.org/10.3390/ijms22179554

Zhou J, Peng Z, Long J (2015) Gene targeting by the TAL effector *PthXo2* reveals cryptic resistance gene for bacterial blight of rice. Plant J 82(4):632–643. https://doi.org/10.1111/tpj.12838

Zhou H, He M, Li J et al (2016) Development of commercial thermo-sensitive genic male sterile rice accelerates hybrid rice breeding using the CRISPR/Cas9-mediated TMS5 editing system. Sci Rep 6(1):37395. https://doi.org/10.1038/srep37395

Zhou X, Liao H, Chern M et al (2018) Loss of function of a rice TPR-domain RNA-binding protein confers broad-spectrum disease resistance. Proc Natl Acad Sci 115(12):3174–3179. https://doi.org/10.1073/pnas.1705927115

Zhou J, Xin X, He Y et al (2019) Multiplex QTL editing of grain-related genes improves yield in elite rice varieties. Plant Cell Rep 38:475–485. https://doi.org/10.1007/s00299-018-2340-3

Zhou Y, Xu S, Jiang N et al (2022) Engineering of rice varieties with enhanced resistances to both blast and bacterial blight diseases *via* CRISPR/Cas9. Plant Biotechnol J 20(5):876–885. https://doi.org/10.1111/pbi.13766

Zhu Y, Lin Y, Chen S et al (2019) CRISPR/Cas9-mediated functional recovery of the recessive rc allele to develop red rice. Plant Biotechnol J 17(11):2096–2105. https://doi.org/10.1111/pbi.13125

Zhu Z, Yin J, Chern M et al (2020) New insights into bsr-d1-mediated broad-spectrum resistance to rice blast. Mol Plant Pathol 21(7):951–960. https://doi.org/10.1111/mpp.12941

Zong Y, Wang Y, Li C et al (2017) Precise base editing in rice, wheat and maize with a Cas9-cytidine deaminase fusion. Nat Biotechnol 35(5):438–440. https://doi.org/10.1038/nbt.3811

Zuris JA, Thompson DB, Shu Y et al (2015) Cationic lipid-mediated delivery of proteins enables efficient protein-based genome editing *in vitro* and *in vivo*. Nat Biotechnol 33(1):73–80. https://doi.org/10.1038/nbt.3081

High-Throughput Phenotyping Enabled Rice Improvement

10

B. S. Harshitha, Aavula Naveen, H. A. Bhargavi,
P. S. Basavaraj, and M. Karthik Kumar

Abstract

Rice is an important cereal crop that feeds over 50% of the global population, making its production significant for global food security. High-throughput phenotyping (HTP) has significantly advanced agricultural research and breeding programs. Traditional methods, while valuable, need help with scalability, precision, and efficiency. However, high-throughput phenotyping has revolutionized this field by facilitating swift, noninvasive, and extensive evaluation of various traits at multiple time intervals. Technologies like drones and sensor networks have allowed precise monitoring of growth, disease resistance, and stress tolerance. This unprecedented data capture helps decode the intricate relationship between genetics and the environment, speeding up trait selection and cultivar development. By integrating these technologies, researchers and breeders can rapidly identify and evaluate traits critical in the face of climate change. Advanced imaging detects early stress signs, aiding in real-time genetic response assessment. This knowledge helps breed resilient rice varieties, ensuring food security amid climatic uncertainties. High-throughput phenotyping synergy with traditional methods opens new doors in creating rice varieties with superior traits. As

B. S. Harshitha
Central Sericultural Research and Training Institute, Berhampore, West Bengal, India

A. Naveen
Division of Genetics, ICAR-Indian Agricultural Research Institute, New Delhi, India

H. A. Bhargavi (✉)
ICAR-Indian Institute of Horticultural Research, Bengaluru, Karnataka, India

P. S. Basavaraj
ICAR-National Institute of Abiotic Stress Management, Baramati, Maharashtra, India

M. Karthik Kumar
Central Sericultural Research and Training Institute, Mysuru, Karnataka, India

A. Singh et al. (eds.), *Climate-Smart Rice Breeding*,
https://doi.org/10.1007/978-981-97-7098-4_10

the global population grows and climate unpredictability looms, these innovations promise sustainable rice production and resilient food systems.

Keywords

High-throughput phenotyping · Rice breeding · Remote sensing · Resilience · Adaptability

10.1 Introduction

The development of high-yielding, input-responsive elite varieties in rice and wheat led to the "Green Revolution," notably boosting their production alongside that of other cereals (Evenson and Gollin 2003). However, recent studies have projected that by 2050, production must be twice the current productivity with an enhanced rate from 1.3% to 2.4% per year to feed the ever-increasing population (Ray et al. 2012). Achieving this amid diminishing arable land, water scarcity, and heightened climatic variability necessitates a shift. Hence, enhancement in productivity must come from crop improvement with optimized agronomic practices. Therefore, a paradigm shift is necessary to integrate and adopt new technologies to advance the rate of genetic gain.

Although traditional breeding methods such as introduction, hybridization-based approaches (pedigree, bulk, and back cross), and selection-driven methods (pure line, mass, and progeny) have retained their significance because of their substantial historical contributions (Chimmili et al. 2022), these methods are time-consuming, laborious, and tedious. Further, these methods heavily depend on selection based on phenotype. It is widely acknowledged that the phenotype of a plant is influenced by the interaction among its genotype, environment, and management practices (phenotype = G + E + (G × E) × M) (Basavaraj and Rane 2020). As a result, selection processes are nearly deceived (Lamichhane and Thapa 2022). Consequently, intentional efforts have been undertaken recently to use trait-based genomic methodologies in addition to traditional crop development techniques (Basavaraj et al. 2023).

Understanding the molecular mechanisms underlying important traits involves dissecting genetic pathways, which requires precise phenotyping of plant populations. However, screening and evaluating such a large population using traditional methods can be costly, laborious, and time-consuming. The expression of a trait is influenced by the interaction of genotype, environment, and management strategies (G × E × M), which ultimately determines the phenotype. These interactions shape the presence and expression of alleles or genes associated with specific structural and functional traits (Dwivedi et al. 2020). Consequently, the interplay between structural and functional phenotypes impacts plant performance, affecting traits such as morphology, growth attributes, and economic yield (Houle et al. 2010). Although next-generation sequencing and chip genotyping techniques have made obtaining genotypic data faster and cheaper, phenotyping remains challenging. Destructive methods, manual assessments, reduced precision, and time constraints hinder our understanding of the genetic basis of complex traits. To address these challenges and improve marker-assisted breeding, reliable, precise, automated, and high-throughput phenotyping technologies are essential (Lorenz et al. 2011). High-throughput plant phenotyping tools offer promise in bridging the gap between genotype and phenotype, aiding in the identification of marker–trait associations, quantitative trait loci (QTL) mapping, single-nucleotide polymorphism (SNP) detection, and candidate gene identification.

Biotic factors such as insects, diseases, and weeds, along with abiotic stresses like drought, heat, salinity, waterlogging, and mineral toxicity, have significantly impacted rice crop growth, development, and, ultimately, economic yield across different growth stages. Given the severity of these challenges, researchers are working to develop rice cultivars that can withstand these stresses (Raza 2021). Hence, the development of stress-resilient rice varieties/hybrids necessitates the screening and identification of suitable donors and their subsequent transfer into desirable genetic backgrounds. This process relies on evaluating morphophysiological traits, yield attributes, and quality traits across numerous plants. Therefore, in plant breeding, accurate phenotyping is crucial. This chapter aims to explore high-throughput phenotyping techniques available to rice researchers, offering options to enhance precision while reducing costs and labor, thereby facilitating rapid genetic improvement.

10.2 Plant Phenotype and Phenotyping

A phenotype refers to the observable features or traits of an individual, which arise from the genotype, environment, and interaction component. "Pheno" means "observe" and comes from the same root as the word "phenomenon." Hence, it represents an observable characteristic of an individual, encompassing various attributes such as shape, color, size, growth rate, number of grains, seed coat patterns, plant height, biomass, and more. The measurement of plant phenotype can be based on morphological, physiological, biochemical, and molecular characteristics in either a qualitative or quantitative manner. Different parameters are measured to study these traits (Kumar et al. 2015). The phenotype develops in an individual through interaction with the environment, starting from the early stages of life to the end. The phenotype changes as the environment changes, influencing the genotype, thereby posing challenges for quantitative measurement (Houle et al. 2010).

The assessment of expressed traits influenced by genetic makeup and environmental changes is known as phenotyping. It started with human civilization. Plant species that gave the highest yield have been considered the best individuals and were used for plant domestication (Diamond and Renfrew 1997). Subsequently, the selection of superior individuals relied on the examination of their phenotypic expressions across diverse environmental contexts. These selected individuals were then incorporated into plant breeding aimed at developing novel and improved genotypes (Pearson et al. 2007).

10.3 Scope for Phenotyping Rice Response

Breeding programs commonly aim to enhance agricultural yield while preserving quality and building resilience to shifting climates. This can be accomplished by evaluating a large population of plants for various traits across the length of the crop cycle, as detailed by many breeders across the globe. However, complexity arises as trials need to be conducted in various environments, necessitating repeated experiments. The conventional methods of phenotyping are expensive, require a lot of labor, and can be destructive, possibly affecting the precision of the results. The integration of automated HTP systems with artificial intelligence (AI) has effectively addressed the challenges linked with crop phenotyping (Gill et al. 2022). High-throughput, automated, reliable, and versatile phenotypic technologies are becoming increasingly valued as critical tools for breeding programs looking to enhance genetic gain quickly.

Utilizing sophisticated tools ranging from ground-based imaging to aerial phenotyping and remote sensing, high-throughput phenotyping enables detailed capture and analysis of plant characteristics. In the quest for future breeding advancements, the establishment of efficient field-based HTP remains a notable bottleneck, particularly when optimizing yield potential and stress resilience. However, the landscape is witnessing notable progress in sensor technologies, aeronautics, and high-performance computing, providing a glimmer of hope for overcoming these challenges. The rising focus on cost-effective solutions for remote sensing, alongside advancements in data management, has the potential to enable widespread adoption of these technologies among breeders. This could enhance the efficiency and speed of the breeding process (Crossa et al. 2021).

10.3.1 Phenotyping for Yield Attributing Traits

Grain-related traits in rice play a crucial role in determining both yield potential and quality. Nevertheless, traditional methods for analyzing these traits typically entail manual counting or the use of specialized seed evaluation devices, resulting in significant labor and financial expenses. The conventional approach for evaluating rice yield-related traits also faces challenges such as threshing issues, complex measurement procedures, and low efficiency. Fortunately, advancements in computer applications and machine learning (ML) have enabled researchers to explore automated methods for obtaining rice grain-related features.

To tackle these obstacles, a high-throughput rice phenotyping facility (HRPF) was established, designed to capture factors associated with morphology, biomass, and yield across the entire rice growth cycle, even postharvest (Yang et al. 2014a). Numerous investigations have employed high-throughput phenotyping to assess a range of traits, viz., canopy ground cover, leaf area, days to 50% flowering, crop nitrogen status, tiller characteristics, plant color attributes, plant architecture, texture traits, height accumulation, plant canopy width, canopy height, crop biomass, and yield (Tanger et al. 2017a; Wu et al. 2019; Huang et al. 2022; Theerawitaya et al. 2022).

A recent development introduced a high-throughput phenotyping approach for acquiring rice grain characteristics directly from the panicle, eliminating the need for grain threshing or branch separation (Lu et al. 2023). Visible light scanning imaging technology easily captures individual rice panicle color images. Utilizing the cascaded Faster R-CNN model and an improved Pix2Pix model allows for the detection, segmentation, and retrieval of each rice grain directly from the panicle. Based on the processing findings, the designed image processing pipeline automatically calculates 15 rice grain traits. High-throughput phenotyping techniques for measuring productivity or yield traits have been mentioned in Table 10.1.

10.3.2 Biotic Stress

Pests and diseases have a notable impact on rice crop yields. When paddy crops face these biotic stresses, observable changes occur in the color, shape, texture, and other characteristics of leaves, stems, roots, and grains. The adoption of machine learning in high-throughput stress phenotyping offers benefits to various stakeholders, including researchers, extension specialists, and farmers. This integration enables accurate, precise, and efficient screening of different biotic stresses (Singh et al. 2016). This approach facilitates prompt gene discovery and the adoption of innovative selection techniques for complex traits like biotic and abiotic stressors and yield (Singh et al. 2016). The application of machine learning in high-throughput phenotyping further enhances our comprehension of pathogen–plant interactions and

Table 10.1 High-throughput phenotyping for productivity or yield traits

Phenotyping technique	Trait	Reference
X-ray computed tomography (CT)	Tillers	Yang et al. (2011)
HRPF	Biomass, morphology, and yield. Postharvest grain morphological parameters	Yang et al. (2014b)
Multispectral reflectance and ultrasonic sensors	Normalized difference vegetation index (NDVI), normalized difference red edge index (NDRE), Chla, canopy temperature depression (CTD), and canopy height	Tanger et al. (2017b)
Micro-CT-red–green–blue (RGB) phenotyping	Tiller growth traits, height accumulation traits, plant architecture traits, digital biomass, plant color traits, texture traits, and tiller traits	Wu et al. (2019)
An automatic and intelligent system based on threshed panicle compensation	Yield traits	Huang et al. (2022)
RGB phenotyping	Plant canopy width, canopy height, and leaf area	Theerawitaya et al. (2022)
Visible light scanner	Panicle and grain traits	Lu et al. (2023)

plant responses to various stresses (Kuska et al. 2015). Table 10.2 outlines high-throughput phenotyping techniques specifically designed for biotic stress.

10.3.3 Diseases

Rice crop is susceptible to many diseases like blast, brown spots, narrow brown leaf spots, sheath rot, sheath blight, and false smut. Diseases in rice can be detected by examining factors such as color, texture, morphology, photosynthetic rate, transpiration rate, and spectral vegetation indices. Different diseases may require another kind of phenotypic examination. A near-infrared hyperspectral imaging system was devised for early identification of rice blast disease in seedlings, achieving overall classification accuracy (infected and healthy leaves) of around 92% (Yang et al. 2012).

Using deep convolutional neural network (CNN) techniques, a novel method for identifying rice diseases was introduced (Lu et al. 2017). Using a dataset of 500 natural images from field trials, the system was trained to distinguish ten common rice diseases including bacterial leaf blight, bakanae disease, rice sheath rot, bacterial wilt, false smut, seedling blight, sheath blight, brown spot, blast, and bacterial sheath rot using CNNs (Lu et al. 2017).

10.3.4 Insect Pests

Insect pests like grasshoppers, gall midge, stem borers, and sucking pests cause severe reduction in yield. They also act as carriers of many pathogens causing plant diseases. Various methods can assess insect pests, including insect numbers, surface temperature, gas exchange, chlorophyll fluorescence spectrum, behavior, carotenoid concentration, spectral vegetation indices, and other physiological characteristics. For example, to address the issue of manual counting in the case of the rice bug (*Leptocorisa chinensis* Dallas), Fukatsu et al. (2012) devised a field monitoring system. Employing a high-resolution web camera and a remote management setup, this system integrates image processing and wireless sensor networks to establish a remote pheromone trap system for continuous monitoring of rice bugs in the field, enhancing efficiency and automation.

Table 10.2 High-throughput phenotyping for biotic stresses in rice

Trait	Phenotyping technique	Reference
Rice bug infestation in the field	Field monitoring support system	Fukatsu et al. (2012)
Blast, bakanae disease, bacterial leaf blight, bacterial wilt, and brown spot	RGB imaging	Lu et al. (2017)

10.3.5 Abiotic Stress

Abiotic stresses encompass challenging environmental conditions like heat, chilling, freezing, drought, salinity, submergence, nutrient deficiencies, and heavy metal contamination. Plant reactions to these stresses can be intricate, with traits exhibiting positive, negative, or neutral effects depending on the specific stress. Consequently, phenotyping rice for abiotic stress resistance often proves to be a complex endeavor. Yet, sophisticated crop stress phenotyping methods provide accurate assessments of how plants respond to diverse abiotic stresses throughout their growth cycles. This is essential for breeding plants that are resilient to abiotic stresses. In the case of rice, phenotyping for diverse abiotic stresses involves assessing water use efficiency, plant water loss rate, leaf rolling, stay-green properties, transpiration rate, plant temperature, photosynthesis efficiency, plant water content, and other relevant parameters.

Drought stress resulting from reduced water availability and cellular dehydration can be assessed using canopy temperature to indicate water status and canopy stomatal conductance. Kim et al. (2020) investigated drought stress tolerance traits in rice (*Oryza sativa*) using the PlantScreen™ robotic XYZ system. The study employed RGB imaging to estimate total growth area, canopy color, and density, alongside near-infrared imaging for assessing rice water content. Infrared technology was utilized to measure canopy temperature depression and fluorescence-based cameras to determine the photosynthetic efficiency. The DroughtSpotter apparatus was employed to assess water use efficiency, plant water loss rate, and transpiration. The objective of this research was to develop HTP techniques to hasten breeding cycles and facilitate functional genetic studies on drought stress tolerance. Table 10.3 outlines various high-throughput phenotyping techniques for measuring abiotic stress responses.

10.3.6 Heat Tolerance

Increasing temperatures and extreme heat events pose significant threats to rice production, with the timing of heat stress playing a critical role in affecting yields (Li et al. 2023). For instance, elevated night-time temperatures can increase dark

Table 10.3 High-throughput phenotyping for abiotic stress in rice

Trait	Phenotyping technique	Reference
Salinity tolerance traits	LemnaTec 3D scanalyzer system (fluorescence images and visible red–green–blue (RGB))	Robson et al. (2023)
Drought tolerance traits	PlantScreen™ robotic XYZ system	Kim et al. (2020)
Photosynthetic heat tolerance	High-throughput chlorophyll fluorescence-based screen	Xu et al. (2021)
NPK deficiencies	Dynamic indices based on leaf morphology and color	Song et al. (2021)

respiration rates, leading to the greater use of photo-assimilates that could otherwise support reproductive growth (Li et al. 2023). Daytime heat stress significantly affects productivity, especially during anthesis or pre-anthesis stages. In the study by Robson et al. (2023), an efficient screening method for assessing photosynthetic heat tolerance in plants was developed. This approach leverages chlorophyll fluorescence for daily evaluation of hundreds of plants. It analyzes how maximum PSII efficiency responds to increasing temperatures, providing data to establish models for the PSII–temperature relationship. These models yield coefficients that reflect traits related to photosynthetic heat tolerance. Notably, these traits demonstrate high heritability in a wide array of African and Asian rice cultivars, highlighting significant genetic diversity crucial for enhancing breeding efforts.

10.3.7 Salinity Tolerance

Soil salinity poses a significant challenge in many regions where rice is cultivated, adversely affecting both plant growth and yield. Developing rice varieties tolerant to salt stress requires effective screening methods to identify suitable genotypes for these areas. Various techniques have been employed to assess salinity tolerance in rice, focusing on morphophysiological traits such as shoot weight, shoot sodium ion (Na^+) concentration, shoot Na^+/K^+ ratio, plant survival rate, leaf damage, leaf area index, and root bypass flow (Robson et al. 2023).

Traditional methods for assessing plant biomass are often destructive, complicating the measurement of dynamic growth responses to salt application and the collection of seeds from evaluated plants. However, recent advancements in image-based phenomics offer a noninvasive approach to monitor plant responses to salinity stress over time, enabling the measurement of aboveground biomass without the need for whole-plant harvest (Hairmansis et al. 2014). The LemnaTec 3D analyzer, an HTP system for image acquisition and analysis utilizing fluorescence imaging and visible red–green–blue (RGB), was employed to evaluate the rice cultivars 'Fatmawati' and 'IR64' under varying salt stress conditions to study salinity tolerance traits (Robson et al. 2023).

10.3.8 Crop Nutrient Deficiencies

Crop nutrient deficiencies can reduce chlorophyll content, slow development, and tissue necrosis and increase disease vulnerability (Jansen et al. 2014). Modern, high-throughput phenotyping methods like RGB, multispectral, and hyperspectral sensors allow quick identification of these symptoms. Sun et al. (2018) employed dynamic analysis to pinpoint deficiencies in nitrogen, phosphorus, and potassium. They observed distinct leaf responses to these deficiencies in incomplete and fully expanded leaves. The effectiveness of dynamic characteristics for identification was assessed across various growth stages.

10.3.9 Role of Spectral Indices in Rice Plant Phenotyping Under Stress Conditions

Images captured by different phenomic tools need to be interpreted, and spectral indices (SI) play a crucial role in extracting and decoding the information within these images. Spectral indices, particularly vegetation indices (VIs), provide valuable insights into plant phenotype, architecture, stress levels, and biomass. These indices are varied, each targeting specific features within the images. Healthy plants reflect more infrared radiation (IR) compared to stressed plants, making near-infrared radiation (NIR) or IR bands useful for assessing abiotic and biotic stresses. Commonly adopted indices for phenotyping plants under stress include a variety of measures to assess different aspects of plant health and vigor. The triangular greenness index and green leaf index measure vegetation greenness, while the enhanced vegetation index (EVI) and normalized difference vegetation index (NDVI) assess overall plant health and vigor. The visible atmospherically resistant index (VARI) accounts for atmospheric conditions affecting visible light reflectance. The normalized green red difference index (NGRDI) and green normalized difference vegetation index (GNDVI) indicate chlorophyll content and overall plant health. The triangular vegetation index (TVI) evaluates canopy structure and biomass. Chlorophyll content and photosynthetic activity are quantified by the chlorophyll vegetation index (CVI) and chlorophyll index—green (CIgreen). The chlorophyll index—red edge (CIRE) is sensitive to chlorophyll concentration and plant stress, while the triangular chlorophyll index (TCI) specifically measures chlorophyll content. The normalized difference red edge index (NDRE) is used to monitor chlorophyll content and plant health, and the MERIS total chlorophyll index (MTCI) assesses total chlorophyll content in vegetation. These indices provide valuable insights into plant health, stress levels, and physiological status, aiding in the phenotyping of plants under various stress conditions.

10.3.10 Quality Improvement

Rice grain quality is generally evaluated based on appearance, milling, cooking, and nutritional characteristics. Appearance quality is determined by post-milling traits such as length of the whole grain, width of the grain, grain length-to-width ratio, and endosperm translucency. Cooking and eating quality are assessed by factors such as cooking ease, texture, springiness, stickiness, and chewiness, which are determined by the chemical and physical properties of starch. Protein content (PC) serves as a key indicator of the nutritional quality of rice grains (Li et al. 2023). An increased night respiration rate at the grain filling stage can negatively impact total yield and quality. This results in a reduction in the grain width, weight of 1000 grains, reduced sink strength, lower starch content in the grains, and heightened grain chalkiness (Sun et al. 2018). Table 10.4 outlines high-throughput phenotyping techniques for quality evaluation.

Table 10.4 HTP tools used to measure quality traits in rice

Trait	Phenotyping technique	Reference
Aflatoxin (*Aspergillus flavus*)	Near-infrared spectroscopy	Wang et al. (2022)
High night temperature-induced chalkiness	Gradient-weighted class activation mapping (grad-CAM)-based tool	Kawamura et al. (2003)

An automated system has been developed utilizing near-infrared technology and visible light separation to improve rice quality assessment. This system measures the amount of moisture and protein present in the seed and kernel ratio (Bahuguna et al. 2017). Additionally, a fully automated method for detecting grain chalkiness has been developed using CNNs and gradient-weighted class activation mapping (Grad-CAM). Grad-CAM identifies the chalky areas within a rice and illustrates these areas with a heatmap indicating the level of chalkiness. Subsequently, the CNN model differentiates between non-chalky and chalky grains based on this information.

10.4 Phenotyping Tools for Trait Measurement in Rice

10.4.1 Aboveground Traits

1. Soil plant analysis development (SPAD): SPAD is a handheld, portable, self-calibrating device designed to nondestructively measure chlorophyll content in plant leaves (Sirisomboon et al. 2013). This gives a value by assessing the difference in transmittance between red light (650 nm) and infrared light (940 nm) passing within the leaf (Jhanji and Sekhon 2011). This tool is used to evaluate chlorophyll content linked to the photosynthetic apparatus, assess plant nutrient status (particularly leaf nitrogen content), determine the impact of environmental stress, and ascertain fertilizer requirements.
2. Normalized difference vegetation index (NDVI): NDVI is a straightforward indicator utilized to analyze remote sensing measurements, often from satellites, to determine the presence of live green vegetation in the observed target. NDVI values are calculated based on the reflection of visible and near-infrared light by vegetation (Uddling et al. 2007). This index is used to evaluate nutrient status and fertilizer needs, investigate the impact of disease attacks, and screen germplasm or breeding populations (Table 10.5).
3. Infrared gas analyzer (IRGA): IRGA is an instrument that measures a variety of heteroatomic gas molecules and gaseous hydrocarbons. This device can assess the stomatal conductance, photosynthetic rate, transpiration rate, intercellular CO_2 concentration, and chlorophyll fluorescence.
4. Hyperspectral imaging: It is a nondestructive tool for measuring leaf chlorophyll and nitrogen content and helps assess photosynthetic capability and crop growth status. It identifies each pixel's primary colors (blue, green, and red) and examines a broad spectrum of light. More information about what is imaged is

Table 10.5 Phenotyping tools to measure aboveground-, belowground-, and postharvest-related traits in rice

Trait/parameter	Equipment/mode of measurement	Reference
Seedling vigor, canopy structure, biomass, root architecture, NDVI, and growth dynamics	Laser scanning, green seeker (NDVI), and 2D and 3D imaging using stereo camera	Uddling et al. (2007)
Photosynthetic parameter, quantum yield, and non-photochemical quenching	Fluorescence camera and SPAD meter	Sirisomboon et al. (2013), Jhanji and Sekhon (2011)
Surface temperature	Thermal imaging	Omasa et al. (2007)
Leaf and canopy water status and water content	Near-infrared, thermal, and hyperspectral camera	Chen et al. (2021), Omasa et al. (2007)
Water content, nutrient status, photosynthetic efficiency, senescence, and pollen viability	RGB and multispectral camera	Jahnke et al. (2016), Cao et al. (2021)
Water stress and canopy temperature	Long-wave, infrared, and thermal camera	Zheng et al. (2018)
PSII activity and chlorophyll fluorescence	PAM fluorescence and hyperspectral radiometer	Gerhards et al. (2019)
Stomatal conductance	Handled porometer and SPAD meter	Nedbal and Whitmarsh (2004)
Root system architecture and root biomass	Shovelomics, magnet resonance imaging, radio detective ranging, and positron emission	Omasa et al. (2007), Burridge et al. (2016), McGrail et al. (2020)
Postharvest-related parameters such as spikelet number, filled spikelets, thousand seed weight, and length and width of the grain	P-TRAP, PANorama, phenoSeeder, and R-ray CT	Pflugfelder et al. (2015), Duan et al. (2011), Al-Tam et al. (2013), Yuan et al. (2016)
Hormone concentration, metabolite estimation, enzyme activity, and carbon isotope discrimination	Mass spectrometry, calorimetry, and hyperspectral imaging	van Dusschoten et al. (2016)

obtained by dividing the light that strikes each pixel into a wide range of spectral bands (Chen et al. 2021).

5. Fluorescence imaging: It is a noninvasive technique used to analyze leaf photosynthesis in situ. This method can uncover various internal leaf components, such as spatial differences due to nutritional status, physiological conditions, developmental stages, pigment distribution, and optical properties. It is widely used for HTP screening and the early sensing of both abiotic and biotic stress factors in plants (Ruszczak et al. 2022).
6. 3D Lidar: This technology provides insights into photosynthesis, transpiration, pigments, stomatal behavior, and plant responses to environmental stress. 3D images with integrated data can be obtained by utilizing computer graphics techniques (Nedbal and Whitmarsh 2004). 3D Lidar is also useful for measuring

width, root volume, plant height, stem angle, stem angle trajectory, root surface area, projected canopy area, and leaf emergence time (Omasa et al. 2007).

7. Thermal imaging: Thermal imaging has high spatial resolution, fast and accurate readings, and the ability to track the transpiration of leaves and canopy. The surface temperature of plants and leaves is a good indicator of transpiration, stomatal conductance in leaves, and water use efficiency (WUE) (Omasa et al. 2007).

10.4.2 Belowground Traits

1. Shovelomics: It is a simple and inexpensive tool to study root systems. Estimating the growth, development, and architecture of the primary roots close to the base of the shoot is the fundamental concept behind analyzing the root crown of field crops. Root crowns can be assessed immediately in the field or cleaned and then preserved for a more thorough examination in the laboratory (Kolhar and Jagtap 2023). Shovelomics can determine several important root architecture factors, including root angle and number of crown roots (Burridge et al. 2016).
2. Magnet resonance imaging (MRI): MRI measures root structure; strong root-to-soil contrast allows MRI to identify a more significant fraction of roots (Omasa et al. 2007). The root length, root mass, root tip length, root diameter, number of tips, growth angles, and geographic dispersion are the root properties that can be nondestructively observed over time (Atkinson et al. 2019).
3. Positron emission: Positron emission tomography is a highly accurate and non-invasive digital imaging method that measures changes in plant metabolism and transport pathways (van Dusschoten et al. 2016). It is a noninvasive tool to measure the functional and structural properties of roots in the soil (McGrail et al. 2020).

10.4.3 Postharvest Traits

1. Seed evaluation accelerator: This tool is used for rapid measurement of yield-related traits, such as the total spikelet count, filled spikelet number, 1000-grain weight, grain width, and grain length (Pflugfelder et al. 2015).
2. P-TRAP: In a lot less time, P-TRAP generated trustworthy results. Compared to human operators, the software gathers much more data and is incredibly practical. It is a novel open-source program that uses numerical images to identify a panicle's structure and its seeds automatically. The program finds and counts the grains, measures the grain shape parameters, and processes a number of characteristics associated with panicle structure. P-TRAP provides effective and user-friendly outcomes (Duan et al. 2011).
3. PANorama: PANorama is an open-source phenotyping tool designed to measure multiple branching and architectural traits from images simultaneously. In order to facilitate precise panicle phenotyping, PANorama automatically removes

skeletons from photos, divides axes into internodes, and thresholds out structures like awns. Compared to current panicle imaging platforms, PANorama is more accurate and efficient, and because of its modular implementation, it can measure a variety of organs from different plant species and quantify rice panicle and grain traits without threshing (Al-Tam et al. 2013).

4. PhenoSeeder: PhenoSeeder is a tool designed to handle and analyze individual seeds with varying sizes. It includes an automated robot and an easily expandable sensor setup. The system identifies three primary biometric properties for each seed: mass (used to calculate seed density), three-dimensional volumetric data, and two-dimensional projection data. Each seed receives a unique identifier, allowing it to be phenotyped, sowed, sorted, or separately stored for further analysis, such as in seed-to-plant tracking pipelines (Crowell et al. 2014).

10.5 Phenomics Platforms

Over 33% of the global population consumes rice, which provides up to 80% of the caloric needs of most Asians (Yang et al. 2020). Rice, owing to its diploid nature and having one of the smallest genomes among economically vital crops, facilitated the sequencing of its genome during the early phases of 2002 (Kearney 2010). Several landraces, wild relatives, and cultivars of rice have undergone sequencing, offering fresh perspectives on the crop's genome evolution and expanding our understanding of potential new genes for rice breeding initiatives.

Researchers are concerned that findings derived from controlled phenotyping platforms may not accurately depict the performance of complex traits under actual field conditions, particularly in environments with high environmental variability (Sun et al. 2017; Atieno et al. 2017). Furthermore, compared to the benefits realized thus far, the high-throughput platform's extensive phenotyping could be more manageable and cost-effective. At the very least, the data produced by these tools leads to big data and data management issues, making it easier to reach a valid conclusion when making decisions with a thorough understanding of machine learning algorithms (Singh et al. 2016). Despite these obstacles, several phenomics platforms, instruments, and sensors have been created, and a fair number of people are using and improving them to reverse this genetic advancement's stagnation (Duan et al. 2018; Ashourloo et al. 2014). Phenomics has a great chance to catch up to genomics in the next decade or two. This will enable the gathering of copious amounts of data, the interpretation of hitherto unknown traits, and the drawing of reliable conclusions from them.

Multiple sensors on various ground-based and aerial platforms are utilized to measure different plant traits precisely, quickly, and accurately across various growth stages. According to Atieno et al. (Sun et al. 2017) and Duan et al. (Atieno et al. 2017), advancements in plant imaging sensors encompass remote sensing and sophisticated autonomous vehicles equipped with RGB, hyperspectral, 3D laser scanning, near- and far-infrared fluorescence, light detection and ranging, thermal, and spectro-radiometry imaging. Examples of advanced automated platforms include unmanned aerial vehicles (UAVs), mobile carts, and ground robots. These vehicles can collect real-time data from multiple plots several times a day throughout the entire season, providing extensive datasets for plant breeders. To derive valuable insights that can be combined with genomic approaches for optimal selection, the longitudinal data produced by these sensors necessitates expertise in statistics, data science, and machine learning (Kearney 2010).

10.5.1 Phenotyping Platforms for Controlled Environment

High-throughput plant phenotyping (HTPP) in greenhouses involve capturing plant responses to specific environments in an increased throughput, noninvasive, repeatable, rapid, and accurate. This process includes precise control of external environmental factors that drive the growth and development. Model crop plants are often used for these studies. HTPP enables the analysis of plant structure, physiology, and biochemical traits to uncover adaptive mechanisms related to environmental cues and clarify their genetic control. Depending on sensor and plant movement and the platform's mechanical design, indoor HTPP systems can be categorized as bench-top or conveyor types.

10.5.2 Phenotyping Platforms Under Field Conditions

The phenotype of a field crop is complex, shaped by a multitude of elements including weather patterns, biotic stress, abiotic stress, and soil attributes. These factors also differ significantly from those in a greenhouse or growth room. Although image-based phenotyping platforms are increasingly prevalent in controlled settings globally, the most of the breeding work takes place in the field, with minimal or no selection occurring in green/glass house environments.

10.5.2.1 Ground-Based Platforms

Close-range phenotyping using ground-based HTPP can provide data with higher resolution compared to aerial remote sensing. The collection and analysis of time-series phenotypic data, along with the examination of dynamic responses and time-dependent traits, are convenient methods. However, large-scale phenotyping tasks are not appropriate for the ground-based HTPP.

10.5.2.2 Aerial Platforms

Aerial HT3Ps include satellite and manned and unmanned aerial platforms (MAP and UAP, respectively), depending on the imaging distance used. To perform the flight task with a UAP, a ground operator usually needs to follow a preplanned, designated route. Onboard sensors automatically capture crop images. Nonetheless, MAP necessitates a dedicated individual to pilot the aircraft, with another person manually capturing crop photos during flight. Both UAP and MAP typically employ global navigation satellite systems. However, for accurate geo-referencing of the gathered data and to fine-tune its spatial resolution, these systems often require the utilization of ground control points and calibration boards.

10.6 Data Management and Analysis Tools

The widespread use of HTP platforms in agriculture has led to a surge in data volume, diversity, and reliability gathered at various intervals, resulting in significant big data challenges. The capacity to effectively analyze and comprehend this data is increasingly pivotal for advancing new tools and discoveries. Until recently, storing vast amounts of data posed significant challenges. However, improved storage capacities now enable efficient archiving of such data, potentially allowing for future use. The primary obstacle hindering the widespread adoption of HTP in most large-scale agricultural operations is the difficulty in analyzing and interpreting the information. Machine learning and deep learning (DL) techniques are valuable tools in computer science, offering significant potential for increased utilization in agriculture, particularly in monitoring plant stresses. It plays a crucial role in analyzing and understanding plants' complex biological characteristics and behaviors. Numerous studies showcased the utilization of ML and DL approaches in analyzing images to identify different forms and degrees of biotic and abiotic stress in rice (Dobbels and Lorenz 2019; Anami et al. 2020). Various machine learning techniques have also been employed to detect seedling growth stages and estimate yield in rice cultivation (Deng et al. 2021; Tan et al. 2022). Managing and analyzing high-throughput phenotyping data in rice involves specialized tools and platforms designed to efficiently handle large volumes of data. Here are some essential tools and approaches commonly used in this field:

1. Phenotyping databases: Databases specifically tailored for phenotypic data management are essential. These databases organize and store diverse datasets related to rice phenotyping. Examples include the following:
   (a) TERRA-REF: A platform that collects, processes, and shares high-resolution images and sensor data for plant phenotyping.
   (b) PhenomeOne: A database for storing and analyzing plant phenotypic and genotypic data.
   (c) The Oryza Tag Line (OTL) database: Rice functional genomics resource.
2. Image analysis software: Specialized software tools assist in processing and analyzing images captured in high-throughput phenotyping experiments. These tools include the following:
   (a) Trait Capture: Software that extracts specific traits or features from images, such as leaf area, color, shape, or disease symptoms.
   (b) PlantCV: A Python-based package for plant image analysis, providing tools for image processing, feature extraction, and trait measurement.
3. Statistical analysis tools: Statistical software and programming languages like R or Python with libraries such as Pandas, NumPy, and SciPy are commonly used for statistical analysis and modeling of phenotypic data. These tools help in:

(a) Descriptive statistics: Calculating means, standard deviations, and distributions of phenotypic traits.
(b) Correlation analysis: Identifying relationships between different traits.
(c) Regression analysis: Predicting plant performance based on phenotypic data and environmental factors.

4. Machine learning frameworks: Machine learning techniques are vital for analyzing complex phenotypic data. Frameworks such as TensorFlow, PyTorch, or scikit-learn in Python enable the implementation of various machine learning algorithms for phenotype classification, regression, or clustering tasks.
5. Ontologies and annotation tools: Ontologies provide standardized vocabularies for describing plant traits, facilitating data integration, and sharing across different research groups. Tools like Plant Ontology (PO) and annotation software help standardize and annotate phenotypic data.
6. Data integration platforms: Platforms that integrate different data types (genomic, phenotypic, environmental) are crucial for holistic analysis. Integration enables researchers to correlate genetic information with phenotypic traits, leading to insights into genotype–phenotype relationships.

Effective utilization of these tools and platforms aids researchers in managing, analyzing, and deriving meaningful insights from high-throughput phenotyping data in rice, ultimately contributing to advancements in cultivar development and agricultural practices.

10.7 Integration of HTP and Genomic Strategies in Rice for Climate Resilience

Connecting the phenotypes of plants with their genotypes is crucial for improving crops and advancing plant genomics. As we have successfully sequenced the entire genomes of various crops, the field of crop functional genomics now relies high-throughput and big data analysis techniques. However, a significant challenge remains: obtaining large-scale phenotypic data vital for crop improvement and genomics experiments (Yuan et al. 2016). Fortunately, advance technologies offer solutions to this challenge, paving the way for more advanced methods of collecting and processing large-scale phenotyping data. Mechanized phenotyping platforms in the lab and the field have significantly benefited many genetic and genomic studies (Fig. 10.1). The implementation of mechanized phenotyping platforms in both controlled and real-world settings has significantly amplified the capacity and efficacy of genetic and genomic investigations across this spectrum of research objectives:

1. Genetic dissection of complex traits.
2. Gene/QTL discovery.
3. Identification of gene function.
4. For genomic selection.
5. G × E studies.

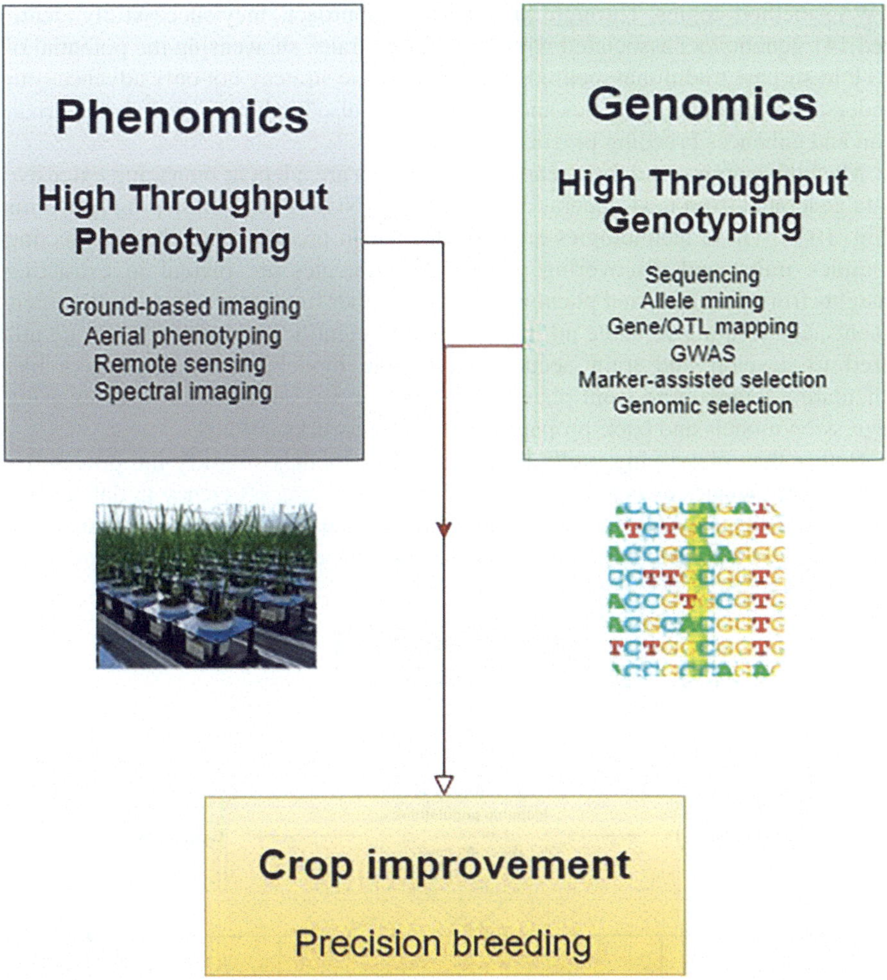

Fig. 10.1 Integration of genomics and phenomics in rice for enhancing climate resilience

6. Linkage mapping, QTL mapping, and QTL cloning.
7. Association mapping.
8. Crop genetic diversity and allele mining.
9. Gene tagging.
10. Marker-assisted breeding.

Recently, researchers have increasingly turned to genome-wide association studies (GWAS) in rice breeding to uncover the intricate relationships between genes and phenotypic traits. For GWAS to be effective, it is essential to gather comprehensive phenotypic and genomic marker data from large populations (Pham et al. 2022). Yang et al. (2014a) utilized a sophisticated approach combining the HRPF with

GWAS methodologies. Through this integrated approach, they successfully identified 141 genetic loci associated with 15 different traits, showcasing the potential of HTP to surpass traditional methods. This innovative strategy not only advances our understanding of plant genetics and genomics but also facilitates gene characterization and enhances breeding practices in rice.

Machine learning and deep learning algorithms are adept at managing extensive data generated from next-generation sequencing (NGS) and phenotyping platforms (Fig. 10.2). These technologies are instrumental in precision breeding, dissecting complex traits, and discovering genes. ML strategies are pivotal in extracting insights from genomics and phenomics data, whether from single or multiple experiments, contributing valuable information to the scientific community. They are utilized to develop and train accurate predictive models, facilitating tasks like calculating temperature from plant leaf images or assessing abiotic stresses using regression models and back-propagation neural networks (BPNN).

Rather than classifying multiple diseases, ML models simplify the process by categorizing images as healthy or non-healthy, enhancing efficiency in data analysis. Connecting extensive datasets from genomics and phenomics poses challenges, requiring statistical models to handle numerous correlated predictors. Wang et al.

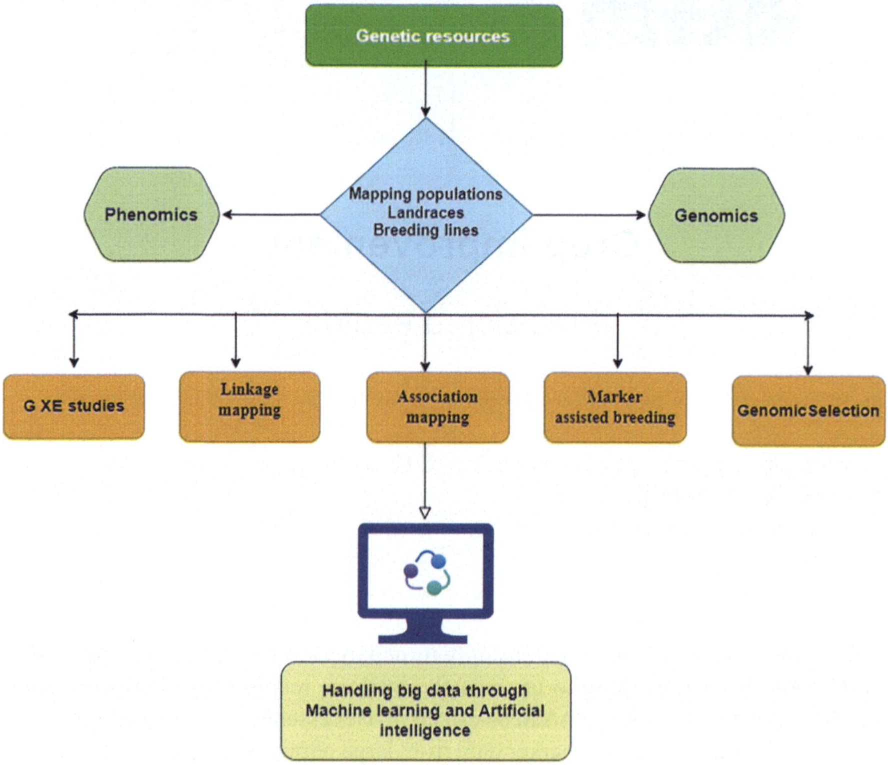

Fig. 10.2 Integration of machine learning with genomics and phenomics

(2022) showcased the efficacy of CNNs and Grad-CAM in accurately detecting chalkiness induced by high night temperatures in rice. Furthermore, Jiang et al. (2020) (Guo et al. 2018) utilized a novel algorithm that combines deep learning with support vector machine to identify four diseases in rice, whereas Lu et al. (2017) successfully distinguished ten prevalent diseases using deep CNNs in rice by utilizing images of diseased and healthy leaves and stems obtained from experimental fields.

10.8 Conclusion

High-throughput phenotyping has revolutionized rice breeding, offering scalability and precision vital for feeding a growing population amid climate uncertainty. By integrating remote sensing, drones, and advanced imaging, it swiftly decodes genetic–environmental interactions, expediting trait selection. These technologies enable real-time stress detection, aiding in the development of climate-smart varieties that are crucial for food security. High-throughput phenotyping synergizes with genomics, swiftly identifying genetic markers for superior traits. The convergence of these methodologies bridges the genotype–phenotype gap, expediting resilient variety development. As the global population surges and climate volatility intensifies, this innovation promises sustainable rice production. The integration of high-throughput phenotyping and traditional methods propels precision breeding, ensuring resilient food systems in a changing world.

References

Al-Tam F, Adam H, Anjos AD, Lorieux M, Larmande P, Ghesquière A, Shahbazkia HR (2013) P-TRAP: a panicle trait phenotyping tool. BMC Plant Biol 13:1–14

Anami BS, Malvade NN, Palaiah S (2020) Deep learning approach for recognition and classification of yield affecting paddy crop stresses using field images. Artif Intell Agric 4:12–20

Ashourloo D, Mobasheri MR, Huete A (2014) Evaluating the effect of different wheat rust disease symptoms on vegetation indices using hyperspectral measurements. Remote Sens 6(6):5107–5123

Atieno J, Li Y, Langridge P, Dowling K, Brien C, Berger B et al (2017) Exploring genetic variation for salinity tolerance in chickpea using image-based phenotyping. Sci Rep 7(1):1300

Atkinson JA, Pound MP, Bennett MJ, Wells DM (2019) Uncovering the hidden half of plants using new advances in root phenotyping. Curr Opin Biotechnol 55:1–8

Bahuguna RN, Solis CA, Shi W, Jagadish KS (2017) Post-flowering night respiration and altered sink activity account for high night temperature-induced grain yield and quality loss in rice (*Oryza sativa* L.). Physiol Plant 159(1):59–73

Basavaraj PS, Rane J (2020) Avenues to realize potential of phenomics to accelerate crop breeding for heat tolerance. Plant Physiol Rep 25(4):594–610

Basavaraj PS, Rane J, Prathibha MD, Boraiah KM, Kumar M (2023) Advances in high-throughput phenotyping of maize (*Zea mays* L.) for climate resilience. In: Maize improvement: current advances in yield, quality, and stress tolerance under changing climatic scenarios. Springer International Publishing, Cham, pp 259–288

Burridge J, Jochua CN, Bucksch A, Lynch JP (2016) Legume shovelomics: high—throughput phenotyping of common bean (*Phaseolus vulgaris* L.) and cowpea (*Vigna unguiculata* subsp, *unguiculata*) root architecture in the field. Field Crop Res 192:21–32
Cao X, Liu Y, Yu R, Han D, Su B (2021) A comparison of UAV RGB and multispectral imaging in phenotyping for stay green of wheat population. Remote Sens 13(24):5173
Chen JJ, Zhen S, Sun Y (2021) Estimating leaf chlorophyll content of buffaloberry using normalized difference vegetation index sensors. HortTechnology 31(3):297–303
Chimmili SR, Kanneboina S, Hanjagi PS, Basavaraj PS, Sakhare AS, Senguttuvel P, Kumar S, Kota S (2022) Integrating advanced molecular, genomic, and speed breeding methods for genetic improvement of stress tolerance in rice. In: Next-generation plant breeding approaches for stress resilience in cereal crops. Singapore, Springer Nature Singapore, pp 263–283
Crossa J, Fritsche-Neto R, Montesinos-Lopez OA, Costa-Neto G, Dreisigacker S, Montesinos-Lopez A, Bentley AR (2021) The modern plant breeding triangle: optimizing the use of genomics, phenomics, and enviromics data. Front Plant Sci 12:651480
Crowell S, Falcão AX, Shah A, Wilson Z, Greenberg AJ, McCouch SR (2014) High-resolution inflorescence phenotyping using a novel image-analysis pipeline PANorama. Plant Physiol 165(2):479–495
Deng R, Tao M, Xing H, Yang X, Liu C, Liao K, Qi L (2021) Automatic diagnosis of rice diseases using deep learning. Front Plant Sci 12:701038
Diamond J, Renfrew C (1997) Guns, germs, and steel: the fates of human societies. Nature 386(6623):339–339
Dobbels AA, Lorenz AJ (2019) Soybean iron deficiency chlorosis high-throughput phenotyping using an unmanned aircraft system. Plant Methods 15:1–9
Duan L, Yang W, Huang C, Liu Q (2011) A novel machine-vision-based facility for the automatic evaluation of yield-related traits in rice. Plant Methods 7:1–13
Duan L, Han J, Guo Z, Tu H, Yang P, Zhang D et al (2018) Novel digital features discriminate between drought resistant and drought sensitive rice under controlled and field conditions. Front Plant Sci 9:492
Dwivedi SL, Goldman I, Ceccarelli S, Ortiz R (2020) Advanced analytics, phenomics and biotechnology approaches to enhance genetic gains in plant breeding. Adv Agron 162:89–142
Evenson RE, Gollin D (2003) Assessing the impact of the green revolution, 1960 to 2000. Science 300(5620):758–762
Fukatsu T, Watanabe T, Hu H, Yoichi H, Hirafuji M (2012) Field monitoring support system for the occurrence of *Leptocorisa chinensis* Dallas (Hemiptera: Alydidae) using synthetic attractants, field servers, and image analysis. Comput Electron Agric 80:8–16
Gerhards M, Schlerf M, Mallick K, Udelhoven T (2019) Challenges and future perspectives of multi-/hyperspectral thermal infrared remote sensing for crop water-stress detection: a review. Remote Sens 11(10):1240
Gill T, Gill SK, Saini DK, Chopra Y, de Koff JP, Sandhu KS (2022) A comprehensive review of high throughput phenotyping and machine learning for plant stress phenotyping. Phenomics 2(3):156–183
Guo Z, Yang W, Chang Y, Ma X, Tu H, Xiong F, Xiong L (2018) Genome-wide association studies of image traits reveal genetic architecture of drought resistance in rice. Mol Plant 11(6):789–805
Hairmansis A, Berger B, Tester M, Roy SJ (2014) Image-based phenotyping for non-destructive screening of different salinity tolerance traits in rice. Rice 7:1–10
Houle D, Govindaraju DR, Omholt S (2010) Phenomics: the next challenge. Nat Rev Genet 11(12):855–866
Huang C, Li W, Zhang Z, Hua X, Yang J, Ye J, Yang W (2022) An intelligent rice yield trait evaluation system based on threshed panicle compensation. Front Plant Sci 13:900408
Jahnke S, Roussel J, Hombach T, Kochs J, Fischbach A, Huber G, Scharr H (2016) Pheno seeder-a robot system for automated handling and phenotyping of individual seeds. Plant Physiol 172(3):1358–1370
Jansen M, Pinto F, Nagel KA, van Dusschoten D, Fiorani F, Rascher U et al (2014) Non-invasive phenotyping methodologies enable the accurate characterization of growth and performance of

shoots and roots. In: Genomics of plant genetic resources: volume 1. Managing, sequencing and mining genetic resources, pp 173–206
Jhanji S, Sekhon NK (2011) Chlorophyll meter (SPAD-502), an effective and reliable tool for estimating chlorophyll content-how. Natl Acad Sci Lett 34(11–12):407–412
Jiang F, Lu Y, Chen Y, Cai D, Li G (2020) Image recognition of four rice leaf diseases based on deep learning and support vector machine. Comput Electron Agric 179:105824
Kawamura S, Natsuga M, Takekura K, Itoh K (2003) Development of an automatic rice-quality inspection system. Comput Electron Agric 40(1–3):115–126
Kearney J (2010) Food consumption trends and drivers. Philos Trans Royal Soc B: Biol Sci 365(1554):2793–2807
Kim SL, Kim N, Lee H, Lee E, Cheon KS, Kim M, Kim KH (2020) High-throughput phenotyping platform for analyzing drought tolerance in rice. Planta 252(3):38
Kolhar S, Jagtap J (2023) Plant trait estimation and classification studies in plant phenotyping using machine vision—a review. Inform Process Agric 10(1):114–135
Kumar J, Pratap A, Kumar S (2015) Plant phenomics: an overview. In: Phenomics in crop plants: trends, options and limitations, pp 1–10
Kuska M, Wahabzada M, Leucker M, Dehne HW, Kersting K, Oerke EC et al (2015) Hyperspectral phenotyping on the microscopic scale: towards automated characterization of plant-pathogen interactions. Plant Methods 11:1–15
Lamichhane S, Thapa S (2022) Advances from conventional to modern plant breeding methodologies. Plant Breed Biotechnol 10(1):1–14
Li L, Zheng X, Wang J, Zhang X, He X, Xiong L, Hu Z (2023) Joint analysis of phenotype-effect-generation identifies loci associated with grain quality traits in rice hybrids. Nat Commun 14(1):3930
Lorenz AJ, Chao S, Asoro FG, Heffner EL, Hayashi T, Iwata H, Smith KP, Sorrells ME, Jannink JL (2011) Genomic selection in plant breeding: knowledge and prospects. Adv Agron 110:77–123
Lu Y, Yi S, Zeng N, Liu Y, Zhang Y (2017) Identification of rice diseases using deep convolutional neural networks. Neurocomputing 267:378–384
Lu Y, Wang J, Fu L, Yu L, Liu Q (2023) High-throughput and separating-free phenotyping method for on-panicle rice grains based on deep learning. Front Plant Sci 14:1219584
McGrail RK, Van Sanford DA, McNear DH Jr (2020) Trait-based root phenotyping as a necessary tool for crop selection and improvement. Agronomy 10(9):1328
Nedbal L, Whitmarsh J (2004) Chlorophyll fluorescence imaging of leaves and fruits. In: Chlorophyll a fluorescence: a signature of photosynthesis. Springer Netherlands, Dordrecht, pp 389–407
Omasa K, Hosoi F, Konishi A (2007) 3D lidar imaging for detecting and understanding plant responses and canopy structure. J Exp Bot 58(4):881–898
Pearson CH, Ernst SM, Barbarick KA, Hatfield JL, Peterson GA, Buxton DR (2007) Agronomy journal turns one hundred. In: Moser LE (ed) The American Society of Agronomy: 100 years of history. ASA, Madison, WI, pp 59–68
Pflugfelder D, Metzner R, Chlubek A, Koller R, Bühler J, Postma JA et al (2015) Noninvasive 3D root imaging. In: Plant 2030 status seminar 2015 (No. FZJ-2015-02187). Pflanzenwissenschaften
Pham HT, Awange J, Kuhn M (2022) Evaluation of three feature dimension reduction techniques for machine learning-based crop yield prediction models. Sensors 22(17):6609
Ray DK, Ramankutty N, Mueller ND, West PC, Foley JA (2012) Recent patterns of crop yield growth and stagnation. Nat Commun 3(1):1293
Raza A (2021) Eco-physiological and biochemical responses of rapeseed (*Brassica napus* L.) to abiotic stresses: consequences and mitigation strategies. J Plant Growth Regul 40(4):1368–1388
Robson JK, Ferguson JN, McAusland L, Atkinson JA, Tranchant-Dubreuil C, Cubry P et al (2023) Chlorophyll fluorescence-based high-throughput phenotyping facilitates the genetic dissection of photosynthetic heat tolerance in African (*Oryza glaberrima*) and Asian (*Oryza sativa*) rice. J Exp Bot 74(17):5181–5197

Ruszczak B, Wijata AM, Nalepa J (2022) Unbiasing the estimation of chlorophyll from hyperspectral images: a benchmark dataset, validation procedure and baseline results. Remote Sens 14(21):5526

Singh A, Ganapathysubramanian B, Singh AK, Sarkar S (2016) Machine learning for high-throughput stress phenotyping in plants. Trends Plant Sci 21(2):110–124

Sirisomboon CD, Putthang REDDY, Sirisomboon PETTR (2013) Application of near infrared spectroscopy to detect aflatoxigenic fungal contamination in rice. Food Control 33(1):207–214

Song P, Wang J, Guo X, Yang W, Zhao C (2021) High-throughput phenotyping: breaking through the bottleneck in future crop breeding. Crop J 9(3):633–645

Sun C, Hu Z, Zheng T, Lu K, Zhao Y, Wang W et al (2017) RPAN: rice pan-genome browser for ~3000 rice genomes. Nucleic Acids Res 45(2):597–605

Sun Y, Zhu S, Yang X, Weston MV, Wang K, Shen Z, Chen L (2018) Nitrogen diagnosis based on dynamic characteristics of rice leaf image. PLoS One 13(4):e0196298

Tan S, Liu J, Lu H, Lan M, Yu J, Liao G, Ma X (2022) Machine learning approaches for rice seedling growth stages detection. Front Plant Sci 13:914771

Tanger P, Klassen S, Mojica JP, Lovell JT, Moyers BT, Baraoidan M, Naredo MEB, McNally KL, Poland J, Bush DR, Leung H (2017a) Field-based high throughput phenotyping rapidly identifies genomic regions controlling yield components in rice. Sci Rep 7(1):42839

Tanger P, Klassen S, Mojica JP, Lovell JT, Moyers BT, Baraoidan M, McKay JK (2017b) Field-based high throughput phenotyping rapidly identifies genomic regions controlling yield components in rice. Sci Rep 7(1):42839

Theerawitaya C, Chutteang C, Arunyanark A, Malumpong C, Kwangern N, Rachsapa N, Toojinda T (2022) Combining high-throughput phenotyping with overall growth measurements of indica rice (*Oryza sativa* L spp. *indica*) cultivars over the whole life cycle. Agric Nat Resour 56(4):713–724

Uddling J, Gelang-Alfredsson J, Piikki K, Pleijel H (2007) Evaluating the relationship between leaf chlorophyll concentration and SPAD-502 chlorophyll meter readings. Photosynth Res 91:37–46

van Dusschoten D, Metzner R, Kochs J, Postma JA, Pflugfelder D, Bühler J, Jahnke S (2016) Quantitative 3D analysis of plant roots growing in soil using magnetic resonance imaging. Plant Physiol 170(3):1176–1188

Wang C, Caragea D, Kodadinne Narayana N, Hein NT, Bheemanahalli R, Somayanda IM, Jagadish SK (2022) Deep learning based high-throughput phenotyping of chalkiness in rice exposed to high night temperature. Plant Methods 18(1):9

Wu D, Guo Z, Ye J, Feng H, Liu J, Chen G, Yang W (2019) Combining high-throughput micro-CT-RGB phenotyping and genome-wide association study to dissect the genetic architecture of tiller growth in rice. J Exp Bot 70(2):545–561

Xu J, Misra G, Sreenivasulu N, Henry A (2021) What happens at night? Physiological mechanisms related to maintaining grain yield under high night temperature in rice. Plant Cell Environ 44(7):2245–2261

Yang W, Xu X, Duan L, Luo Q, Chen S, Zeng S, Liu Q (2011) High-throughput measurement of rice tillers using a conveyor equipped with x-ray computed tomography. Rev Sci Instrum 82(2):025102

Yang Y, Chai R, He Y (2012) Early detection of rice blast (Pyricularia) at seedling stage in Nipponbare rice variety using near-infrared hyper-spectral image. Afr J Biotechnol 11(26):6809–6817

Yang W, Guo Z, Huang C, Duan L, Chen G, Jiang N, Fang W, Feng H, Xie W, Lian X, Wang G (2014a) Combining high-throughput phenotyping and genome-wide association studies to reveal natural genetic variation in rice. Nat Commun 5(1):5087

Yang W, Guo Z, Huang C, Duan L, Chen G, Jiang N, Xiong L (2014b) Combining high-throughput phenotyping and genome-wide association studies to reveal natural genetic variation in rice. Nat Commun 5(1):5087

Yang W, Feng H, Zhang X, Zhang J, Doonan JH, Batchelor WD et al (2020) Crop phenomics and high-throughput phenotyping: past decades, current challenges, and future perspectives. Mol Plant 13(2):187–214

Yuan Z, Cao Q, Zhang K, Ata-Ul-Karim ST, Tian Y, Zhu Y et al (2016) Optimal leaf positions for SPAD meter measurement in rice. Front Plant Sci 7:719

Zheng H, Cheng T, Li D, Zhou X, Yao X, Tian Y et al (2018) Evaluation of RGB, color-infrared and multispectral images acquired from unmanned aerial systems for the estimation of nitrogen accumulation in rice. Remote Sens 10(6):824

Artificial Intelligence and Machine Learning for Rice Improvement

11

Nitesh Kumar Sharma, Abhishek Anand,
Neeraj Budhlakoti, Dwijesh Chandra Mishra,
and Girish Kumar Jha

Abstract

Food security poses a major challenge to developing and underdeveloped nations worldwide because of rapid growth of the world population, necessitating increased food production. This escalating demand exerts pressure on agricultural lands as farmers and breeders strive to boost food grain output. Ninety percent of the world's rice production and consumption takes place in Asia, where rice is a vital staple for 1.3 billion people. In developing countries, many farmers still rely on traditional farming techniques that prove inadequate in meeting the escalating demand for food grains. These outdated methods often involve excessive use of harmful pesticides and chemical fertilizers, leading to deteriorating effects on soil microorganisms and flora, resulting in soil degradation and reduced fertility. The forthcoming agriculture hinges on embracing automation to ensure food security for growing populations. This chapter delves into the assessment and exploration of various solutions, methodologies, and perspectives related to artificial intelligence, machine learning, big data applications, and high-end computing. These innovations present viable paths for data-intensive, transdisciplinary agro-technology research. The urgent requirement for diverse automation approaches is evident in addressing challenges related to rice and other crop cultivation, farm management, agricultural practices, irrigation, fertilization, weeding, harvesting, marketing, food distribution, pesticide management, drainage control, emissions reduction, and environmental impact. Implementing automated farming practices not only enhances soil health and fertility but also streamlines agricultural operations. By leveraging advanced technologies and innovative solutions, the agricultural sector is poised to enter a

N. K. Sharma (✉) · A. Anand · N. Budhlakoti · D. C. Mishra · G. K. Jha
Division of Agricultural Bioinformatics, ICAR-Indian Agricultural Statistics Research Institute, New Delhi, India

A. Singh et al. (eds.), *Climate-Smart Rice Breeding*,
https://doi.org/10.1007/978-981-97-7098-4_11

new era of data-driven exploration, paving the way for sustainable and efficient food production systems.

Keywords

Agriculture · Artificial intelligence · Food · Machine learning · Rice

Abbreviations

AI	Artificial intelligence
CNN	Convolutional neural network
ICT	Information and communication technologies
IoT	Internet of things
kNN	k-Nearest neighbors
ML	Machine learning
RL	Reinforcement learning
SVM	Support vector machine

11.1 Introduction

In the pursuit of knowledge, humanity has made significant strides in scientific and technological advancements over the years. This progress has been driven by widespread industrialization, mechanization, and urbanization across the globe, leading to a surge in food insecurity due to the escalating competition for resources like labor, energy, capital, land, and water (Godfray et al. 2010). Consequently, agricultural lands are under mounting pressure to produce higher yields per unit area, exacerbating the influence of climate change and the greenhouse effect, which, in turn, disrupt crop yields and supply chains (Kesavan and Swaminathan 2006; Suzuki et al. 2014). To navigate these challenges, effective agricultural policies must be devised to mitigate natural calamities like floods, droughts, salinity, and cold spells (Timsina and Connor 2001).

In the contemporary landscape, the spectre of hunger and malnutrition persists into the twenty-first century, driven by flawed agricultural practices, inadequate food distribution, and a lack of farming expertise (Petersen 2003; Fogel 2004). Many developing nations grapple with subpar farm management practices, resulting in food scarcity, malnutrition, and mismanagement, thus indicating that traditional farming methods may be insufficient to meet future demands (Long et al. 2015; Ahmar et al. 2020). Furthermore, poor farming techniques contribute to environmental degradation through the excessive use of harmful pesticides and chemical fertilizers, jeopardizing soil fertility and agricultural sustainability (Brown and Jameton 2000; Carvalho 2006; Chowdhury et al. 2008). Agricultural industries also face multifaceted challenges such as pest management, escalating costs of inputs like pesticides and fertilizers, unresolved biotic and abiotic stresses, and inadequate water conservation measures (Spielman and Von Grebmer 2004; Dwivedy 2011).

To address these pressing issues, modern agricultural practices incorporating automation and artificial intelligence (AI) have been introduced. These innovative strategies improve millions of workers' economic prospects while simultaneously increasing food output. AI-driven technologies safeguard crop yields against diverse threats like climate change and population growth, revolutionizing farming through automated irrigation, weed control, and spraying using sensors, robots, and drones (Talaviya et al. 2020). Leveraging machine learning (ML) algorithms, these technologies enable precision farming by processing vast amounts of data collected from various sources, thus optimizing decision-making processes, and ensuring judicious of agricultural inputs at the same time (Donde et al. 2021).

In essence, the convergence of artificial intelligence with agriculture marks the dawn of a new era, akin to the Green Revolution. This pattern shift toward digital agriculture emphasizes information-intensive strategies to bolster crop processing without compromising environmental integrity. High-throughput sensors play a pivotal role in understanding dynamic climate and weather patterns, empowering farmers to make cognizant decisions to enhance productivity while diminishing environmental impact (Araus and Cairns 2014; Holman et al. 2016; Davis et al. 2016; Van Evert et al. 2017; Liakos et al. 2018).

11.2 Artificial Intelligence (AI)

The goal of artificial intelligence (AI) is to build computer systems that are capable of activities that typically require human intelligence, such as learning from past experiences, comprehending spoken language, discerning patterns, and making judgments. Many subfields are included in this topic, including robotics, expert systems, image recognition, natural language processing, and machine learning, all of which are aimed at enabling machines to replicate human intelligence and decision-making capabilities. Its integration has been transformative, expanding our ability to use computation to change the surroundings (Ahir et al. 2020). AI algorithms facilitate self-adaptation, self-learning, and self-improvement in decision-making efficiency. For instance, in 2019, Plessen introduced a system for optimizing harvest preparation by integrating crop allocation and vehicle routing, thereby broadening its applicability beyond manufacturing industries (Plessen 2021). Through depiction from diverse disciplines such as psychology, computer science, linguistics, and biology, AI finds remarkable application, notably in agriculture, where pairing farm management systems with AI-driven automation yields substantial benefits. The synergy between AI-based automation and farm management promises enhanced efficiency and the creation of adaptive systems, empowering decision-making processes with reliable data insights (Ahir et al. 2020).

These days, big data technologies combined with AI and deep learning (DL) open up a world of possibilities for data-intensive analyses and agricultural applications in transdisciplinary agri-technology (Liakos et al. 2018). Concurrently, the swift adoption of emerging machine engineering innovations in manufacturing marks a significant stride toward sustainable and intelligent production facilities, promising advancements for human welfare (Cioffi et al. 2020). Notably, intensive research endeavors in AI have resulted various techniques and algorithms that enhance our understanding of nature, including the integration of ML approaches in sustainable manufacturing practices (Liakos et al. 2018). This drive toward AI and ML integration in high-tech industries heralds the onset of a smart factory revolution.

In the agricultural domain, rice stands out as a vital staple crop, supporting the sustenance of over 1.3 billion people worldwide. Consequently, AI and ML applications in rice cultivation hold immense potential and significance. This chapter explores the implementation of AI and ML techniques in optimizing rice and other crop yields, spanning areas such as crop management, disease detection, pest control, water and soil management, and fertilizer optimization (Liakos et al. 2018). Through the deployment of AI and ML in farm management, there is a transition toward real-time data processing, facilitated by high-throughput sensors, enabling swift decision-making and offering invaluable insights for farmers (Liakos et al. 2018). Thus, AI and ML technologies are not just theoretical constructs but practical tools driving agricultural innovation and bolstering decision-making processes in real time, thereby enhancing farm productivity and sustainability.

11.3 Machine Learning (ML)

Machine learning (ML) is part of artificial intelligence (AI) and involves a range of methods used to create computer programs that use statistical models and algorithms to analyze data and generate predictions or decisions (Fig. 11.1). ML is typically characterized by its methodological diversity, involving various algorithms that undergo a learning process from a designated dataset, often referred as training dataset, to perform specific tasks. These datasets are composed of a series of

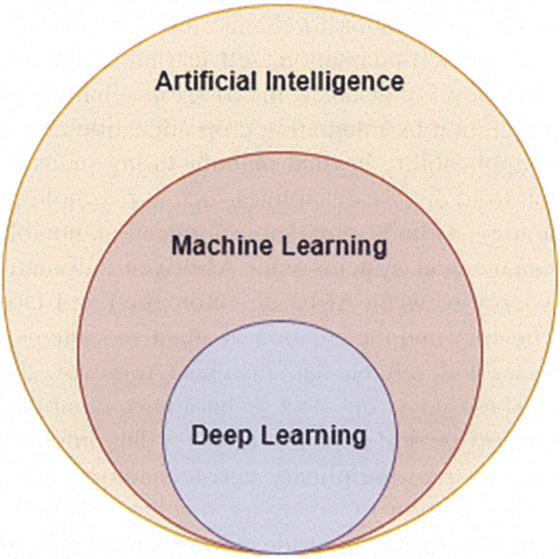

Fig. 11.1 Hierarchy of AI, ML, and DL

instances, each of which uses a set of features and variables to represent a particular item or attribute. ML functions can be represented in various formats, such as binary (0 or 1), numeric, ordinal (such as A or B), or nominal (enumeration), and are evaluated based on their ability to perform tasks with minimal resource consumption. To evaluate the efficiency of ML models and algorithms, several computational and analytical techniques are used. After the learning phase, a trained model can be applied to categorize, predict, or group new instances, known as a testing dataset, based on the knowledge gained during training. The four primary types of machine learning (ML) techniques and algorithms are supervised learning, unsupervised learning, semi-supervised learning, and reinforcement learning (Fig. 11.2).

Supervised learning entails training a model with labeled input data along with corresponding output. In this approach, the ML algorithm is trained to identify specific features, allowing the model to learn and recognize underlying patterns and relationships. Through iterative training, the model becomes proficient at generating accurate outcomes when presented with new, unlabeled data. In unsupervised learning, the learning algorithm operates without predefined labels, akin to humans learning independently. Here, the objective is to discern significant patterns within the data. The machine must deduce these patterns from an unlabeled dataset itself, lacking guidance on how to categorize each data point and instead inferring its own criteria for classification. Reinforcement learning stands apart from other machine learning approaches. Likewise, semi-supervised learning involves training the algorithm using a mixture of both labeled and unlabeled data. It falls somewhere in between supervised learning, which depends only on labeled training data, and unsupervised learning, which completely lacks labeled training data. Reinforcement learning (RL) supports automation by learning from the environment in which it is present. In RL, a computer program interacts with an ever-changing environment, aiming to achieve specific objectives, like driving a vehicle or competing in a game. Throughout the training process, the agent earns rewards when completing tasks, as defined by a reward function. Through reinforcement learning, the agent can rapidly acquire strategies to surpass human performance levels.

The advancement of AI and ML, bolstered by big data and high-performance computing technologies, has catalyzed breakthroughs across numerous scientific domains, including agriculture, aquaculture, bioinformatics, climatology,

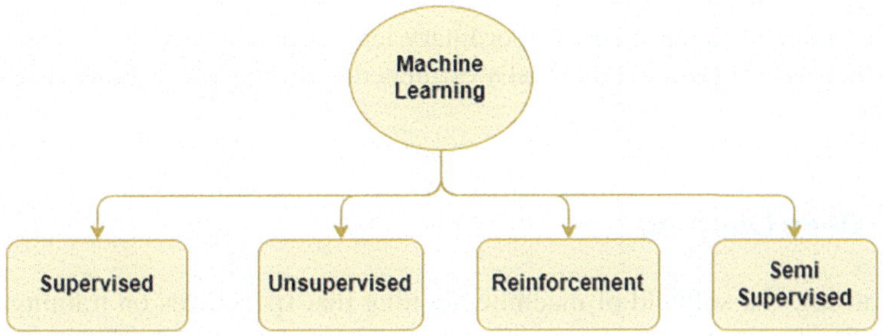

Fig. 11.2 Classification of ML

economics, food safety, medicine, meteorology, and robotics (Liakos et al. 2018). This chapter comprehensively investigates the integration of AI and ML in agricultural practices, highlighting several ML models and their distinguishing characteristics. Adopting AI and ML in agriculture has the potential to bring about revolutionary changes, especially in terms of guaranteeing food security for both developed and developing countries (Table 11.1).

11.4 AI and ML Algorithms

AI is a branch of computer science that analyzes and interprets data using various techniques such as machine and deep learning in order to replicate human intelligence. These algorithms establish dynamic connections between input and output variables, enabling them to provide predictions. These predictions offer solutions to a wide range of simple and complex scenarios (Jung et al. 2021). AI technologies are now pervasive in our daily lives, from facial recognition apps on mobile devices to autonomous vehicles. While numerous industries have witnessed significant productivity enhancements through AI systems and ML capabilities, the agricultural sector is also undergoing a digital transformation facilitated by AI. AI introduces numerous applications in agriculture, like pest and disease forewarning, disease identification, automatic weeding and harvesting, and optimizing agricultural inputs through precision farming. By leveraging data from ML, AI assists farmers in tasks such as irrigation, choosing crops, planting, harvesting, rotating them, controlling pests, and planting them—all contributing to the future of one of the oldest enterprises (Shadrin et al. 2019).

11.4.1 Regression Algorithms

Regression stands as a fundamental and extensively employed method in statistics and machine learning. By revealing the relationship between characteristics in the data and an observable continuous-valued response, it makes predictions easier. Regression is a versatile tool utilized in various domains, such as predicting stock prices and explaining gene regulatory networks. Some of the primary regression algorithms are linear regression (Hope 2020), logistic regression, stepwise regression (Jennrich and Sampson 1968), ordinary least squares regression, multivariate adaptive regression splines, and locally estimated scatterplot smoothing (Zhang and Goh 2016).

11.4.2 Deep Learning

Deep learning is a subfield of machine learning that specializes on training multilayer artificial neural networks to generate complex data models. These networks can learn features independently from unprocessed input, making them particularly

Table 11.1 List of major machine learning algorithms

Class of algorithms	Algorithms	References
Regression algorithms	1. Linear regression 2. Logistic regression 3. Stepwise regression 4. Ordinary least squares regression (OLSR) 5. Multivariate adaptive regression splines (MARS) 6. Locally estimated scatterplot smoothing (LOESS)	Hope (2020); Jennrich and Sampson (1968); Zhang and Goh (2016)
Deep learning algorithms	1. Convolutional neural networks (CNN) 2. Recurrent neural networks (RNNs) 3. Long short-term memory networks (LSTMs) 4. Stacked auto-encoders, deep Boltzmann machine (DBM) 5. Deep belief networks (DBN)	Wu (2017); Schmidt (2019); Woodbridge et al. (2016); Salakhutdinov and Hinton (2009); Mohamed et al. (2012)
Bayesian algorithms	1. Naive Bayes 2. Gaussian naive Bayes 3. Multinomial naive Bayes 4. Averaged one-dependence estimators (AODE) 5. Bayesian belief network (BBN) 6. Bayesian network (BN)	Jiang and Zhang (2006); Grover (2012); Gandhi et al. (2016); Bensi et al. (2013)
Neural network	1. Artificial neural networks (ANN) 2. Multilayer perceptions (MLP) 3. Back-propagation 4. Stochastic gradient descent 5. Hopfield network 6. Radial basis function network (RBFN)	Gershenson (2003); Taud and Mas (2018); Ramsauer et al. (2020); Ghosh and Nag (2001)

(continued)

Table 11.1 (continued)

Class of algorithms	Algorithms	References
Instance-based algorithms	1. k-Nearest neighbor (kNN) 2. Learning vector quantization (LVQ) 3. Self-organizing map (SOM) 4. Locally weighted learning (LWL) 5. Support vector machines (SVM)	Peterson (2009); Nova and Estévez (2014); Vesanto and Alhoniemi (2000); Englert (2012); Jakkula (2006)
Regularization	1. Ridge regression 2. Least absolute shrinkage and selection operator (LASSO) 3. Elastic net 4. Least-angle regression (LARS)	Kukreja et al. (2006); Zhang and Zamar (2014)
Decision tree	1. Chi-squared automatic interaction detection (CHAID) 2. Classification and regression tree (CART) 3. Iterative Dichotomiser 3 (ID3) 4. C4.5 and C5.0 5. Conditional decision trees 6. Decision stump 7. M5	Lewis (2000); Khedr et al. (2016); Quinlan (2014); Pandya and Pandya (2015); Binkert et al. (2006)
Clustering	1. k-Means 2. k-Medians 3. Expectation maximization (EM) 4. Hierarchical clustering 5. Minimum spanning tree 6. BIRCH 7. Fuzzy C-means 8. Fuzzy K-modes 9. Mini batch K-means 10. DBSCAN 11. Fuzzy clustering 12. Mean-shift 13. OPTICS algorithm	Ahuja et al. (2020)

(continued)

Table 11.1 (continued)

Class of algorithms	Algorithms	References
Dimensionality reduction	1. Principal component analysis (PCA) 2. Principal component regression (PCR) 3. Partial least squares regression (PLSR) 4. Sammon mapping 5. Multidimensional scaling (MDS) 6. Projection pursuit 7. Linear discriminant analysis (LDA) 8. Independent component analysis (ICA) 9. Nonnegative matrix factorization (NMF) 10. Regularized discriminant analysis (RDA) 11. Mixture discriminant analysis (MDA) 12. Partial least squares dimension analysis (PLSDA) 13. Quadratic discriminant analysis (QDA) 14. Canonical correlation analysis (CCA) 15. Flexible discriminant analysis (FDA) 16. Diffusion map	Kherif and Latypova (2020); Fleming and Garen (2022); Wang et al. (2022a); Buja et al. (2008); Tharwat et al. (2017); Bingham et al. (2015); Lopes et al. (2015); Ye and Wang (2006); Gkalelis et al. (2012); Andrew et al. (2013); Reddy et al. (2020)
Ensemble algorithms	1. Boosting 2. AdaBoost 3. Bootstrapped aggregation (bagging) 4. Weighted average (blending) 5. Stacked generalization (stacking) 6. Gradient boosting machines (GBM) 7. Gradient-boosted regression trees (GBRT) 8. Random forest	Dietterich (2002)
Association rule learning	1. Apriori algorithm 2. Eclat algorithm 3. FP-growth algorithm	Hegland (2007); Shawkat et al. (2022)

(continued)

Table 11.1 (continued)

Class of algorithms	Algorithms	References
Rule-based system	1. Ripper algorithm 2. Repeated incremental pruning to produce 3. Cubist 4. OneR 5. ZeroR	Liu et al. (2015)
Reinforcement learning	1. Q-learning 2. Deep Q-networks (DQN) 3. Policy gradient methods 4. Actor-critic methods 5. Deep deterministic policy gradient (DDPG) 6. Proximal policy optimization (PPO) 7. Trust region policy optimization (TRPO) 8. Asynchronous advantage actor-critic (A3C) 9. Dueling DQN 10. Rainbow	Watkins and Dayan (1992); Tiong et al. (2020); Schulman et al. (2015); Muhati and Rawat (2021); François-Lavet et al. (2018)

(continued)

Table 11.1 (continued)

Class of algorithms	Algorithms	References
Other ML algorithms	1. ALOPEX 2. CN2 method 3. Dynamic time warping 4. FastICA 5. Forward–backward technique 6. GeneRec 7. Leabra 8. Linde–Buzo–Gray method 9. Local outlier factor 10. Logic learning machine 11. LogitBoost 12. Markov chain Monte Carlo 13. Prefrontal cortex basal ganglia working memory 14. Weighted majority approach (WMA) 15. Rprop 16. Sparse PCA 17. Structured kNN 18. T-distributed stochastic neighbor embedding 19. Wake–sleep technique	Unnikrishnan and Venugopal (1994); Liu et al. (2020); Alhejazi and Mohammad (2022); Das and Behera (2017)

effective at tasks such as image identification, natural language processing, and audio recognition. Deep learning has significantly impacted numerous sectors, including computer vision, healthcare, finance, and self-driving cars, due to its capacity to manage vast quantities of data and identify intricate patterns. Deep learning models are based on neural network architectures such as transformers, recurrent neural networks, convolutional neural networks, deep neural networks, and deep belief networks. They can be supervised, semi-supervised, or unsupervised. These models can achieve high levels of accuracy, often outperforming humans. However, because of their intricacy, deep learning models are frequently viewed as "black boxes" that are difficult to understand. To address this, a new field called explainable AI provides methods for trying to translate the behavior of deep learning models into understandable words for humans. When exploring deep learning algorithms, it is essential to familiarize oneself with the following prominent ones: CNN (Wu 2017); recurrent neural networks (Schmidt 2019); long short-term memory networks (Woodbridge et al. 2016); stacked auto-encoders, deep Boltzmann machine (Salakhutdinov and Hinton 2009); and deep belief networks (Mohamed et al. 2012).

11.4.3 Bayesian Algorithms

A group of algorithms operating on the shared principle that each pair of classified features is independent of each other serves as the basis for Bayesian machine learning. These algorithms are rooted in Bayes' theorem, a method for calculating the likelihood of an event happening in light of the occurrence of another event. Delving into the essential types of Bayesian algorithms, there are some major algorithms like naive Bayes, Gaussian naive Bayes, multinomial naive Bayes, averaged one-dependence estimators (Jiang and Zhang 2006), Bayesian belief network (Grover 2012), and Bayesian network (Gandhi et al. 2016; Bensi et al. 2013).

11.4.4 Neural Network

Artificial neural networks, or ANNs for short, are networks of artificial neurons that integrate numerous inputs to produce a single output, emulating the structure of the human brain. Input and output nodes, coupled by one or more tiers of hidden nodes, make up an artificial neural network (ANN). Through activation functions, which activate or deactivate the hidden nodes based on the input data, information is transferred from input nodes to hidden nodes. To process the input, the hidden layers use weighting functions. When a node or a group of nodes in the hidden layer passes a predetermined threshold, a signal is transmitted to one or more nodes in the output layer. Various types of ANN algorithms include multiple layer perceptron (Gershenson 2003; Taud and Mas 2018), back-propagation, stochastic gradient descent, Hopfield network (Ramsauer et al. 2020), and radial basis function network (Ghosh and Nag 2001).

11.4.5 Instance-Based Algorithms

In order for this supervised machine learning algorithm to work, instances from training data are compared to instances that are currently kept in memory. It is termed instance-based as it relies on instances generated during the training process. Below are some of the widely recognized instance-based algorithms: kNN (Peterson 2009), learning vector quantization (Nova and Estévez 2014), self-organizing map (Vesanto and Alhoniemi 2000), locally weighted learning (Englert 2012), and SVM (Jakkula 2006).

11.4.6 Regularization

Regularization is a technique that entails making minor modifications to the learning algorithm to improve the model's capacity to generalize effectively, thereby enhancing its performance when confronted with new data. To combat overfitting in data, these methods or algorithms are combined with classification or regression algorithms. By fine-tuning these algorithms, it becomes possible to walk out a balance between training the model effectively and ensuring its predictive accuracy. Here are the most commonly utilized regularization algorithms: ridge regression, least absolute shrinkage and selection operator (Kukreja et al. 2006), elastic net, and least-angle regression (Zhang and Zamar 2014).

11.4.7 Decision Tree

The decision tree algorithm is a supervised learning technique that is applied to both regression and classification applications. By learning basic decision rules from past data, or training data, a decision tree model is to be constructed that can predict the class or value of the target variable. In a decision tree, one begins at the root of the tree and compares the values of the root attribute with the attribute of the record in order to forecast the class label for a record. In light of this assessment, one advances to the subsequent node by following the branch that corresponds to that value. This process continues until a leaf node is reached, which contains the predicted class label. Decision trees are a popular choice for their simplicity and interpretability, as they can be visualized and easily understood. However, they can be prone to overfitting and require techniques such as pruning or ensemble methods like random forests to mitigate this issue. The following are some widely recognized decision tree algorithms: chi-squared automatic interaction detection, classification and regression tree (Lewis 2000), Iterative Dichotomiser 3 (Khedr et al. 2016), C4.5 (Quinlan 2014) and C5.0 (Pandya and Pandya 2015), conditional decision trees, decision stump, and M5 (Binkert et al. 2006).

11.4.8 Clustering

Clustering refers to the process of organizing a collection of similar items into groups. Clustering is a technique in machine learning that involves grouping data points. With a given dataset, clustering algorithms classify each data point into distinct groups. Some of the clustering algorithms commonly used in ML include k-Means, k-Medians, expectation maximization, hierarchical clustering, minimum spanning tree, BIRCH, fuzzy C-means, fuzzy K-modes, mini batch K-means, DBSCAN, fuzzy clustering, mean-shift, and OPTICS algorithm (Ahuja et al. 2020).

11.4.9 Dimensionality Reduction

Dimensionality reduction serves as an unsupervised learning technique that aids in reducing the dimensions of data. This ML technique is instrumental in mitigating overfitting in trained models and alleviating high variance in training set, thereby enabling improved predictions on test set. Below are some of the most prominent and extensively used dimensionality reduction algorithms: principal component analysis (Kherif and Latypova 2020), principal component regression (Fleming and Garen 2022), partial least squares regression (Wang et al. 2022a), Sammon mapping, multidimensional scaling (Buja et al. 2008), projection pursuit, linear discriminant analysis (Tharwat et al. 2017), independent component analysis (Bingham et al. 2015), nonnegative matrix factorization (Lopes et al. 2015), regularized discriminant analysis (Ye and Wang 2006), mixture discriminant analysis (Gkalelis et al. 2012), partial least squares dimension analysis, quadratic discriminant analysis, canonical correlation analysis (Andrew et al. 2013), flexible discriminant analysis, and diffusion map (Reddy et al. 2020).

11.4.10 Ensemble Algorithms

Multiple learning algorithms are used in ensemble approaches to produce better predicting performance compared to any individual learning algorithm alone. They generally outperform single models by amalgamating various models to enhance accuracy. If one is unfamiliar with the types of ensemble algorithms, here is a list one should be acquainted with: boosting, AdaBoost, bootstrapped aggregation (bagging), weighted average (blending), stacked generalization (stacking), gradient boosting machines, gradient-boosted regression trees, and random forest (Dietterich 2002).

11.4.11 Association Rule Learning

Association rule learning is a rule-driven ML approach aimed at uncovering noteworthy relationships among variables within extensive databases. Its purpose is to unearth robust rules found in databases using various measures of significance. Association rules, which are based on if/then statements, reveal relationships between independent data that is kept in databases, relational databases, or other information repositories. These guidelines are used to identify connections between frequently used things in tandem. Notably, three association rule learning algorithms include Apriori algorithm (Hegland 2007), Eclat algorithm, and FP-growth algorithm (Shawkat et al. 2022).

11.4.12 Rule-Based System

Rule-based machine learning employs learning algorithms to autonomously discern valuable rules, obviating the need for human intervention to manually devise rules and compile a rule set based on prior domain knowledge. The following algorithms fall into this category: Ripper algorithm, repeated incremental pruning to produce, Cubist, OneR, and ZeroR (Liu et al. 2015).

11.4.13 Reinforcement Learning

Reinforcement learning algorithms are techniques used to train agents to mark sequential decisions in a situation to maximize cumulative rewards. Some notable reinforcement learning algorithms include Q-learning (Watkins and Dayan 1992), deep Q-networks, policy gradient methods, actor-critic methods, deep deterministic policy gradient (Tiong et al. 2020), proximal policy optimization, trust region policy optimization (Schulman et al. 2015), asynchronous advantage actor-critic (Muhati and Rawat 2021), dueling DQN, and rainbow. These algorithms vary in complexity, applicability, and performance, and the choice depends on the specific problem domain and requirements (François-Lavet et al. 2018).

11.4.14 Other ML Algorithms

Other machine learning algorithms include feature selection methods, algorithm performance assessment, evaluation metrics, optimization techniques, ALOPEX (Unnikrishnan and Venugopal 1994), CN2 method, dynamic time warping, FastICA (Liu et al. 2020), forward–backward technique, GeneRec, Leabra, Linde–Buzo–Gray method, local outlier factor, logic learning machine, LogitBoost, Markov chain Monte Carlo, prefrontal cortex basal ganglia working memory, weighted majority approach (Alhejazi and Mohammad 2022), Rprop, sparse PCA, structured kNN, T-distributed stochastic neighbor embedding, and wake–sleep technique (Das and Behera 2017).

11.5 AI and ML Applications in Agriculture

AI might completely transform agriculture by offering advisory services, data analytics, Internet of things (IoT), and camera and sensor applications. By examining a variety of data sources, including temperature, crop performance, soil conditions, and weather patterns, AI can enhance predictive insights in agriculture (Fig. 11.3). However, there are concerns involved with the use of AI in agriculture, such as interoperability, safety, data reliability, and unanticipated socio-ecological consequences. Using ML models, AI can maximize crop production; yet, this could also result in an excessive dependence on technology and unforeseen repercussions. AI

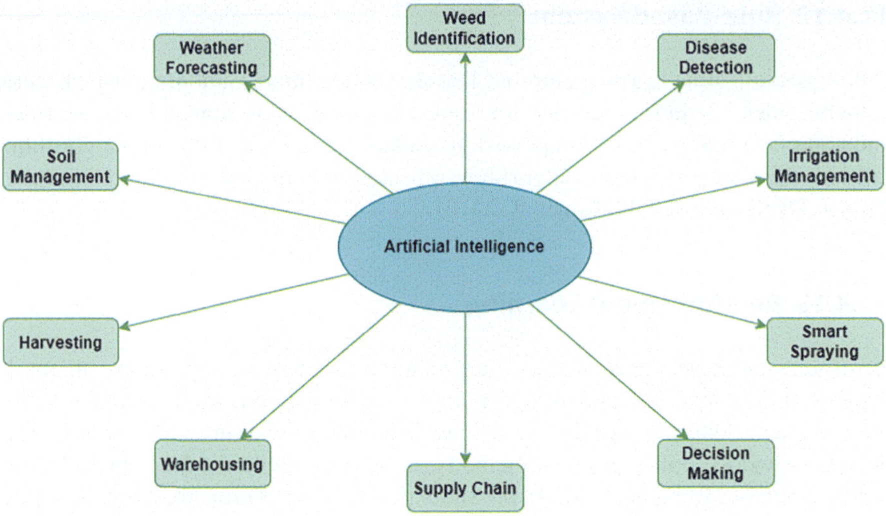

Fig. 11.3 Parameters monitored by AI

can help increase crop productivity and management by identifying plant illnesses fast and using agrochemicals more effectively. Rapid plant phenotyping, soil analysis, weather forecasting, agricultural monitoring, and yield calculation are all possible with machine learning approaches. To improve the productivity and efficiency of their farming techniques, many farmers are implementing AI, IoT, and other technological advancements.

11.5.1 Plant Disease Prediction

AI can detect and predict plant diseases and propose effective pest management strategies. It aids in predicting the optimal combination of agronomic inputs and determining the most suitable schedules for watering and when to apply nutrients and pesticides. Field management using high-resolution photos taken by devices such as drones and helicopters enables real-time prediction. During cultivation, predictive analytics involve creating yield maps and field maps to identify areas requiring varying levels of water, fertilizer, or pesticides. With the help of cognitive solutions, farmers can get important information on local insect infestations, seed varieties that are suitable, weather forecasts, and soil conditions.

Rice diseases pose significant threats to crop yield, and accurate diagnosis is crucial to mitigate these risks. However, conventional methods for diagnosing rice diseases lack both precision and efficiency, often requiring specialized equipment. An automated diagnostic approach can be developed and used in a smartphone application. Leveraging deep learning techniques, the model may be trained on a substantial dataset comprising images representing common diseases that affect rice, such as brown spot, bacterial stripe disease, sheath blight, fake smut, neck

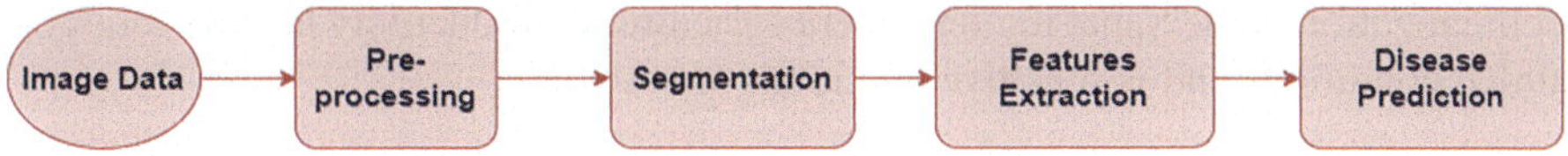

Fig. 11.4 Methodology of disease prediction using ML

blast, and leaf blast. Subsequently, the model is rigorously validated using a distinct set of images (Deng et al. 2021).

Image processing technology could be used to detect diseases in rice plant. Rice leaf diseases can be automatically identified through image processing methods. Image preprocessing steps has to be integrated into the model for leaf disease detection. Feature extraction can be done by utilizing techniques like the discrete wavelet transform, scale-invariant feature transformation, and gray level co-occurrence matrix. These extracted features underwent classification using various classifiers, including KNN, NN, SVM, and Naive Bayes, to distinguish between healthy and diseased plants (Gayathri Devi and Neelamegam 2019; Vasavi et al. 2022) (Fig. 11.4).

CNN techniques can be employed for more accurate detection of various rice diseases by extracting features. Deep-level features extracted through CNN techniques are superior and more distinct compared to conventional manually crafted features. Therefore, the accuracy of the models significantly surpasses traditional machine learning models. Identity models can combine with domain expertise on rice diseases to accomplish automated rice disease diagnosis. This way includes both qualitative and quantitative analyses to assess the efficiency of the CNN-based approach. Utilizing CNN with SVM and CNN with the activation function softmax enhances the reliability of agricultural disease identification (Liang et al. 2019; Lu et al. 2017; Zhou et al. 2019). AI can also be used for detecting paddy leaf blast disease and categorizing its severity of disease in three stages: spreading, worst stage, and infection (Bakar et al. 2018).

11.5.2 Weather Forecasting

The utilization of AI technology facilitates the prediction of weather patterns and various agricultural conditions, including crop cycles, soil quality, groundwater levels, and the identification of plant diseases, all of which are crucial aspects of agriculture. Monitoring the health of crops is enabled through the deployment of soil and plant sensors, as well as the utilization of multispectral imagery captured by satellites or drones (Singh and Kaur 2022; Tzachor et al. 2022). Variables including average, maximum, and minimum sun energy, soil temperature, average relative humidity, and air temperature can be used for predicting rice disease occurrences using short-term environmental data to mitigate significant losses (Liu et al. 2021a). A tool known as YIELDCAST (Traore et al. 2020), which forecasts rice yield responses, holds significant value as a decision support mechanism for climate acclimatization in the Sahel region, where agricultural productivity heavily relies on rainfed conditions and is sensitive to climate stressors.

11.5.3 Crop and Soil Monitoring

Crop and soil monitoring is one area where artificial intelligence (AI) is being used. AI-powered applications may process and analyze data gathered from drones, IoT devices, and satellite photos of fields to find the best solutions. Machine learning algorithms provide quick and simple data analysis, while AI-enabled applications can detect diseases, pests, and deficiencies in the soil. Applications for mobile agriculture let farmers analyze crop data, manage activities, keep an eye on weather patterns, and run their farms more profitably. The goal of optimizing the use of pesticides and herbicides is to protect food safety while improving agricultural sustainability and productivity. Instead of following a preset timetable, AI systems can track current insect and weed activity and modify the administration of pesticides and herbicides based on real-time data. By analyzing satellite or drone images and identifying patterns in pest activity, AI can predict pest outbreaks and alert farmworkers to potential threats, enabling them to take preventive measures without negotiating crop health or resorting to excessive pesticide use (Lowe et al. 2022; Weng et al. 2019; Shelake et al. 2021; Maraveas et al. 2021). Deep learning and IoT have emerged as prominent areas of research across various disciplines. With combination of deep learning and IoT, now paddy fields can be monitored remotely. Two strategies can be implemented for image identification: transfer learning and deep feature extraction. AI-based prototype not only controls temperature and humidity but also monitors nitrogen status and diseases, providing superior functionality compared to existing solutions (Sethy et al. 2021). A unique IoT gadget with machine learning capabilities was just unveiled; it is intended to track soil nutrients and provide accurate crop advice. Together with nutrient levels, the device collects data in real time on temperature, wetness, humidity, and composition of the soil. Then, using machine learning algorithms, the data is analyzed to provide customized recommendations. These recommendations include plans for high-yielding crop varieties, appropriate fertilizer types, and quantities based on crop needs and soil nutrient levels (Islam et al. 2023).

Precision farming contributes to increased output by accurately determining the essential activities to be taken at each step of the agricultural cycle. This include predicting the weather, assessing the quality of the soil, suggesting appropriate crops for planting, and figuring out how much fertilizer and insecticide to use. Modern technologies like IoT, data mining, data analytics, and machine learning are used in precision farming to collect data, train systems, and generate predictions. By harnessing these technologies, precision farming streamlines processes, reduces manual labor, and enhances productivity. In light of the challenges faced by farmers, such as crop failure due to inadequate rainfall and soil infertility, the proposed approach offers insights into managing crops and harvesting intelligently. It serves as a guide for implementing smart farming practices (Durai and Shamili 2022).

11.5.4 Growth Rate Prediction in Rice

Predicting the growth rate of rice is essential for achieving precision agriculture. By using growth data from several rice cultivars grown in various climate zones throughout critical growth periods, the rice growth rate may be estimated. Different algorithms like regression algorithm, ANN, and gene expression programming can be used for predicting rice growth rate based on ambient temperature from Gr observations of different cultivation instances (Traore et al. 2020; Ferreira 2002; Liu et al. 2021b). The identification of rice seedling growth stages is essential for efficient crop management, including temperature control, fertilization, irrigation, cultivation, and disease management. Accurate staging ensures healthy seedlings for mechanical transplanting and increased yield. However, traditional methods of assessing growth stages are manual, time-consuming, and subjective. Using machine learning algorithms on unmanned aerial vehicle (UAV) images offers a high-throughput and noninvasive alternative for large-scale fields (Tan et al. 2022). Smart farming has gained significant popularity, and machine learning models like random forest (RF) can be used to predict rice growth stages for beginners. The experimental setup includes a high-definition (HD) smart camera for image and video data collection, along with IoT devices such as a weather monitoring station, flow meter, and milometer. Various image processing techniques, including object detection, object classification, instance segmentation, excess green index, and modified excess green index, can be used to estimate paddy height and canopy cover or green coverage (Sheng et al. 2022).

11.5.5 Crop Yield Prediction in Rice

Agricultural management experts require straightforward and precise estimation methods to anticipate rice yields during the planning phase. Enhancing farmers' predictive capabilities regarding crop productivity under diverse climatic conditions can be instrumental in guiding crucial decisions related to agronomy and crop selection. Rice production yields can be forecasted by leveraging neural networks and by exploring the influencing factors. Parameters like precipitation, minimum and maximum temperatures, average temperature, reference crop evapotranspiration, as well as area, production, and yield for the Kharif season can be considered as input (Gandhi and Petkar 2016; Ji et al. 2007).

In Thailand, the current scenario depicts rapid changes in the planted area, climate, and rainfall, resulting in unstable Thai rice production. This unpredictability necessitates dependable rice response models for effective decision-making processes. Local authorities and farmers alike seek straightforward and precise estimation methods to forecast rice yield during the planning phase, and this can be done by creating a decision support system utilizing ANN with fine-tuning parameter values (Jabjone and Wannasang 2014).

Bayesian networks are utilized for forecasting rice crop yield as well. Factors such as area; production; yield; reference crop evapotranspiration; minimum, average, and maximum temperatures; as well as precipitation from the previous season are taken into account as significant variables in the prediction of rice yield (Gandhi et al. 2016).

11.5.6 Self-driving Tractors and Robotics

Agriculture faces a significant challenge with labor shortages, but solutions are emerging through the integration of AI, self-driving tractors, and IoT. These machineries not only offer precision and reduced errors but also present cost-effective alternatives. Precision farming, enabled by the confluence of AI, self-driving tractors, and IoT, holds the key to addressing these difficulties. Additionally, robotics is becoming more and more popular, especially for jobs like harvesting fruits and vegetables and trimming lettuce. While robots offer certain advantages over human labor, such as increased accuracy and longer working hours, they are still less common on farms. Agriculture generates vast amounts of data, but its complexity makes analysis challenging. However, AI advancements have the potential to revolutionize the industry by providing deep insights and predictive analytics. By leveraging AI and ML techniques, businesses can quickly adapt to new information and make informed decisions in a complex environment (Jiayu et al. 2015; Sharma et al. 2022; Santangeli et al. 2020; Songol et al. 2021).

11.5.7 Weed Identification

Crop yields are greatly impacted by weed suppression; hence, accurate identification of weed species is crucial for successful management techniques including hoeing and targeted herbicide use. However, the use of AI approaches in weed management has been hampered by the scarcity of field weed information. The Weed25 (Wang et al. 2022b) dataset comprises 14,035 images representing 25 different weed species, including both monocots and dicots, along with images capturing various growth stages of weeds. This dataset aims to fill the gap of unavailability and facilitate the development of weed identification models. Common deep learning detection techniques such as YOLOv3, YOLOv5, and Faster R-CNN, among others, can be used to train weed recognition models utilizing weed picture datasets (Wang et al. 2022b).

Deep learning models can also be used to identify weeds in rice crop photos, with the goal of providing accurate, real-time, low-cost machine identification for broad practical application. In order to accomplish this, online crawling and on-site photography can be used to collect a dataset of pictures of rice crops and different types of weeds. Weed detection is, however, significantly hampered by the difficulty of crop and weed overlap. A RetinaNet-based model known as WeedDet was developed to tackle this problem. To efficiently manage the complex information in the photos, WeedDet improves the backbone, feature pyramid network, and detecting head's feature extraction capabilities. In order to improve detection speed, the feature pyramid and detection head structure are also adjusted (Peng et al. 2022).

In modern agriculture, machine learning technology is increasingly employed for weed detection, a crucial task that poses challenges to the industry. Weed detection systems rely on image patches captured from crops, soils, and weeds. Algorithms are trained to identify weeds and crops within predefined plant regions in each image, pinpointing the location of weeds within rows and measuring the distances between them. The input data for these systems typically consists of four-channel NIR + RGB images or regular RGB images, depending on the type of sensor deployed on agricultural robots. Machine learning methods are then applied to process this input data and identify weeds. To achieve accurate and precise weed detection, models are trained using extensive datasets containing images of weeds and crops, often focusing on specific crops such as rice, with the potential for later application to various other crops (Nagothu et al. 2023).

11.5.8 Decision-Making

The agricultural sector is increasingly leveraging AI technology to improve decision-making processes. With the growing volume of data available, agricultural decision-making is becoming more data-driven. This change has been assisted by innovations such as extensive use of sensors, greater access to satellite imaging, lower costs of data logging equipment, increased drone usage, and enhanced access to data archives, particularly in irrigation management. Automation has made labor-intensive operations that were previously laborious more efficient. AI-enabled devices are capable of analyzing previous weather trends and data on soil quality, offering insightful analysis that maximizes crop productivity overall (Taberkit et al. 2021). Precision farming, which employs precise and controllable techniques, replaces the repetitive and labor-intensive aspects of traditional agriculture. This includes technologies like systems for precise location, automated steering, sensors, remote sensing, geomapping, and integrated electronic communication. These technologies not only automate tasks but also offer guidance to farmers on various aspects such as crop rotation, long-term sustainability, profitability, and efficiency. Given the complexity and various stages involved in agriculture, AI complements these technologies by addressing both routine and intricate tasks. By integrating AI with other technologies, agricultural data can be collected, processed, and analyzed on digital platforms to formulate optimal strategies and initiate necessary actions. To make informed agricultural decisions, factors such as temperature, weather, water usage, soil conditions, and other farm-related data can be examined (Leal Filho et al. 2022; Banthia and Chaudaki 2022; Hemming et al. 2019).

11.6 Challenges of Machine Learning and AI in Rice

Various government agencies have accepted the significant role of AI in transferring knowledge and modern agricultural skills to farmers. The main variables impacting the application of AI/ML and the difficulties in educating farmers has been categorized into four main groups: socio-economic, cultural, technical, and infrastructural support. However, smallholder farmers have not fully benefited from AI due to several factors, including low literacy, expense, the digital divide levels.

The digital divide is a major barrier, particularly in rural and developing areas, where physical access and the ability to utilize innovations are limited. This results in missed opportunities for citizens to fully leverage IoT or AI services provided by the government or private for agricultural development. Initiatives aimed at improving agricultural livelihoods are hindered by the lack of access and usability among rural populations. Furthermore, gender disparities exacerbate the digital divide in rural areas (Panganiban 2019). Women in rural regions face significant challenges in accessing and utilizing modern computerized techniques. Data from the International Telecommunication Union (ITU) revealed that women's Internet usage worldwide was 12% lower than men's, with even larger disparities in Africa, where it was 25% lower. Overall, addressing the digital and gender divides is crucial for ensuring equitable access to IoT resources and maximizing their impending benefits for agricultural development among smallholder farmers (Treinen and van der Elstraeten 2018).

Various factors influencing the adoption of AI-based information and communication technologies (ICTs) in small-scale farmers' agribusiness have been identified. Their inferential statistics highlighted the cost of technology as a major obstacle restricting the adoption of AI-based techniques in farming practices (Awojide and Akintelu 2018). Similarly, some socio-economic conditions such as the high cost associated with broadcast equipment, expensive charges for television and radio presentations, and the elevated costs of accessing electricity connectivity affect the utilization of ICT by agricultural extension officers (Tata and McNamara 2016).

Smallholder farmers' attitudes toward innovation can have a big impact on how well they are able to use tools to support their farming operations. Farmers may be discouraged from adopting new technologies due to a variety of reasons, including low self-esteem, fear of technology, and a sense of inferiority toward sophisticated technological instruments. This lack of confidence and resistance to adopting new technology often results from limited knowledge, understanding, and skills. Consequently, farmers may encounter difficulties in successfully integrating new technologies into their farming practices, leading to feelings of frustration and reluctance to continue trying (Pignatti et al. 2015). Similarly, even among computer-literate farmers, technophobia exists, hindering the adoption and usage of Information Technology Enabled Systems (ITES). This reluctance stems from confusion and a fear of failure associated with implementing new technologies (Boniface et al. 2019).

Some research have looked into the difficulties experienced by dairy farmers employing machinery in both plain and hilly parts of India. They found a number of factors, including illiteracy, that are causing these problems. Many farmers fail to understand how to use computers, Internet services, and email for communication, and many are only able to speak in their local tongue (Singh et al. 2015). This lack of literacy led to a lack of awareness about the technology, making its adoption unattainable for many farmers (Balraj and Pavalam 2012).

Among the many challenges identified by the respondents in utilizing AI/ML is inadequate infrastructure, particularly the lack of power supply. This limitation significantly hindered the implementation of modern initiatives in rural areas of India (Dhaka and Chayal 2010). Regarding the infrastructure assistance area, one of the biggest issues facing Sudanese farmers is the absence of electricity. The study also looked at how energy availability affects the spread of agricultural knowledge (Musa et al. 2014).

11.7 Conclusions and Future Perspective

This chapter evaluated several review articles on the application of AI/ML in contemporary and future agriculture. When combined with solid data, AI/ML can convert conventional farm management systems for agriculture into useful AI platforms that provide in-the-moment observations, recommendations, and decision-making for next actions. Many scientists have noted that ML-based algorithms are the most commonly used techniques, particularly in identifying diseases in agriculture. These algorithms also have diverse applications in rice cultivation practices, such as predicting grain yield, identifying pests, managing irrigation and fertilizer, distinguishing rice species and varieties, assessing grain quality, and detecting weeds. Furthermore, ML and DL models are widely adopted in soil management categories and various crop management tasks using image patterns such as spectral, hyperspectral, and near-infrared data. Statistical data analysis is essential for predicting the progress of different crop management techniques, enabling forecasts without requiring additional data integration from other tools. However, implementing ML algorithms during big data collection may require extra effort, which can limit their widespread use. Despite this, applications of machine learning in soil, water, and livestock management remain consistent, with ANNs and SVMs being the most commonly used ML versions. The use of various ML algorithms can lead to the development of integrated and relevant tools, promising even wider adoption in the future. However, current implementations often lack direct integration with the decision-making process compared to other fields. Integrating automatic dataset recording and ML-based processing can enhance decision-making processes, leading to knowledge-based agriculture aimed at improving agricultural yield and the effectiveness of biological products. In conclusion, the additional deployment of various ML algorithms with strong data has the potential to raise farm management systems to real AI in agriculture, offering precise feedback and recommendations for enhanced decision-making and action planning.

References

Ahir K, Govani K, Gajera R, Shah M (2020) Application on virtual reality for enhanced education learning, military training and sports. Augment Hum Res 5:1–9

Ahmar S, Gill RA, Jung K-H et al (2020) Conventional and molecular techniques from simple breeding to speed breeding in crop plants: recent advances and future outlook. Int J Mol Sci 21:2590

Ahuja R, Chug A, Gupta S et al (2020) Classification and clustering algorithms of machine learning with their applications. In: Nature-inspired computation in data mining and machine learning, pp 225–248

Alhejazi MM, Mohammad RMA (2022) Enhancing the blockchain voting process in IoT using a novel blockchain weighted majority consensus algorithm (WMCA). Inf Secur J Glob Perspect 31:125–143

Andrew G, Arora R, Bilmes J, Livescu K (2013) Deep canonical correlation analysis. In: International conference on machine learning. PMLR, pp 1247–1255

Araus JL, Cairns JE (2014) Field high-throughput phenotyping: the new crop breeding frontier. Trends Plant Sci 19:52–61

Awojide S, Akintelu SO (2018) Empirical investigation of factors affecting information and communication technologies (icts) in agric-business among small scale farmers in Esan Community, Edo State, Nigeria. J Res Market 9:713–722. issn:2292-9355

Bakar MNA, Abdullah AH, Rahim NA et al (2018) Rice leaf blast disease detection using multilevel colour image thresholding. J Telecommun Electron Comp Eng 10:1–6

Balraj PL, Pavalam SM (2012) Integrating ICT in agriculture for knowledge-based economy. Rwanda J 27:44–56

Banthia V, Chaudaki G (2022) The study on use of artificial intelligence in agriculture. J Adv Res Appl Artif Intell Neural Netw 5:18–22

Bensi M, Der Kiureghian A, Straub D (2013) Efficient Bayesian network modeling of systems. Reliab Eng Syst Saf 112:200–213

Bingham E, Kaski S, Laaksonen J, Lampinen J (2015) Advances in independent component analysis and learning machines. Academic Press

Binkert NL, Dreslinski RG, Hsu LR et al (2006) The M5 simulator: modeling networked systems. IEEE Micro 26:52–60

Boniface PJ, Jose AM, Husain AS (2019) Constraints faced by farmers and agricultural extensionists in using selected information technology enabled systems for agriculture. J Soc Sci 58:7–12

Brown KH, Jameton AL (2000) Public health implications of urban agriculture. J Public Health Policy 21:20–39

Buja A, Swayne DF, Littman ML et al (2008) Data visualization with multidimensional scaling. J Comput Graph Stat 17:444–472

Carvalho FP (2006) Agriculture, pesticides, food security and food safety. Environ Sci Pol 9:685–692

Chowdhury A, Pradhan S, Saha M, Sanyal N (2008) Impact of pesticides on soil microbiological parameters and possible bioremediation strategies. Indian J Microbiol 48:114–127

Cioffi R, Travaglioni M, Piscitelli G et al (2020) Artificial intelligence and machine learning applications in smart production: progress, trends, and directions. Sustainability 12:492

Das K, Behera RN (2017) A survey on machine learning: concept, algorithms and applications. Int J Innov Res Comp Commun Eng 5:1301–1309

Davis KF, Gephart JA, Emery KA et al (2016) Meeting future food demand with current agricultural resources. Glob Environ Chang 39:125–132

Deng R, Tao M, Xing H et al (2021) Automatic diagnosis of rice diseases using deep learning. Front Plant Sci 12:701038

Dhaka BL, Chayal K (2010) Farmers' experience with ICTs on transfer of technology in changing agri-rural environment. Indian Res J Extens Educ 10:114–118

Dietterich TG (2002) Ensemble learning. In: The handbook of brain theory and neural networks, vol 2, pp 110–125

Donde R, Gouda G, Sabarinathan S, et al (2021) Artificial intelligence and machine learning in rice research. In: Applications of bioinformatics in rice research, pp 239–275
Durai SKS, Shamili MD (2022) Smart farming using machine learning and deep learning techniques. Decis Anal J 3:100041
Dwivedy N (2011) Challenges faced by the agriculture sector in developing countries with special reference to India. Int J Rural Stud 18:1–6
Englert P (2012) Locally weighted learning. In: Seminar class on autonomous learning systems. Citeseer, pp 1–9
Ferreira C (2002) Gene expression programming in problem solving. In: Soft computing and industry: recent applications. Springer, pp 635–653
Fleming SW, Garen DC (2022) Simplified cross-validation in principal component regression (PCR) and PCR-like machine learning for water supply forecasting. JAWRA J Am Water Resour Assoc 58:517–524
Fogel RW (2004) The escape from hunger and premature death, 1700–2100: Europe, America, and the third world. Cambridge University Press, Cambridge
François-Lavet V, Henderson P, Islam R et al (2018) An introduction to deep reinforcement learning. In: Foundations and trends® in machine learning, vol 11, pp 219–354
Gandhi N, Petkar O, Armstrong LJ (2016) Rice crop yield prediction using artificial neural networks. In: 2016 IEEE technological innovations in ICT for agriculture and rural development (TIAR). IEEE, pp 105–110
Gandhi N, Armstrong LJ, Petkar O (2016) Predicting rice crop yield using Bayesian networks. In: 2016 international conference on advances in computing, communications and informatics (ICACCI). IEEE, pp 795–799
Gayathri Devi T, Neelamegam P (2019) Image processing based rice plant leaves diseases in Thanjavur, Tamilnadu. Clust Comput 22:13415–13428
Gershenson C (2003) Artificial neural networks for beginners. arXiv preprint cs/0308031
Ghosh J, Nag A (2001) An overview of radial basis function networks. In: Radial basis function networks 2: new advances in design, pp 1–36
Gkalelis N, Mezaris V, Kompatsiaris I, Stathaki T (2012) Mixture subclass discriminant analysis link to restricted Gaussian model and other generalizations. IEEE Trans Neural Netw Learn Syst 24:8–21
Godfray HCJ, Beddington JR, Crute IR et al (2010) Food security: the challenge of feeding 9 billion people. Science (1979) 327:812–818
Grover J (2012) A literature review of Bayes' theorem and Bayesian belief networks (BBN). In: Strategic economic decision-making: using Bayesian belief networks to solve complex problems, pp 11–27
Hegland M (2007) The apriori algorithm—a tutorial. In: Mathematics and computation in imaging science and information processing, pp 209–262
Hemming S, de Zwart F, Elings A et al (2019) Remote control of greenhouse vegetable production with artificial intelligence—greenhouse climate, irrigation, and crop production. Sensors 19:1807
Holman FH, Riche AB, Michalski A et al (2016) High throughput field phenotyping of wheat plant height and growth rate in field plot trials using UAV based remote sensing. Remote Sens 8:1031
Hope TMH (2020) Linear regression. In: Machine learning. Elsevier, pp 67–81
Islam MR, Oliullah K, Kabir MM et al (2023) Machine learning enabled IoT system for soil nutrients monitoring and crop recommendation. J Agric Food Res 14:100880
Jabjone S, Wannasang S (2014) Decision support system using artificial neural network to predict rice production in Phimai District, Thailand. Int J Comp Electric Eng 6:162–166
Jakkula V (2006) Tutorial on support vector machine (svm). School of EECS, Washington State University 37:3
Jennrich RI, Sampson PF (1968) Application of stepwise regression to non-linear estimation. Technometrics 10:63–72
Ji B, Sun Y, Yang S, Wan J (2007) Artificial neural networks for rice yield prediction in mountainous regions. J Agric Sci 145:249–261

Jiang L, Zhang H (2006) Weightily averaged one-dependence estimators. In: PRICAI 2006: trends in artificial intelligence: 9th Pacific rim international conference on artificial intelligence Guilin, China, August 7–11, 2006 proceedings 9. Springer, pp 970–974

Jiayu Z, Shiwei X, Zhemin L et al (2015) Application of intelligence information fusion technology in agriculture monitoring and early-warning research. In: 2015 international conference on control, automation and robotics. IEEE, pp 114–117

Jung J, Maeda M, Chang A et al (2021) The potential of remote sensing and artificial intelligence as tools to improve the resilience of agriculture production systems. Curr Opin Biotechnol 70:15–22

Kesavan PC, Swaminathan MS (2006) Managing extreme natural disasters in coastal areas. Philos Trans R Soc A Math Phys Eng Sci 364:2191–2216

Khedr AE, Idrees AM, El Seddawy AI (2016) Enhancing iterative Dichotomiser 3 algorithm for classification decision tree. Wiley Interdiscip Rev Data Min Knowl Discov 6:70–79

Kherif F, Latypova A (2020) Principal component analysis. In: Machine learning. Elsevier, pp 209–225

Kukreja SL, Löfberg J, Brenner MJ (2006) A least absolute shrinkage and selection operator (LASSO) for nonlinear system identification. IFAC proceedings volumes 39:814–819

Leal Filho W, Wall T, Mucova SAR et al (2022) Deploying artificial intelligence for climate change adaptation. Technol Forecast Soc Change 180:121662

Lewis RJ (2000) An introduction to classification and regression tree (CART) analysis. In: Annual meeting of the society for academic emergency medicine in San Francisco, California. Citeseer

Liakos KG, Busato P, Moshou D et al (2018) Machine learning in agriculture: a review. Sensors 18:2674

Liang W, Zhang H, Zhang G, Cao H (2019) Rice blast disease recognition using a deep convolutional neural network. Sci Rep 9:1–10

Liu H, Gegov A, Cocea M (2015) Rule based systems for big data: a machine learning approach. Springer

Liu J, Song H, Sun H, Zhao H (2020) High-precision identification of power quality disturbances under strong noise environment based on FastICA and random forest. IEEE Trans Industr Inform 17:377–387

Liu L-W, Hsieh S-H, Lin S-J et al (2021a) Rice blast (*Magnaporthe oryzae*) occurrence prediction and the key factor sensitivity analysis by machine learning. Agronomy 11:771

Liu L-W, Ma X, Wang Y-M et al (2021b) Using artificial intelligence algorithms to predict rice (Oryza sativa L.) growth rate for precision agriculture. Comput Electron Agric 187:106286

Long SP, Marshall-Colon A, Zhu X-G (2015) Meeting the global food demand of the future by engineering crop photosynthesis and yield potential. Cell 161:56–66

Lopes N, Ribeiro B, Lopes N, Ribeiro B (2015) Non-negative matrix factorization (NMF). In: Machine learning for adaptive many-core machines—a practical approach, pp 127–154

Lowe M, Qin R, Mao X (2022) A review on machine learning, artificial intelligence, and smart technology in water treatment and monitoring. Water (Basel) 14:1384

Lu Y, Yi S, Zeng N et al (2017) Identification of rice diseases using deep convolutional neural networks. Neurocomputing 267:378–384

Maraveas C, Loukatos D, Bartzanas T, Arvanitis KG (2021) Applications of artificial intelligence in fire safety of agricultural structures. Appl Sci 11:7716

Mohamed A, Hinton G, Penn G (2012) Understanding how deep belief networks perform acoustic modelling. In: 2012 IEEE international conference on acoustics, speech and signal processing (ICASSP). IEEE, pp 4273–4276

Muhati E, Rawat DB (2021) Asynchronous advantage actor-critic (a3c) learning for cognitive network security. In: 2021 third IEEE international conference on trust, privacy and security in intelligent systems and applications (TPS-ISA). IEEE, pp 106–113

Musa NS, Githeko JM, El-Siddig K (2014) The adoption and use of ICT by small scale farmers in Gezira State, Sudan

Nagothu SK, Anitha G, Siranthini B, et al (2023) Weed detection in agriculture crop using unmanned aerial vehicle and machine learning. Mater Today Proc

Nova D, Estévez PA (2014) A review of learning vector quantization classifiers. Neural Comput Appl 25:511–524
Pandya R, Pandya J (2015) C5.0 algorithm to improved decision tree with feature selection and reduced error pruning. Int J Comput Appl 117:18–21
Panganiban GGF (2019) E-governance in agriculture: digital tools enabling Filipino farmers. J Asian Public Policy 12:51–70
Peng H, Li Z, Zhou Z, Shao Y (2022) Weed detection in paddy field using an improved RetinaNet network. Comput Electron Agric 199:107179
Petersen PE (2003) The World Oral Health Report 2003: continuous improvement of oral health in the 21st century—the approach of the WHO Global Oral Health Programme. Community Dent Oral Epidemiol 31:3–24
Peterson LE (2009) K-nearest neighbor. Scholarpedia 4:1883
Pignatti E, Carli G, Canavari M (2015) What really matters? A qualitative analysis on the adoption of innovations in agriculture. J Agric Inform 6
Plessen MG (2021) Freeform path fitting for the minimisation of the number of transitions between headland path and interior lanes within agricultural fields. Artif Intell Agric 5:233–239
Quinlan JR (2014) C4.5: programs for machine learning. Elsevier
Ramsauer H, Schäfl B, Lehner J et al (2020) Hopfield networks is all you need. arXiv preprint arXiv:200802217
Reddy GT, Reddy MPK, Lakshmanna K et al (2020) Analysis of dimensionality reduction techniques on big data. IEEE Access 8:54776–54788
Salakhutdinov R, Hinton G (2009) Deep Boltzmann machines. In: Artificial intelligence and statistics. PMLR, pp 448–455
Santangeli A, Chen Y, Kluen E et al (2020) Integrating drone-borne thermal imaging with artificial intelligence to locate bird nests on agricultural land. Sci Rep 10:10993
Schmidt RM (2019) Recurrent neural networks (RNNS): a gentle introduction and overview. arXiv preprint arXiv:191205911
Schulman J, Levine S, Abbeel P et al (2015) Trust region policy optimization. In: International conference on machine learning. PMLR, pp 1889–1897
Sethy PK, Behera SK, Kannan N et al (2021) Smart paddy field monitoring system using deep learning and IoT. Concurr Eng 29:16–24
Shadrin D, Menshchikov A, Somov A et al (2019) Enabling precision agriculture through embedded sensing with artificial intelligence. IEEE Trans Instrum Meas 69:4103–4113
Sharma P, Vimal A, Vishvakarma R et al (2022) Deciphering the blackbox of omics approaches and artificial intelligence in food waste transformation and mitigation. Int J Food Microbiol 372:109691
Shawkat M, Badawi M, El-ghamrawy S et al (2022) An optimized FP-growth algorithm for discovery of association rules. J Supercomput 78:5479–5506
Shelake S, Sutar S, Salunkher A et al (2021) Design and implementation of artificial intelligence powered agriculture multipurpose robot. Int J Res Eng Sci Manage 4:165–167
Sheng RT-C, Huang Y-H, Chan P-C et al (2022) Rice growth stage classification via RF-based machine learning and image processing. Agriculture 12:2137
Singh P, Kaur A (2022) A systematic review of artificial intelligence in agriculture. In: Deep learning for sustainable agriculture, pp 57–80
Singh P, Bardhan D, Tripathi SC (2015) Constraints faced in using modern ICT tools: a study of dairy cooperative societies in Uttarakhand. J Rural Dev 34:149–166
Songol M, Awuor F, Maake B (2021) Adoption of artificial intelligence in agriculture in the developing nations: a review. J Lang Technol Entrepreneurship Africa 12:208–229
Spielman DJ, Von Grebmer K (2004) Public-private partnerships in agricultural research: an analysis of challenges facing industry and the Consultative Group on International Agricultural Research. International Food Policy Research Institute, Washington, DC
Suzuki N, Rivero RM, Shulaev V et al (2014) Abiotic and biotic stress combinations. New Phytol 203:32–43

Taberkit AM, Kechida A, Bouguettaya A (2021) Algerian perspectives for UAV-based remote sensing technologies and artificial intelligence in precision agriculture. In: Proceedings of the 4th international conference on networking, information systems & security, pp 1–9
Talaviya T, Shah D, Patel N et al (2020) Implementation of artificial intelligence in agriculture for optimisation of irrigation and application of pesticides and herbicides. Artif Intell Agric 4:58–73
Tan S, Liu J, Lu H et al (2022) Machine learning approaches for rice seedling growth stages detection. Front Plant Sci 13:914771
Tata JS, McNamara PE (2016) Social factors that influence use of ICT in agricultural extension in southern Africa. Agriculture 6:15
Taud H, Mas J-F (2018) Multilayer perceptron (MLP). In: Geomatic approaches for modeling land change scenarios, pp 451–455
Tharwat A, Gaber T, Ibrahim A, Hassanien AE (2017) Linear discriminant analysis: a detailed tutorial. AI Commun 30:169–190
Timsina J, Connor DJ (2001) Productivity and management of rice–wheat cropping systems: issues and challenges. Field Crop Res 69:93–132
Tiong T, Saad I, Teo KTK, Bin Lago H (2020) Deep reinforcement learning with robust deep deterministic policy gradient. In: 2020 2nd international conference on electrical, control and instrumentation engineering (ICECIE). IEEE, pp 1–5
Traore S, Zhang L, Guven A, Fipps G (2020) Rice yield response forecasting tool (YIELDCAST) for supporting climate change adaptation decision in Sahel. Agric Water Manag 239:106242
Treinen S, van der Elstraeten A (2018) Gender and ICTs: mainstreaming gender in the use of information and communication technologies (ICTs) for agriculture and rural development. Food and Agriculture Organization of the United Nations (FAO), Rome
Tzachor A, Devare M, King B et al (2022) Responsible artificial intelligence in agriculture requires systemic understanding of risks and externalities. Nat Mach Intell 4:104–109
Unnikrishnan KP, Venugopal KP (1994) Alopex: a correlation-based learning algorithm for feedforward and recurrent neural networks. Neural Comput 6:469–490
Van Evert FK, Fountas S, Jakovetic D et al (2017) Big data for weed control and crop protection. Weed Res 57:218–233
Vasavi P, Punitha A, Rao TVN (2022) Crop leaf disease detection and classification using machine learning and deep learning algorithms by visual symptoms: a review. Int J Electric Comp Eng 12:2079
Vesanto J, Alhoniemi E (2000) Clustering of the self-organizing map. IEEE Trans Neural Netw 11:586–600
Wang H, Chu X, Chen P et al (2022a) Partial least squares regression residual extreme learning machine (PLSRR-ELM) calibration algorithm applied in fast determination of gasoline octane number with near-infrared spectroscopy. Fuel 309:122224
Wang P, Tang Y, Luo F et al (2022b) Weed25: a deep learning dataset for weed identification. Front Plant Sci 13:1053329
Watkins CJCH, Dayan P (1992) Q-learning. Mach Learn 8:279–292
Weng S, Zhu W, Zhang X et al (2019) Recent advances in Raman technology with applications in agriculture, food and biosystems: a review. Artif Intell Agric 3:1–10
Woodbridge J, Anderson HS, Ahuja A, Grant D (2016) Predicting domain generation algorithms with long short-term memory networks. arXiv preprint arXiv:161100791
Wu J (2017) Introduction to convolutional neural networks. National Key Lab for Novel Software Technology Nanjing University China 5, pp 495
Ye J, Wang T (2006) Regularized discriminant analysis for high dimensional, low sample size data. In: Proceedings of the 12th ACM SIGKDD international conference on knowledge discovery and data mining, pp 454–463
Zhang W, Goh ATC (2016) Multivariate adaptive regression splines and neural network models for prediction of pile drivability. Geosci Front 7:45–52
Zhang H, Zamar RH (2014) Least angle regression for model selection. Wiley Interdiscip Rev Comput Stat 6:116–123
Zhou G, Zhang W, Chen A et al (2019) Rapid detection of rice disease based on FCM-KM and faster R-CNN fusion. IEEE Access 7:143190–143206

Fast-Forward Breeding in Rice

12

Tushar Arun Mohanty, Puja Mandal,
Dharmalingam Kumaresan,
and Manonmani Swaminathan

Abstract

Rice, a fundamental crop and vital staple food worldwide, stands at the junction of numerous global challenges. With the planet undergoing rapid environmental changes and the global population steadily increasing, there is an urgency to harness the full potential of rice production—in terms of both quantity and quality. Sustainable expansion of crop production systems is necessary to provide food for the growing human population. Progress in genome sequencing technologies, combined with efficient methods for identifying desirable traits, accelerates the availability of advantageous genetic elements for breeding and research. Fast-forward breeding (FFB) in rice introduces a groundbreaking approach that expedites the traditional breeding process by integrating cutting-edge technologies and innovative methods. FFB in rice represents a revolutionary method that employs genomics, molecular biology, and data-driven techniques to expedite the development of improved rice varieties with enhanced characteristics, such as increased yields, resistance to environmental stress, and improved nutritional content. The adoption of FFB has the potential to tackle global food security challenges by delivering more robust and productive rice crops, although it necessitates responsible consideration of regulatory and ethical concerns in its implementation. This chapter explores the innovative concept of "Fast Forward Breeding" in rice breeding and showcases examples where this approach has resulted in the development of high-yielding, disease-resistant, and climate-resilient rice varieties. Additionally, it addresses the critical significance of main-

T. A. Mohanty (✉) · P. Mandal
Department of Genetics and Plant Breeding, Centre for Plant Breeding and Genetics, Tamil Nadu Agricultural University, Coimbatore, Tamil Nadu, India

D. Kumaresan · M. Swaminathan
Department of Rice, Centre for Plant Breeding and Genetics, Tamil Nadu Agricultural University, Coimbatore, Tamil Nadu, India

A. Singh et al. (eds.), *Climate-Smart Rice Breeding*,
https://doi.org/10.1007/978-981-97-7098-4_12

taining genetic diversity and preserving landraces in the pursuit of sustainable rice breeding.

Keywords

Food security · High-throughput technology · Rice · Smart breeding · Speed breeding

12.1 Introduction

Rice, scientifically known as *Oryza sativa* L., is a globally significant cereal crop with a history of cultivation spanning 8200–13,500 years (Molina et al. 2011). It constitutes over 20% of the total calorie intake for more than half of the world's population, with over 90% of its production and consumption concentrated in Asia. Asia, which had historically faced food shortages, achieved self-sufficiency in rice production by the early 1980s, mainly due to the adoption of high-yielding dwarf plant varieties introduced in the mid-1960s. Looking ahead, it is estimated that by 2050, the global demand for rice will be 70% higher than current production levels. Meeting this substantial demand sustainably, especially considering the limited availability of resources like arable land, irrigation water, and genetic diversity, poses a significant challenge. This situation underscores the need to discover new sources of genetic material and innovative breeding and selection strategies to develop rice varieties capable of meeting this demand. While systematic breeding for the improvement of Asian rice (*Oryza sativa* L.) began more than a century ago, the past 60 years have witnessed rapid advancements, marked by notable achievements in applied and mission-oriented basic research.

Rice breeding holds a pivotal role in ensuring food security and sustainability. Rice breeding programs are crucial for developing varieties that can adapt to various environmental conditions, combat diseases and pests, and withstand climate-related challenges such as droughts and floods. These initiatives not only enhance farmers' economic stability by increasing yields and income but also address malnutrition concerns by improving the nutritional content of rice through biofortification. Additionally, by advocating for resource-efficient and environmentally friendly practices, rice breeding helps reduce the environmental impact of rice production. It also plays a role in conserving the biodiversity of rice, ensuring its adaptability and enhancement in the future.

Fast-forward breeding in rice represents a transformative approach that harnesses cutting-edge techniques and technology to accelerate the development of new rice varieties. In a world grappling with the critical challenge of feeding a growing global population while facing the threats of climate change, this innovative methodology is poised to play a pivotal role in ensuring food security and sustainable agriculture. Traditional rice breeding methods have historically been time-consuming and labor-intensive, often taking several years to develop and release new varieties. However, the urgency of the current global agricultural landscape demands a more agile and efficient process. Fast-forward breeding in rice combines the power of genomics, marker-assisted selection, and other advanced biotechnological tools to streamline the breeding cycle. This approach not only significantly reduces the time required to develop improved rice strains but also enhances the precision and effectiveness of breeding efforts.

12.2 Traditional Rice Breeding Methods

Rice is cultivated on six continents, excluding Antarctica. Two main rice species are *Oryza glaberrima* (African rice) and *Oryza sativa* L. (Asian rice, with *japonica* and *indica* subspecies). Cultivated rice is diploid ($2n = 2x = 24$) with 12 basic chromosomes (Brar and Khush 2006). There are 21 wild rice varieties. Rice is a vital global food crop, particularly in Asia. Plant breeding is a coevolutionary process between humans and edible plants, resulting in changes in both. Traditional rice breeding methods, used for centuries, have enhanced characteristics like yield, disease resistance, and adaptability. These methods involve selection and crossbreeding, not genetic modification. Key steps and techniques in traditional rice breeding include the following.

12.2.1 Mass Selection

Mass selection, by altering gene frequencies, enhances cultivar performance by removing undesired plants or cultivating desirable ones. Effective for purity maintenance, resistance introduction, and regional adaptation, it relies on highly heritable traits (Brown and Caligari 2008). The resulting cultivar is genetically diverse, adaptable, and stable, requiring just one generation per cycle. Cost-effective and swift, it suits large populations. Notably, rice varieties like IR8, the first high-yielding modern rice cultivar released by the International Rice Research Institute (IRRI) in 1966, marked the beginning of the "green revolution" in Asia. It features semidwarf stature, prolific tillering, stiff culms, upright leaves, insensitivity to photoperiod, high nitrogen responsiveness, and a high harvest index as compared to traditional cultivars (Peng et al. 2008).

12.2.2 Heterosis Breeding

Heterosis is the superior performance of hybrids over parental individuals, resulting from crossing different genotypes with diverse traits. It is generally associated with the heterozygous loci. The challenges in hybrid breeding are (1) improving two populations for agronomic adaptation while maintaining genetic distance to express strong heterosis and (2) efficient seed production without excessive costs (Breseghello and Coelho 2013). In China, rice hybrids with a 10–15% yield advantage over inbred varieties cover 45% of rice cultivation. Notable hybrids like Hunan,

Jiangxi, and Sichuan were developed for yield and pest resistance (Li et al. 2007). In India, "Sudhir" was developed from "FR13A" × "Biraj" crosse for submergence tolerance and improved rice yield (Khan et al. 2015). The evaluation of hybrids over 32 years showed 10% greater yield than local cultivars in multi-site experiments with an increased level of productivity per day among the early-maturing and mid-early-maturing hybrid genotypes (Muralidharan et al. 2022). APRH-1, the first hybrid variety released by APRRI in 1993–1994, marked the beginning of 117 developed rice hybrids. These hybrids, spanning public and private sectors, cover 3.0 million hectares, about 7.0% of India's total rice acreage (Rout et al. 2020). Notably, CORH 1, CORH 3, and CORH 4 are three-line hybrids from the Paddy Breeding Station, TNAU, Coimbatore.

12.2.3 Pedigree Breeding

The pedigree method is the primary approach in rice breeding (Peng et al. 2008). In this method, individual plants are selected from the segregating F2 generation, followed by selfing and character selection up to the F_6 generation. Notable varieties like IR64 were developed using this method. IR64, introduced in the Philippines in 1985, replaced IR36 in most regions due to its broad adaptability, early maturity, and improved quality. It has become a standard for high-quality rice and is widely used in research studies as a representative *indica* variety (Mackill and Khush 2018). Tamil Nadu Agricultural University, Coimbatore, India, released CO52, a medium-duration fine grain rice, suitable for late samba/thaladi season. It was developed through pedigree breeding, combining quality, yield, and blast resistance (Pi54) (Robin et al. 2019). ADT 53, a short-duration variety for Kuruvai/Kodai/Navarai season, was also released in 2019, evolved from the cross ADT 43/JGL 384 through pedigree breeding (Pushpa et al. 2020).

12.2.4 Ideotype Breeding

The ideotype breeding approach enhances the pedigree method by modifying simpler traits to boost quantitative traits like yield, particularly beneficial in smaller breeding programs (Donald 1968). Changing the crop's ideotype may involve exploring beyond elite germplasm, sometimes requiring backcrossing for improvement and addressing challenges from wild relatives through a pre-breeding phase. In the quest for higher rice productivity, scientists at IRRI designed the "new plant type," with few tillers, large panicles, sturdy stems, thick leaves, and rich green color (Bin Rahman and Zhang 2023). However, it failed to outperform traditional modern plant types due to reduced canopy biomass and incomplete grain filling. A second generation with more *indica* genotype showed some improvement but still fell short of the best checks. Tamil Nadu Agricultural University, Coimbatore, introduced CO(R) 50, a medium-duration rice variety suitable for irrigated zones in Southern India. It is a high-yielding variety with new plant type characteristics and attributes by registering higher crop growth rate and leaf area and relative growth rate and better light transmission ratio, spikelet fertility and absolute productive tillers, dry matter production, and harvest index, resulting from the cross between CO 43 and ADT 38, maturing in 130–135 days (Rajeswari et al. 2015).

12.2.5 Pure Line Selection

Pure line cultivars, characterized by identical alleles at all loci, provide phenotypic uniformity but have a narrow genetic base (Allard 1999). Suited for applications requiring trait uniformity, like mechanized farming, this method involves repeated selfing until no segregation is observed in the progeny (Poehlman and Sleper 1995). While it might be suitable for traits with low heritability, pure line selection may lack yield stability across diverse conditions. In self-pollinating species like rice and wheat, landraces are mixtures of pure lines with occasional heterozygous individuals from cross-pollination. While inbred progenies from selection may excel under specific conditions, they exhibit less stability under stress and lack long-term adaptability. Initial exploitation of germplasm can be done using pure line selection. Pure line selections from local types in Bihar resulted in the identification of submergence-tolerant improved varieties BR 14, BR 15, and BR 46 (Rao and Babu n.d.). The Department of Rice, formerly known as the "Paddy Breeding Station" at Tamil Nadu Agricultural University, Coimbatore, India, contributed 23 pure lines, including CO 1–CO 13, CO 18–CO 23, CO 27, and CO 32.

12.2.6 Single-Seed Descent

Single-seed descent accelerates breeding programs by randomly selecting one seed from desirable F_2 plants for the next generation (Allard 1960). This method, extended through F_3, rapidly achieves homozygosity while postponing selection, allowing the breeder to advance numerous F_2 plants through multiple generations. It is ideal for self-pollinated small grains and legumes like soybean, especially in early generations that can be conducted in a small space or greenhouse (Tigchelaat and Casali 1976).

12.2.7 Challenges and Limitations of Traditional Breeding

Traditional captive propagation for biodiversity preservation faces recognized limitations. Time-consuming, it requires generations to achieve desired traits, leading to unpredictable outcomes influenced by complex gene–environment interactions. Limited genetic variation, inbreeding risks, and ethical concerns arise. Additionally, traditional breeding is labor-intensive and imprecise in introducing specific genetic changes and may have unintended environmental impacts. Modern biotechnology methods like genetic engineering offer more precise control but come with their own challenges and ethical considerations.

12.2.8 The Need for Speed Breeding in Rice

Conventional breeding methods typically involve selecting parent genotypes, making crosses, selecting and advancing progenies with desired traits, screening for performance in multiple environments, and registering and distributing cultivars to growers. However, these procedures can take over a decade to develop and release improved varieties. The slow pace is due to the extended time required for field selection, the inherent nature of crop cycles, and the need for skill and resources in phenotyping.

The necessity for speed breeding (SB) in rice is driven by several factors: a growing global population, the need for food security, and the urgency imposed by climate change. It can significantly reduce the duration of each crop breeding cycle. Speed breeding, in particular, manipulates environmental conditions to accelerate flowering and seed set, enabling rapid generation advancement. This approach integrates various selection methods and can yield three to nine generations per year compared to the one to two generations achieved with conventional methods. Speed breeding facilitates the rapid development and release of new cultivars and aligns well with marker-assisted selection and high-throughput phenotyping for multiple trait selection. It also streamlines the traditional, time-consuming breeding process, enhancing crop sustainability and genetic diversity. This approach is crucial for meeting the evolving needs of agriculture, economy, and global food security.

12.3 Introduction to Speed Breeding

The urgent need to enhance the current rate of genetic gain in essential food crops to ensure global food security was emphasized by Lin et al. (2016). The improvement of genetic gain relies on crop breeding pipelines for the rapid delivery of enhanced crop varieties. The breeder's equation suggests that accelerating plant breeding involves improving factors that influence genetic gain over time, with a crucial emphasis on reducing breeding cycle time.

Conventional crop variety development involves selecting parental genotypes with desired traits, followed by crosses and a series of selections to release candidate cultivars meeting market demands. The conventional breeding procedures, while effective, can take more time to develop and release an improved variety, especially without an integrated pre-breeding program. Resources and time are heavily invested; field selection processes, limited by the inherent nature of crop cycles, contribute to delays, especially in common cereal and legume crops with a cycle of 3–6 months per generation.

Modern technologies such as doubled haploid breeding and speed breeding offer ways to reduce the duration of each crop breeding cycle (Dwivedi et al. 2015; Hickey et al. 2019). Speed breeding, characterized by the manipulation of environmental conditions to expedite flowering and seed set, can accomplish approximately three to nine generations per year, a notable improvement compared to the one to two generations achievable through conventional methods (Ghosh et al. 2018; Ochatt et al. 2002). This accelerated development facilitates the swift introduction of new cultivars, synergizing effectively with marker-assisted selection and high-throughput phenotyping techniques for the simultaneous selection of multiple traits.

Speed breeding methodologies provide a mechanism to accelerate diverse breeding objectives, encompassing the creation of crosses, generation of mapping populations, and assessment of agronomic traits. This entails cultivating plants in controlled environmental (CE) settings, where researchers adjust parameters like day/night temperature, light spectrum and intensity, and photoperiod duration. The objective is to minimize the duration to floral initiation and expedite processes such as embryo development and seed maturation.

In speed breeding, considerable attention is directed toward manipulating light conditions. Plants react to alterations in both the duration and quality of light by expediting their flowering time. The utilization of artificial electric lamps to accelerate plant growth and development has a well-established history. Photoperiod extension, especially in long-day species, has been extensively utilized to control the flowering timing. The advent of advanced light-emitting diode (LED) lighting systems has amplified these endeavors, enabling researchers to manipulate the composition of wavelengths. This manipulation induces specific light responses, such as shade avoidance, expediting the transition to flowering.

12.3.1 Significance of Speed Breeding in Rice

To keep pace with the increasing demand for rice, it is crucial to breed high-yielding and improved varieties more rapidly. The success of past yield enhancements can be largely attributed to the widespread adoption of high-yielding varieties by farmers. However, to meet future requirements, surpassing the average annual genetic gain of 1% in rice is essential. Developing new and improved higher-yielding varieties with reduced breeding cycle times becomes a modest yet effective way to increase genetic gain in crop (Collard et al. 2017).

Rice, being an annual crop, is highly sensitive to photoperiod and varying temperature levels during its growth and reproductive stages. The typical time period for rice from seed to seed is around 120–140 days, allowing for a maximum of two generations per year. However, for long-duration photoperiod-sensitive varieties, achieving desirable traits becomes more challenging and time-consuming. Under such circumstances, it may take nearly 10 years to develop a new variety, which is difficult to afford. Speed breeding provides an opportunity to expedite the development of homozygous lines by manipulating photoperiod, temperature, humidity, environmental CO_2, soil moisture, hormone application, and plant density (Wanga et al. 2021).

Integrating speed breeding protocols with modern genomic tools such as marker-assisted selection, genomic selection, and next-generation sequencing technologies would allow the fast-track development of new superior rice varieties.

12.4 Historical Perspective

The emergence of speed breeding as an advanced alternative to rapid generation advancement stems from the concept of inducing early flowering via manipulation of artificial environmental conditions. This technique originated from the concept of using artificial light to grow plants, with early studies dating back to Pfeiffer in 1926 and Siemens in 1880.

Continuous light exposure was found to induce early flowering in various plant species. Inspired by the National Aeronautics and Space Administration (NASA)'s efforts to grow crops in space, speed breeding aims to accelerate improvements in crops by creating controlled environments with extended photoperiods. Initially developed at the University of Queensland, Australia, speed breeding allows researchers to quickly mobilize genetic variation from wild relatives of crops and introduce them into elite varieties (Ghosh et al. 2018; Watson et al. 2018). The approach has been successfully applied to wheat and peanuts, demonstrating its efficiency in screening for disease resistance. Published in *Nature Protocols*, speed breeding protocols, involving artificial conditions like modified temperature and photoperiod, have been successful for various crops, including *Brassica* species, wheat, barley, and chickpea. The method accelerates breeding programs, achieving at least four generations in a single year (Ghosh et al. 2018; Watson et al. 2018; O'Connor et al. 2013).

Scientists in the United Kingdom, including the John Innes Centre and the University of Queensland, have adapted speed breeding for use in vast glasshouses and desktop growth chambers. In a 2016 Norwich experiment, maize and wheat were grown successfully without supplementary light or heat sources. This efficient technique has gained global attention, with significant investments from agricultural firms and research agencies in crops such as rice, pigeon pea, grass pea, and chickpea. Speed breeding offers a streamlined approach to crop improvement, contributing to faster and more resilient agricultural systems (Ghosh et al. 2018; Watson et al. 2019).

The first wheat variety which was developed through speed breeding was "DS Faraday" by Lee Hickey (Singh and Janeja 2021). It has high protein content, is tolerant to preharvest sprouting, and has milling quality. The speed breeding technique has mainly been used for purpose of research, but is now being widely adopted by the industries as well.

Table 12.1 Controlled speed breeding facilities for rice across India

Sr. no.	Center	Location
1	**Speed Breed Facility**, IRRI South Asia Regional Centre	ISARC, Varanasi
2	Accel Breed, PAU-Ludhiana	Punjab
3	ICAR-IARI-New Delhi[a]	New Delhi
4	**Speed Breeding Facility,** Department of Rice, CPBG, Tamil Nadu Agricultural University, Coimbatore	Coimbatore, Tamil Nadu

[a] Under developmental stage

In India, the first speed breeding facility was setup at Varanasi. The team led by Dr. Vikas K Singh has developed protocols to procure four to five rice generations per year in early-, medium-, and late-duration varieties (Anonymous n.d.). However, refinements and validation of protocols are yet to be made. The most recent facility commissioned by Saveer Biotech Limited was established at Tamil Nadu Agricultural University as a part of the ongoing National Agriculture Development Programme (Table 12.1).

12.5 Principles of Speed Breeding

Speed breeding accelerates the development of new crop varieties to address global food demand, climate change, and the need for enhanced traits. Key principles include optimized light conditions with LED technology, controlled temperature for rapid growth, hydroponics or aeroponics for direct nutrient supply, and manipulation of photoperiods for faster seed production. Genetic tools like marker-assisted selection expedite trait identification. Ensuring an optimal growth medium, high plant density for competitive growth, continuous monitoring, and strict trait selection criteria contribute to this transformative process. Speed breeding facilitates multiple plant generations in a short timeframe, potentially revolutionizing plant breeding for resilient and high-yielding crops.

12.5.1 Accelerated Plant Growth

12.5.1.1 Optimized Growing Conditions

- Provide an ideal balance of nutrients, water, and light to support rapid plant growth. This may involve the use of advanced fertilizers, hydroponics, or aeroponics systems to deliver nutrients directly to plants.

12.5.1.2 High-Intensity Lighting

- Implement high-intensity artificial lighting, such as LEDs, to extend the photoperiod and simulate optimal growing conditions. This helps plants photosynthesize more efficiently, promoting faster growth.

12.5.1.3 Genetic Manipulation

- Explore genetic techniques to enhance growth rates. This may include the identification and modification of genes associated with rapid growth without compromising other desirable traits.

12.5.2 Controlled Environment Conditions

12.5.2.1 Temperature Control

- Maintain consistent and optimal temperature conditions throughout the growth cycle. This involves careful regulation of day and night temperatures to create an environment that encourages accelerated plant development.

12.5.2.2 Humidity Management

- Control humidity levels to prevent the risk of diseases and create an environment that supports plant health. Maintaining the right balance of humidity is crucial for optimal growth.

12.5.2.3 Carbon Dioxide Enrichment

- Increase carbon dioxide levels in the growing environment to enhance photosynthesis. Elevated carbon dioxide concentrations can contribute to faster plant growth under controlled conditions.

12.5.3 Efficient Selection and Generation Advancement

12.5.3.1 Genetic Markers and Molecular Tools

- Employ genetic markers and molecular tools to identify and select plants with desirable traits more efficiently. This allows for targeted breeding and reduces the time spent on traditional trial-and-error approaches.

12.5.3.2 High-Throughput Phenotyping

- Use advanced technology for high-throughput phenotyping to rapidly assess and quantify plant traits. Automated systems can quickly analyze large numbers of plants, streamlining the selection process.

12.5.3.3 Accelerated Seed Production

- Manipulate photoperiods to shorten the time plants spend in the flowering phase, facilitating faster seed production. This enables quicker advancement to the next generation in the breeding cycle.

12.5.3.4 Optimized Breeding Schemes

- Develop and implement breeding strategies that maximize the efficiency of selection and generation advancement. This may involve carefully planned crosses and the use of advanced statistical methods to accelerate progress.

These principles collectively contribute to the overarching goal of speed breeding, which is to expedite the development of new plant varieties with improved traits to address evolving agricultural challenges.

12.6 Key Components of Speed Breeding in Rice

12.6.1 Controlled Environment Facilities

12.6.1.1 Growth Chambers

12.6.1.1.1 Temperature Control

The interplay of temperature and photoperiod influences flowering. In speed breeding, a tailored temperature regime is vital. Efficient temperature manipulation, like adjusting it during the photoperiod, accelerates generation time, enhancing breeding efficiency. Temperature can be controlled by using hardware items such as cooling fans, temperature sensors, thermoelectric coolers, and aluminum heat sinks (Ghosh et al. 2018). These need to be regulated and set up according to the requirements of each crop.

12.6.1.1.2 Photoperiod Manipulation

Speed breeding generally adopts a 22-h continuous light followed by a 2-h dark period in a 24-h diurnal cycle. This needs to be standardized for each crop separately. This method maximizes light exposure for accelerated growth. Continuous light optimizes photosynthesis, while brief dark periods benefit overall plant health. Balancing continuous light with intermittent darkness is crucial in speed breeding, aiming for rapid growth and robust plant health.

Lighting is usually provided with SVLs or LEDs, but LED lamps are preferred because of the significant cost savings as well as maintenance and energy consumption savings (Ghosh et al. 2018). The use of LED modules and combination of different LED lights of varying intensities can be useful in different conditions. LED light spectrum with controllable channels may be of potential efficiency. Ecotune, Daypro, Beaglebone, Blue Panel, Blue Rail, and Relumity are examples of different LED modules used (Jähne et al. 2020).

12.6.1.1.3 Humidity Control

Relative humidity directly impacts plant water relations, affecting leaf development, photosynthesis, and pollination. Stomatal movements respond to vapor pressure deficit and air humidity. Managing humidity in controlled environments is challenging, with 60–70% considered optimal. Lower humidity may suit crops adapted to drier conditions. Balancing humidity is crucial for plant health and growth in controlled environments (Ghosh et al. 2018).

Humidity sensor and airflow can be used to control the humidity; i.e., the humidity sensor will trigger a fan which would regulate the humidity inside the growth chambers.

12.6.1.1.4 Carbon Dioxide Control and Enrichment

Speed breeding facilities are equipped with sophisticated environmental control systems that allow for the regulation of various parameters, including CO_2 levels. These systems can automatically adjust CO_2 concentrations based on predefined set points.

Proper ventilation is crucial for controlling CO_2 levels. Facilities may have ventilation systems in place to ensure the exchange of air and prevent the buildup of excess carbon dioxide. This is especially important in closed or semi-closed systems. CO_2 control is often integrated into the overall management of plant growth chambers. These chambers create a controlled environment for plants, and CO_2 levels are adjusted to optimize photosynthesis and plant growth.

The precise methods used can vary depending on the specific design and technology employed by the speed breeding facility.

12.6.1.1.5 Soilless Systems

In the context of speed breeding, hydroponics and aeroponics are often preferred over traditional soil-based cultivation methods. These soilless systems offer several advantages that contribute to the acceleration of plant growth and development in terms of nutrient precision, water and nutrient efficiency, space efficiency, and reduced disease risk.

When integrated with speed breeding, the controlled environments coupled with hydroponics or aeroponics allow for faster seed germination, vegetative growth, and flowering, reducing the overall time required for a generation turnover.

12.6.2 High-Quality Germplasm

12.6.2.1 Germplasm Collection

12.6.2.1.1 Diverse Genetic Pool

The germplasm collection should encompass a wide spectrum of genetic diversity within the rice species. This diversity serves as the foundation for the breeding program, offering a broad range of potential traits and adaptations. A diverse genetic pool provides a reservoir of genes that can confer traits such as resistance to diseases, tolerance to environmental stress, and improved nutritional content.

12.6.2.1.2 Trait Diversity

Beyond genetic diversity, the germplasm collection must include variations in specific traits relevant to the breeding objectives. Traits could include resistance to pests and diseases, high yield, nutritional content, and adaptation to specific environmental conditions. Having a diverse set of traits allows breeders to select and combine characteristics to create rice varieties with multiple desirable attributes.

Integration with Speed Breeding

- **Rapid trait introgression**: The diverse germplasm pool allows for the quick identification of plants with specific traits of interest, which can then be introduced into breeding programs through techniques like crossbreeding.
- **Accelerated variety development**: The combination of diverse germplasm and selective breeding methods contributes to the rapid development of new rice varieties with improved traits. This is essential for addressing challenges such as changing environmental conditions and evolving pest and disease pressures.
- **Customized varieties**: The ability to selectively breed and cross diverse germplasm allows researchers to tailor rice varieties to meet specific needs, such as adaptation to different climates, resistance to emerging pests, or improved nutritional content.

12.6.3 Efficient Data Management

12.6.3.1 Data Collection and Analysis

12.6.3.1.1 High-Throughput Phenotyping

High-throughput phenotyping involves the rapid and automated collection of data on plant traits, growth rates, and responses to environmental conditions. This method allows for the efficient monitoring of a large number of plants, providing detailed information on their development and performance in a short time.

12.6.3.1.2 Remote Sensing

Using technologies such as drones to capture data from the field or growth chambers without physically disturbing the plants. Remote sensing enables nonintrusive, real-time monitoring of plant health, growth, and environmental conditions, contributing to timely decision-making in the breeding process.

12.6.3.1.3 Machine Learning

Using algorithms and statistical models to analyze patterns and correlations within genomic and phenotypic datasets. Machine learning enhances the efficiency of data analysis by automating the identification of complex relationships. It aids in predicting traits, optimizing breeding strategies, and making data-driven decisions.

12.6.3.1.4 Applications of Speed Breeding in Rice till Date

Various generation advancement methods, including the biotron breeding system, simplified biotron breeding system, speed-breeding–modified controlled-biotron breeding, and speed breeding, have been developed to accelerate crop breeding (Table 12.2).

Table 12.2 Applications of speed breeding in rice till date

Generation advancement method	Reduction in crop duration	Features	References
Biotron breeding system	Crop duration reduced to 2 months	Regulation of CO_2 levels, tiller removal, and embryo rescue	Ohnishi et al. (2011)
Biotron breeding system	Crop duration reduced to 3 months	Combination of biotron breeding system with early flowering habit and cytoplasmic male sterility (CMS) system	Ohnishi et al. (2013)
Simplified-biotron breeding system	Crop duration reduced to less than 3 months	Eliminates tiller removal and embryo rescue, controlled CO_2, and day length and root volume	Tanaka et al. (2016)
Speed-breeding–modified controlled-biotron breeding	70 days from germination to flowering	Integration of speed breeding with marker-assisted gene introgression for salinity tolerance	Rana et al. (2019)
Speed breeding	7 days reduction in generation time	Development of speed breeding protocol using LEDs to modify light quality mainly for short-day crops like rice	Jähne et al. (2020)

12.7 Methods and Techniques

Before starting an SB experiment, replicate germplasm growth under standard conditions for your program. Compare plant parameters, operational costs, and quality. Source seeds appropriately, choose soil based on the crop, and consider nutrient needs, especially for small pots. Swiftly address nutrient deficiencies due to rapid growth. Use slow-release fertilizer for maturity, maintain pH, and be aware of calcium deficiency, possibly requiring foliar fertilizer.

In a speed breeding facility dedicated to rice cultivation, the planting procedure is finely tuned to exploit optimal conditions for swift and efficient growth. The process commences with the careful selection of high-quality rice seeds, ensuring they are free from pests and diseases. These seeds may undergo pre-germination to promote uniformity. Subsequently, a suitable growth medium is chosen, with options ranging from traditional soil to soilless mixes or hydroponic systems. The seeds are then sown at the recommended depth and spacing in trays or containers designed for efficient space utilization and optimal root development.

Once planted, the trays are meticulously arranged in growth chambers equipped with controlled environmental conditions specifically tailored to mimic the ideal rice-growing environment. This includes precise control over temperature, humidity, and lighting. Artificial lighting sources, such as LEDs, are often employed to extend the photoperiod and stimulate continuous growth. Day length and duration must also be standardized. Regular monitoring throughout the growth cycle allows for prompt intervention in the face of nutrient deficiencies, pests, diseases, or any other stress factors that may arise.

Selective pruning may be employed strategically to direct the plant's energy toward desired traits, and the maintenance of adequate air circulation within the growth chambers is essential. Harvesting is performed at the optimal stage, with a keen focus on collecting seeds for subsequent planting cycles. The utilization of a sequential planting schedule ensures a continuous supply of rice plants at different growth stages, maximizing the efficiency of the speed breeding process.

Detailed record-keeping, including planting dates, environmental conditions, and observations made during the growth cycle, is paramount for assessing the success of the speed breeding program and informing future iterations. This comprehensive approach enables researchers and breeders to achieve multiple generations of rice in a condensed timeframe, facilitating accelerated progress in rice crop improvement efforts.

The biotron breeding system developed for rice follows a temperature of 30 °C/25 °C and (11 h day/13 h night) system. It also involves the techniques of embryo rescue and tiller removal at appropriate stages for generation advancement (Ohnishi et al. 2011).

12.8 Applications of Speed Breeding

Speed breeding (SB) techniques have not only revolutionized crop improvement but have also made a significant impact in various areas of plant research and development, bringing forth innovative applications in agriculture and biotechnology.

12.8.1 Molecular Breeding

SB has transformed the creation of mapping populations by rapidly cycling through generations, a process traditionally time-consuming. This enables researchers to efficiently produce large and genetically diverse mapping populations. The quick generation turnover aids in identifying genes associated with key traits through quantitative trait locus (QTL) analysis and marker-assisted selection (MAS). For example, SB has been utilized to enhance salinity stress tolerance in rice by developing mapping populations with improved traits.

Although genomic selection (GS) can expedite the breeding cycle by progressing through a specific number of generations without requiring phenotyping, the enhancement of improved varieties could occur at an even more accelerated rate by employing "SpeedGS." This innovative approach combines genomic selection with speed breeding, as detailed in studies by Voss-Fels et al. (2019) and Hickey et al. (2019).

12.8.2 Genetic Modification (GM) for Crop Development

In the realm of GM crops, SB is a valuable tool for biotechnologists. SB's accelerated generation times and controlled environments enable researchers to significantly reduce the time required for creating and testing GM crops, addressing challenges such as pest resistance and disease resilience. The integration of gene editing technologies, such as CRISPR/Cas9, with SB facilitates the development of crops with enhanced yield-related traits and improved nutritional quality.

12.8.3 Evaluation of Landraces

Speed breeding can be efficiently put to use to evaluate various landraces available in rice. There are innumerable varieties of rice landraces across the country and the world. Controlled conditions and standard photoperiod setup, along with the integration of other tools, can be used to screen the germplasm for various individual traits. This can help in easier and faster identification of potential donors, which can be further utilized in breeding programs.

12.8.4 Trait Stacking for Resilient Crop Varieties

SB plays a pivotal role in streamlining the process of trait stacking, where multiple desirable traits are introduced into a single crop variety. The ability to rapidly generate and assess multiple generations allows breeders to efficiently create and evaluate multi-trait combinations. This contributes to the development of adaptable and resilient crop varieties capable of withstanding diverse challenges, including changing climatic conditions and evolving pest pressures.

12.8.5 Enhancing Research Across Disciplines

SB's unique capabilities extend beyond crop improvement and genetic research, finding applications in diverse fields such as ecology, physiology, and agronomy. Researchers can use SB techniques to accelerate experiments, gather data more efficiently, and make informed decisions in areas like plant responses to environmental stressors, cultivation practice optimization, and plant physiology exploration.

12.8.6 Integration with Systems Biology

The integration of SB with systems biology provides a holistic understanding of plant growth processes. Computational modeling allows precise control over growth conditions, facilitating the fine-tuning of environmental factors like light intensity to accelerate plant growth while maintaining optimal conditions. Predictive models enhance the breeding process by estimating the effects of genetic modifications on crop traits, streamlining the design of plants with desired characteristics.

12.8.7 Advances in High-Throughput Phenotyping

SB supports advancements in high-throughput phenotyping, allowing for extensive screening of diverse plant populations. This rapid generation turnover enables the identification of superior genotypes with precision, enhancing the efficiency of breeding programs and accelerating the release of improved crop varieties.

12.8.8 Varietal Protection

Speed breeding can also be utilized for varietal protection as a method of quicker evaluation of varieties for their essential traits, to reduce the overall timeline of variety registration. Jamali et al. (2020) proposed "speed DUS testing," a speed breeding model for DUS testing, integrating speed breeding with DNA markers to extend its utility from varietal development to varietal protection (Jamali et al. 2020).

As we look to the future, the collaborative approach between SB and cutting-edge methodologies holds promise for continued innovation in plant breeding and crop enhancement. Figure 12.1 shows the general overview of speed breeding approach, uses, and generation advancement process in rice.

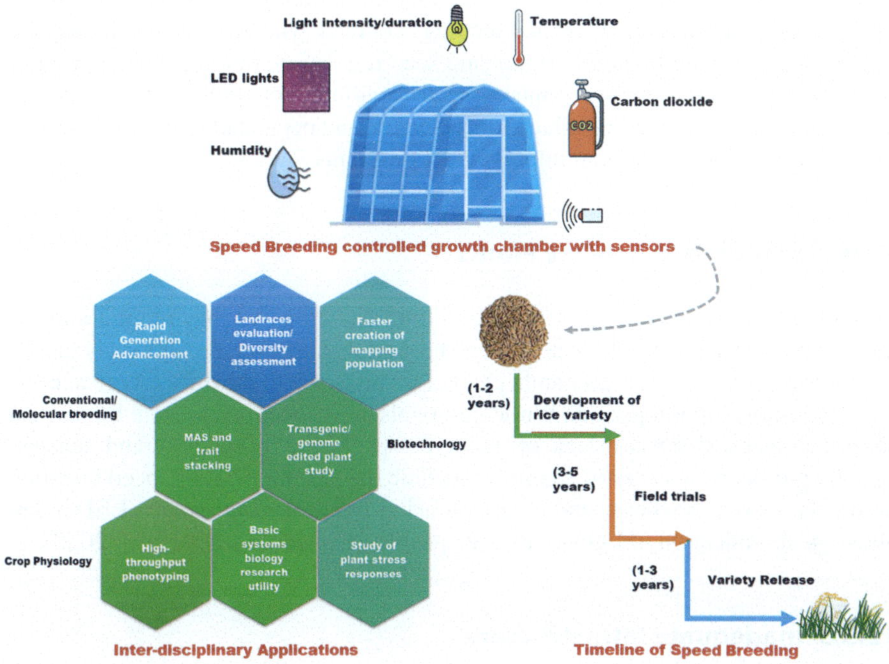

Fig. 12.1 Overview of speed breeding

12.9 Challenges of Speed Breeding

The utilization of speed breeding techniques is a valuable strategy to expedite conventional breeding programs. However, the successful implementation of this technology demands expertise, efficient plant phenomics facilities, suitable infrastructure, and sustained financial support for research and development (Shimelis et al. 2019). Establishing these resources necessitates the recognition of speed breeding approaches as indispensable components of conventional plant breeding, marker-assisted selection, and genetic engineering.

Moreover, the integrated suite of tools requires proficiency in plant breeding and biotechnology, sustained funding, and governmental policy support. While speed breeding methods have the potential to accelerate the release of both conventional and genetically modified crop cultivars, several challenges impede their widespread use.

12.9.1 Lack of Trained Plant Breeders and Breeding Technicians

A significant obstacle to the adoption of speed breeding in the public sector is the shortage of trained and active plant breeders and technicians in developing countries (Shimelis et al. 2019; Morris et al. 2006). Public sector breeding programs are adversely affected by a high turnover of personnel to private seed companies and training institutes offering better remuneration. Additionally, the limited availability of scientists specializing in plant breeding, coupled with insufficient legislative frameworks for plant breeders' rights and seed regulation, hinders the adoption of speed breeding. Developing countries need to adjust their policies to encourage investments in plant breeding education, research, and personnel retention to ensure the viability of long-term crop improvement programs.

12.9.2 Breeding Short-Day Plants

Speed breeding is more straightforward in long-day crops, as they require a maximum photoperiod greater than the critical day length. However, for short-day plants, such as rice, soybean, and amaranth, challenges arise due to the difficulty in achieving the required photoperiod. While protocols have been optimized for certain short-day crops, there is a lack of genotype-specific, crop-specific, and species-specific protocols. Successful examples include the development of speed breeding protocols for rice, soybean, and amaranth using light-emitting diodes (LEDs) that allow the modification of light quality at specific crop stages (Jähne et al. 2020).

12.9.3 Inadequate Infrastructure

Limited infrastructure in developing countries, like India, poses a significant bottleneck for supporting modern plant breeding tools and variety development (Ribaut et al. 2010). The shift toward private companies in plant breeding is attributed to the inadequacy of resources in public plant breeding programs. To overcome this challenge, there is a need for substantial infrastructure development and the training of skilled labor to facilitate fast-track research.

12.9.4 Unreliable Water and Electricity Supplies for Sustainable Operations

The success of speed breeding relies on manipulating environmental factors in indoor growing facilities, requiring reliable water and electricity supplies for cooling, heating, and lighting. The cost of temperature regulation in some regions has been a significant factor, emphasizing the need for sustainable energy sources. Developing countries face challenges related to unreliable supplies of electricity, necessitating innovative solutions such as the use of sustainable solar power. However, achieving water and electricity supplies that are affordable and reliable remains a considerable challenge.

12.9.5 Other Challenges

Continuous photoperiod in speed breeding environments induces early flowering and completes the life cycle before the normal period, but it also presents challenges such as increased risk of pest and disease attacks, potential photo damage to plants, and alterations in biochemical compounds. Addressing these challenges is crucial for the continued success and global adoption of speed breeding technology.

Despite these challenges, speed breeding technology is gaining worldwide recognition as a crucial tool to address issues of food security and climate change.

12.10 Conclusion

The escalating global population, projected to reach nine billion by 2050, intensifies the challenge of increasing food production by 60–80%. Speed breeding, alongside modern breeding technologies, emerges as a pivotal strategy to accelerate crop development. In summary, fast breeding in rice relies on controlled environments with temperature, photoperiod, and humidity control, complemented by LED lighting and soilless systems. This approach expedites breeding processes, enabling swift development of new rice varieties. A robust germplasm collection with genetic diversity and desired traits, alongside selective breeding methods, forms the foundation for addressing agricultural challenges efficiently. Incorporating molecular techniques and genetic markers enhances precision, accelerating the creation of rice varieties with specific traits. Effective data management is crucial for success, facilitating informed decisions and optimizing breeding techniques for the rapid production of superior rice varieties.

Despite the benefits, addressing challenges such as the demand for advanced facilities, a skilled workforce, genotypic variations, and ethical considerations is crucial. Nevertheless, the synergy of speed breeding with other modern breeding tools holds promise for significantly contributing to global food security and sustainability goals, potentially ushering in a second green revolution. The integration of speed breeding with advanced techniques, including plant phenotyping, marker-assisted selection, marker-assisted backcrossing, genomic selection, genome editing, and express editing, offers a pathway to develop climate-smart crop varieties. To ensure success, optimizing speed breeding protocols for diverse crops and providing personnel training are essential. This integrated approach, leveraging the time-constrained nature of speed breeding, proves effective in rapidly advancing generations and enhancing the efficiency of breeding operations, particularly in the context of rice cultivation.

With the anticipated surge in rice demand, the detrimental impacts of climate change, and potential upcoming biotic threats such as new races, pathotypes, or biotypes, rice breeders must contemplate the integration of accelerated breeding methods to contemporize their programs. This strategic approach aims to boost the pace of genetic improvement while concurrently minimizing operational expenses (Atlin et al. 2017).

References

Allard RW (1960) Principles of plant breeding. Wiley, New York

Allard RW (1999) Principles of plant breeding. Wiley, Hoboken

Anonymous. https://www.irri.org/news-and-events/news/speedbreed-crop-breeding-center-built-speed

Atlin GN, Cairns JE, Biswanath D (2017) Rapid breeding and varietal replacement are critical to adaptation of cropping systems in the developing world to climate change. Glob Food Sec 12:31–37. https://doi.org/10.1016/j.gfs.2017.01.008

Bin Rahman AR, Zhang J (2023) Trends in rice research: 2030 and beyond. Food Energy Secur 12(2):e390

Brar DS, Khush, GS (2006) Cytogenetic manipulation and germplasm enhancement of rice (*Oryza sativa L.*). In: Genetic resources, chromosome engineering and crop improvement, vol 2, pp 115–158

Breseghello F, Coelho ASG (2013) Traditional and modern plant breeding methods with examples in rice (*Oryza sativa L.*). J Agric Food Chem 61(35):8277–8286. https://doi.org/10.1021/jf305531j

Brown J, Caligari P (2008) An introduction to plant breeding. Blackwell Publishing, Oxford, p 209

Collard BC, Beredo JC, Lenaerts B, Mendoza R, Santelices R, Lopena V, Verdeprado H, Raghavan C, Gregorio GB, Vial L, Demont M (2017) Revisiting rice breeding methods–evaluating the use of rapid generation advance (RGA) for routine rice breeding. Plant Prod Sci 20(4):337–352

Donald CT (1968) The breeding of crop ideotypes. Euphytica 17:385–403

Dwivedi SL, Britt AB, Tripathi L, Sharma S, Upadhyaya HD, Ortiz R (2015) Haploids: constraints and opportunities in plant breeding. Biotechnol Adv 33(6, Part 1):812–829. https://doi.org/10.1016/j.biotechadv.2015.07.001

Ghosh S, Watson A, Gonzalez-Navarro OE, Ramirez-Gonzalez RH, Yanes L, Mendoza-Suárez M, Simmonds J, Wells R, Rayner T, Green P, Hafeez A (2018) Speed breeding in growth chambers and glasshouses for crop breeding and model plant research. Nat Protoc 13(12):2944–2963. https://doi.org/10.1038/s41596-018-0072-z

Hickey LT, Hafeez AN, Robinson H, Jackson SA, Leal-Bertioli SCM, Tester M, Gao C, Godwin ID, Hayes BJ, Wulff BBH (2019) Breeding crops to feed 10 billion. Nat Biotechnol 37(7):744–754. https://doi.org/10.1038/s41587-019-0152-9

Jähne F, Hahn V, Würschum T, Leiser WL (2020) Speed breeding short-day crops by LED-controlled light schemes. Theor Appl Genet 133(8):2335–2342

Jamali SH, Cockram J, Hickey LT (2020) Is plant variety registration keeping pace with speed breeding techniques? Euphytica 216(8):131

Khan MH, Dar ZA, Dar SA (2015) Breeding strategies for improving rice yield: review. Agric Sci 6(5):467

Li J, Xin Y, Yuan L (2007) Hybrid rice technology development ensuring china's food security. International Food Policy Research Institute, Washington, DC, pp 1–40

Lin Z, Cogan NO, Pembleton LW, Spangenberg GC, Forster JW, Hayes BJ, Daetwyler HD (2016) Genetic gain and inbreeding from genomic selection in a simulated commercial breeding program for perennial ryegrass. Plant Genome 9(1):plantgenome2015-06

Mackill DJ, Khush GS (2018) IR64: a high-quality and high-yielding mega variety. Rice 11:1–11

Molina J, Sikora M, Garud N, Flowers JM, Rubinstein S, Reynolds A, Huang P, Jackson S, Schaal BA, Bustamante CD, Boyko AR, Purugganan MD (2011) Molecular evidence for a single evolutionary origin of domesticated rice. Proc Natl Acad Sci U S A 108(20):8351–8356. https://doi.org/10.1073/pnas.1104686108

Morris M, Edmeades G, Pehu E (2006) The global need for plant breeding capacity: what roles for the public and private sectors? HortScience 41(1):30

Muralidharan K, Prasad GS, Rao CS, Sridhar R, Siddiq EA (2022) Grain yield performance of hybrid rice in relation to inbred cultivars in long-term multi-environment tests in India. Crop Sci 62(3):1133–1148

O'Connor DJ, Wright GC, Dieters MJ, George DL, Hunter MN, Tatnell JR et al (2013) Development and application of speed breeding technologies in a commercial peanut breeding program. Peanut Sci 40(2):107–114. https://doi.org/10.3146/PS12-12.1

Ochatt SJ, Sangwan RS, Marget P, Issoufou Ndong Y, Rancillac M, Perney P (2002) New approaches towards the shortening of generation cycles for faster breeding of protein legumes. Plant Breed 121(5):436–440. https://doi.org/10.1046/j.1439-0523.2002.746803.x

Ohnishi T, Yoshino M, Yamakawa H, Kinoshita T (2011) The biotron breeding system: a rapid and reliable procedure for genetic studies and breeding in rice. Plant Cell Physiol 52(7):1249–1257

Ohnishi T, Yoshino M, Toriyama K, Kinoshita T (2013) Rapid establishment of introgression lines using cytoplasmic male sterility and a restorer gene in *Oryza sativa* cv. *Nipponbare*. Mol Breed 32:831–839

Peng S, Khush GS, Virk P, Tang Q, Zou Y (2008) Progress in ideotype breeding to increase rice yield potential. Field Crop Res 108(1):32–38

Pfeiffer NE (1926) Microchemical and morphological studies of effect of light on plants. Bot Gaz 81:173–195

Poehlman JM, Sleper DA (1995) Methods in plant breeding. In: Breeding field crops, pp 172–174

Pushpa R, Saraswathi R, Sassikumar D, Chandirakala R, Suresh R, Meenakumari B, Chozan K, Suresh S, Karthikeyan A, Ravi V (2020) ADT 53-short duration high yielding fine grain rice variety for Kuruvai/Kodai seasons of Tamil Nadu. Electron J Plant Breed 11(1):103–109

Rajeswari S, Robin S, Pushpam R, Manonmani S, Malarvizhi D, Suresh S, Loganadhan S, Ravichandran V (2015) CO (R) 50 (IET 19321) a new plant type high yielding medium duration rice variety for irrigated ecosystem. Madras Agric J 102:1

Rana MM, Takamatsu T, Baslam M, Kaneko K, Itoh K, Harada N et al (2019) Salt tolerance improvement in rice through efficient SNP marker-assisted selection coupled with speed-breeding. Int J Mol Sci 20(10):2585

Rao LS, Babu VR. Genetic resources and biodiversity in rice. In: Biodiversity, ecosystem services and genetic resources, pp 42

Ribaut JM, De Vicente MC, Delannay X (2010) Molecular breeding in developing countries: challenges and perspectives. Curr Opin Plant Biol 13(2):213–218

Robin S, Pushpam R, Rajeswari S, Amudha K, Jeyaprakash P, Manonmani S, Raveendran M, Rabindran R, Ramanathan A, Suresh S, Soundararajan RP (2019) CO 52 (IET 25487): a highly remunerative medium duration fine grain rice variety. Electron J Plant Breed 10(3):1148–1160

Rout D, Jena D, Singh V, Kumar M, Arsode P, Singh P et al (2020) Hybrid rice research: current status and prospects, vol 2020. IntechOpen, London

Shimelis H, Gwata ET, Laing MD (2019) Crop improvement for agricultural transformation in southern Africa. In: Sikora RA, Terry ER, Vlek PLG, Chitja J (eds) Transforming agriculture in southern Africa, 1st ed. Routledge, pp 97–103. https://doi.org/10.4324/9780429401701

Siemens CW III (1880) On the influence of electric light upon vegetation, and on certain physical principles involved. Proc R Soc Lond B 30:210–219

Singh H, Janeja HS (2021) Speed breeding a ray of hope for the future generation in terms of food security: a review. Plant Arch 21(1):155–158

Tanaka J, Hayashi T, Iwata H (2016) A practical, rapid generation-advancement system for rice breeding using simplified biotron breeding system. Breed Sci 66(4):542–551. https://doi.org/10.1270/jsbbs.15038

Tigchelaat EC, Casali VWD (1976) Single seed descent: applications and merits in breeding self-pollinated crops. In: International symposium on floriculture plant breeding and genetics. Acta Hort 63:85–90

Voss-Fels KP, Herzog E, Dreisigacker S, Sukumaran S, Watson A, Frisch M et al (2019) "SpeedGS" to accelerate genetic gain in spring wheat. In: Applications of genetic and genomic research in cereals. Woodhead Publishing, Sawston, pp 303–327

Wanga MA, Shimelis H, Mashilo J, Laing MD (2021) Opportunities and challenges of speed breeding: a review. Plant Breed 140(2):185–194

Watson A, Ghosh S, Williams MJ et al (2018) Speed breeding is a powerful tool to accelerate crop research and breeding. Nat Plants 4:23–29. https://doi.org/10.1038/s41477-017-0083-8. https://www.biorxiv.org/content/biorxiv/early/2017/07/09/161182

Watson A, Hickey LT, Christopher J, Rutkoski J, Poland J, Hayes BJ (2019) Multivariate genomic selection and potential of rapid indirect selection with speed breeding in spring wheat. Crop Sci 59(5):1945–1959

Application of Next-Generation Sequencing Technology for Rice Improvement

13

Vemula Anjula Krishna, Ritik Digamber Bisane, Amrit Prasad Poudel, Akansha Singh, and Shravan Kumar Singh

Abstract

Rice is one of the most significant food crops worldwide. Population growth and climate change posed great challenges for further rice production. In the past decade, we have observed enormous developments in next-generation sequencing technologies (NGS) and genomics, revolutionizing almost every aspect of crop improvement. The launching of new platforms employing long-read sequencing safeguarded further efficient sequencing of huge and complex genomes. Currently, rice genetic resources and germplasms are first characterized by sequencing platforms, pursued by bioinformatics tools and genomic methodology. These are applied to acquire masses of genomic data from these various resources. These data are further explored for insights into the molecular mechanisms of quantitative and qualitative traits of rice. The deep integration of genomic technologies and NGS into rice study and breeding practices has changed the scheme and framework of rice breeding. Additionally, these genomic data offer molecular knowledge and wide range of applications like rice marker discovery by NGS, wide crosses and alien introgression, transcriptome investigation, gene expression profiling, small non-coding RNA profiling, gene annotation using transcriptome sequence data, phylogenetic and ecological studies, allele mining, epigenetics studies, metagenomics analysis, single-cell genomics, exome sequencing, and whole genome resequencing (WGRS). NGS permits the straight identification of candidate genes through strategies such as MutMap, Genome-Wide Association Studies (GWAS), and QTL-sequencing. Understanding the molecular mechanisms behind rice character expression will

V. A. Krishna · R. D. Bisane · A. P. Poudel · A. Singh · S. K. Singh (✉)
Genetics and Plant Breeding, Institute of Agricultural Sciences, Banaras Hindu University, Varanasi, Uttar Pradesh, India

A. Singh et al. (eds.), *Climate-Smart Rice Breeding*,
https://doi.org/10.1007/978-981-97-7098-4_13

boost rice breeding for developing climate resilient crops and enhancing yield compared to old conventional breeding methods.

Keywords

Rice · Genomics · Sequencing · Breeding · Transcriptome

13.1 Introduction

Over half of all people on Earth receive their staple food from rice (*Oryza sativa*), a significant cereal and model monocot. Rice output has doubled, particularly between the 1960s and 2000s. This increase is mainly due to genetic enhancements to the current rice varieties, such as the introduction of hybrids and semi-dwarf kinds (Mohapatra et al. 2021). In addition to climate change, the earth's human population is anticipated to surge to nearly ten billion by 2060–2070, a vital prerequisite is to boost rice production by half in the approaching decade (Gu et al. 2022). The worldwide rice community continually pursues innovative approaches to cultivate new varieties that possess nutritional value, sustainability, and resilience to climate change (Singh et al. 2023), in addition to increased yield.

The demand for high-quality rice has surged due to recent improvements in global living standards, putting pressure on rice-producing regions across the globe (Jin et al. 2023). The fundamental need to address this pressing need for food demand is exploiting the inherent genetic diversity in the rice genomes of over 500K accessions and their wild relatives stored in germplasm banks (Wing et al. 2018). Next-generation sequencing (NGS) allows for examining nucleotide variants and detecting genetic markers on a large scale. This can help to investigate the genetic foundation of ecologically significant trait variations through quantitative trait loci (QTL) (Ziv et al. 2017).

13.2 Rice Genomics

Rice is ideal for knowing plant genomics because of its compact genome size of approximately 400 Mb, efficacious and proven transformation methods, abundant germplasm, and its similarity in gene sequences with other cereal crops. International Rice Genome Sequencing Project (IRGSP) was founded in 1998 to mine the whole genome (WG) rice sequence correctly and comprehensively. This was achieved by a clone-by-clone shotgun technique, map-based cloning, resulting in the generation of a reference sequence (Matsumoto et al. 2008). Genomics can be coarsely divided into three fragments: structural, quantitative, and functional. In rice, structural genomics offers the genome sequences of numerous accessions that are advantageous for quantitative genomics (measuring the genetic impacts of causative QTLs) and functional genomics (identifying and characterizing genes) (Wang and Han 2022).

13.3 Next-Generation Sequencing Technologies

Approximately 25 years after the structure of DNA discovery in 1953, the initial technique for DNA sequencing was made available to the public (Hu et al. 2021). The development of first-generation DNA sequencing technologies involved the simultaneous use of template DNA, amplification, and gel electrophoresis. However, the time for completion differs and consists of two unique methods: the Sanger method, often referred to as chain termination, and the Maxam–Gilbert method, which employs chemical degradation, are distinct approaches utilized in DNA sequencing (Eren et al. 2022) (Fig. 13.1). In the past, the process of sequencing was difficult and necessitated the use of radioactive reagents. Additional limitations include restricting output data to a single reaction and the labor-intensive process of sequencing diploid/haploid DNA through subcloning. NGS forums can generate genome sequences by utilizing libraries created through the fragmentation and attachment of adapters to RNA/DNA/amplicon. Sequencing strategies adhere to a standard workflow, regardless of the specific sequencing research platform. This workflow involves three main steps:

1. Constructing a library using nucleic acid
2. Operating a sequencing machine to generate sequenced data
3. Interpreting the data using bioinformatics or software

Short-read and long-read sequencing refer to second-generation sequencing technology (SGST) and third-generation sequencing technology (TGST) (Ashraf et al. 2022).

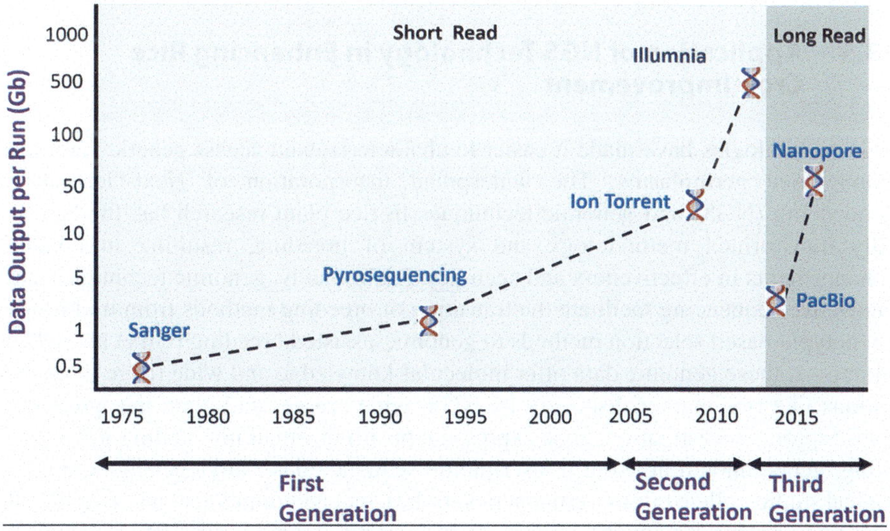

Fig. 13.1 Evolution of sequencing platforms

NGS technologies can be categorized into three groups: (1) sequence by synthesis, (2) sequence by ligation, and (3) single-molecule sequencing. (1) The primary platforms commonly employed for sequence by synthesis are Ion Torrent, PGM Roche-454 GSFLX, and Illumina HiSEq. 2000. (2) The leading platforms frequently employed for sequence by ligation (SBL) include the Life/AB Solid 5500 Series and the Polonator G007. (3) The two primary platforms for single-molecule sequencing that were more commonly utilized are the Helicos BioSciences HeliScope and the Pacific BioSciences PacBio HRS (Satam et al. 2023).

13.4 Bioinformatics Tools for Analyzing NGS Data

The emergence of NGS technologies, with their high throughput, quick run times, and affordable costs, has made it easier to provide sequence data for various applications. However, there are still many difficulties in analyzing such data, which is why several solutions for managing and analyzing data in a particular way have been developed (Gupta and Verma 2019). NGS methods typically yield substantial amounts of sequencing data in a high-throughput manner for each run. The standard procedure for NGS data analysis typically includes the following (Ahmed et al. 2019) steps:

1. Evaluating the quality of the data
2. Aligning the data
3. Identifying variants
4. Annotating the variants
5. Visualizing the data

13.5 Application of NGS Technology in Enhancing Rice Crop Improvement

NGS technologies have made it easier to characterize and access genetic variation among rice germplasms. The widespread incorporation of Next-Generation Sequencing (NGS) and genomic techniques in rice plant research has fundamentally transformed methodology and system of breeding, resulting in notable enhancements in effectiveness and accuracy. Additionally, genomic techniques and innovative sequencing facilitate the transition of breeding methods from traditional phenotypic based selection methods to genomic-assisted breeding (Gu et al. 2022). Moreover, these genomic data offer molecular knowledge and wide range of applications like rice marker discovery by NGS, wide crosses and alien introgression, transcriptome investigation, gene expression profiling, small non-coding RNA profiling, gene annotation using transcriptome sequence data, phylogenetic and ecological studies, allele mining, epigenetics studies, metagenomics analysis, single-cell genomics, exome sequencing, and whole genome resequencing (WGRS) (Fig. 13.2).

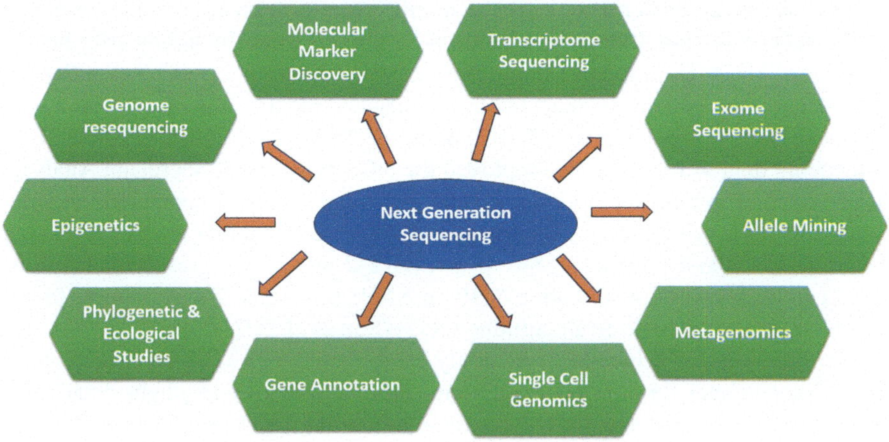

Fig. 13.2 Various application of sequencing techniques in crop improvement

13.5.1 Rice Marker Discovery by NGS

The molecular markers are precise DNA sequences located on a chromosome that are passed down in accordance with Mendelian principles and remain unaffected by external environmental factors. They may or may not encode any particular characteristics. In crop biology, molecular markers are used for fundamental and practical purposes, such as creating linkage maps, mapping QTLs, genotype fingerprinting, parentage analysis, marker-assisted selection (Majhi et al. 2022), genetic diversity research, gene flow, and evolutionary studies (Dida 2022). SSRs are repetitive DNA sequences containing one to six nucleotides arranged in a series throughout the genome. They are widely utilized in improving crop yield and quality (Singh et al. 2022).

SSR markers were first made by the development of genomic libraries, a laborious and time-intensive method that entailed isolating and sequencing clones and harboring SSRs (Homchan et al. 2021). However, with the aid of NGS techniques, such as transcriptome profiling or RNA-Seq, a robust and reliable tool now offers intriguing opportunities in identifying and developing a significant number of SSR markers, being more affordable, quicker, and more accessible than traditional SSR development processes. Using various analytical methods, one can directly sequence the transcriptomic RNA or genomic DNA and search for SSRs (Taheri et al. 2018). Studies on genetic diversity in Thai rice blast isolates demonstrate that creating SSRs using in silico methods enhances polymorphism rates and reduces the need for extensive time and labor. In the rice blast reference genome, 11,072 SSR locations were investigated. The most prevalent patterns of SSRs have AGC/CTG motifs. There were 14 SSR sites with significant polymorphism identified through computer-based analysis of 18 blast sequences. Custom primers were designed for these sites accordingly. Ten different SSR loci were found to be polymorphic with two and seven alleles in Thai rice blast isolates (Ngernmuen et al. 2019).

SNPs are essential for discovering genomic variations, creating genetic linkage maps, and connecting SNPs with phenotypic characteristics, including crop breeding (Korsa 2022). Traditionally, Sanger-expressed sequenced tags (ESTs) or Sanger sequencing amplicons were analyzed to identify and locate SNPs. However, a new technique for the discovery of SNPs is mining DNA/RNA molecules from NGS sequencing data. A recent study employed the NGS-based Bulk Segregant Analysis (BSA) method in rice to pinpoint potential genes that exhibit tolerance to alkaline conditions. Following the sequencing data, it was concluded that *Os02g39884* was the candidate alkali-tolerance gene of *qATGW2-2* due to the presence of a non-synonymous substitution mutation SNP (nsSNP) that caused a glutamine (cAg) residue to be transformed from arginine (cGg) (Sun et al. 2021).

The fundamental principle of diversity array technology (DArT) is to employ restriction enzymes for minimizing genomic intricacy, followed by hybridization to microarrays for simultaneously evaluating numerous markers across the entire genome (Adeboye et al. 2021). The development of NGS techniques has also benefited this methodology, as NGS-DArT markers have superseded array-based DArT markers. Microarray markers, known *as silico* DArT markers, are scored based and dominant on the absence or presence of single allele. DArTseq technology was used to explore Myanmar rice's genetic structure and diversity (Thant et al. 2021).

13.5.2 Wide Crosses and Alien Introgression

Traditional introgression strategies have fruitfully introduced several traits into cultivated rice, but sterility and crossing barriers have restricted their efficiency. Genomic technology will accelerate rice gene discovery from wild ancestors and provide evolutionary understanding to feed future generations (Zhou et al. 2022a). Recent research studies showed 6.8 and 19.3 megabytes of alien-introgressed chromosomal locations in temperate *japonica* and *indica* accession, respectively. Mostly, these segments contained significant sequence variants, QTL, or genes that controlled grain, flowering traits, and resistance (Chen et al. 2020).

13.5.3 Transcriptome Investigation

The sequencing of mRNA extracted from diverse plant cells at different times is known as transcriptome sequencing, which focuses on the transcribed section of the plant genome (Tu et al. 2022). Sequencing information has diverse applications in enhancing the comprehension of crop biology, encompassing gene expression studies, intron splicing, genome rearrangement detection, genome annotation, and identification and quantification of non-coding RNA that govern numerous agronomic traits (Xie et al. 2023). Next-generation sequencing methods with high throughput provide a potent approach to identifying plant transcriptomes. It has been demonstrated that long-read sequencing assists in the identification of new transcript isoforms (different mature mRNA products from the transcription of the same gene).

This method allows the production of whole sequences that can identify splice variants that may play a critical role in regulating gene function. RNA from seeds, leaves, panicles, and roots of *O. sativa ssp. japonica var* Nipponbare was used to create a cDNA library, which was sequenced using PacBio to reveal 14,874 new isoforms (Hasan et al. 2022). RNA-seq in paddy seed detected 14,391 DEGs among germinating and dry seeds (Zhang et al. 2023a). A study using RNA-seq analyzed pattern gene expression in seedlings of two cultivars of *indica* rice, XZX45 (sensitive to cold) and SQSL (resistant to freezing), in response to cold stress (Yan et al. 2023).

13.5.4 Gene Expression Profiling

The rice gene expression atlas, derived from microarray data, showcases distinct and dynamic gene expression patterns across many phases and tissues. This comprehensive resource offers valuable insights for genomic research on rice (Miya et al. 2021). Furthermore, transcriptomics research is anticipated to radically change due to the recent development of RNA-seq as a potent technique for large-scale measurement and quantization of global gene expression (D'Agostino et al. 2022). Integrating RNA-seq or microarray profiling with microdissection enables the identification of transcriptional regulatory mechanisms that govern specific developmental processes in different tissues (Woronik et al. 2022). There are currently numerous gene expression databases available for rice, such as RiceXPro, Transcriptome Encyclopaedia Of Rice (TENOR), and the Rice Expression Database (RED) (Sato et al. 2011; Kawahara et al. 2016; Xia et al. 2017). Transcriptomics helps in identifying genes and provides base for research on eQTL study and hybrid vigor (Zhou et al. 2022b). NGS, bulk segregant, and rice transcriptome analysis showed that purple leaf is due to recessive genes. Two of the 12 differentially expressed unigenes for the recessive purple leaf genes *Os04g0577800* and *Os04g0616400* were downregulated and may be candidates for purple leaf (Gao et al. 2019).

13.5.5 Small Non-coding RNA Profiling

It has been discovered that the sncRNA (Short Non-coding RNA) segment of the transcriptome is involved in numerous significant important biological activities in plant life (Halder et al. 2023). RNAs, such as miRNA, snRNA, rRNA, siRNA, tasiRNA, and tRNA belong to this category. Both miRNAs and siRNAs play an increasingly major role in post-transcriptional modulation of gene expression (Lakhotia et al. 2020). SiRNAs are 21–24 nucleotide transcripts detected mostly in crop plants, while microRNAs are 21 nucleotide transcripts found in both animals and plants. Despite their comparable sizes, siRNAs and miRNAs affect gene activity by blocking translation or mRNA degradation, but their biosynthesis and processes differ (Dong et al. 2022).

MicroRNAs have been reported in other organisms, including viruses, plants, fungi, and animals, since they were first recognized in nematodes (Quesnelle et al. 2023). Traditionally, miRNAs were determined through cloning, followed by sequencing each miRNA, which is time-consuming. The next-generation sequencing technology that is currently being developed has dramatically increased our ability to explore short RNAs (sRNAs) by omitting the cloning phase, making the research of ncRNA easier. This technique's high efficiency and accuracy allow the study of short ncRNAs across the whole genome. Additionally, the tiny read length produced by this technique is consistent with the dimensions of small interfering RNAs and microRNAs (Ratti et al. 2020). Small non-coding RNA characterization based on NGS provides a few merits over microarray approaches, as well as the capability to identify newly recognized miRNA and the potential miRNA editing (Di Bella et al. 2020).

The miRNA landscapes of high-temperature sensitive and tolerant lines in rice differ, and utilizing the short RNA databases from 16 genome-wide sequencing libraries of rice, tolerant cultivars demonstrated successful miRNA-mediated rescue. Under heat stress, roots control miRNAs more sensitively than shoots (Mangrauthia et al. 2018). An analysis of rice genes related to nitrogen transport and/or metabolism was conducted using strand-specific RNA-Seq. This analysis discovered 2588 previously unknown loci that may encode long non-coding RNA (Shin et al. 2018). Jain et al. (2021) analyzed RNA-seq information from the reproductive phase of salt-susceptible (IR-29) and salt-tolerant (Horkuch) rice lines. Correspondingly, they discovered 2103 and 1529 as potential lncRNAs in leaf and root cells (Jain et al. 2021).

RiceNCexp was recently constructed and built upon 274 sRNA-seq and 491 RNA-seq datasets in rice. It included (1) phasiRNAs, (2) lncRNAs, (3) miRNAs, and (4) PHAS genes. It can be obtained by the plant breeder and user at the following website: https://cbi.njau.edu.cn/RiceNCexp/. This dataset is an accessible and user-friendly atlas that provides information on the co-expression of non-coding RNA (ncRNA) in rice (Zhang et al. 2022).

13.5.6 Gene Annotation Using Transcriptome Sequence Data

Insufficient or imprecise annotation can impede the comprehensive study and functional understanding of the molecular processes in rice that contribute to complex agronomic traits (Hasan et al. 2022). Precise genomic annotation is essential for conducting rice research and generating scientific findings. An enormous quantity of sequencing information is produced every year as a result of more affordable NGS platforms.

Currently, protein-coding genes are identified and described using advanced techniques such as full-length cDNA or EST sequencing, which are then aligned with a reference genome. According to FGSTs, the majority of EST sequencing programs are unable to detect 20–40% of atypical or long mRNA (Qian et al. 2014). These NGS approaches offer deeper coverage, making finding rare transcripts simple. It is noteworthy that different transcript expression profiling investigations based on NGS also aid in gene annotation, along with the identification of novel genes, alternative splicing variations, and sequences of exons and introns. Furthermore, SSRs and SNPs can be identified using these EST-NGS data (Glinos et al. 2022). The latest IRGSP sequences, along with manually curated, literature-dependent annotation, are available in the RAP-Database to promote genomics-assisted studies (Rice Annotation Project 2008). At the same time, the RAP databases were working toward the goal of achieving structural and functional annotation through the utilization of pseudomolecules rather than BACs (Ouyang et al. 2007).

Transcriptome anthocyanin biosynthesis analysis in rice revealed the presence of the purple leaves recessive regulatory gene plr4, positioned near the 27.9–31.1 Megabytes segment of the fourth chromosome (Gao et al. 2019). Transcript genome sequencing was subsequently employed to investigate the DEGs and revealed 19 of these exhibited identified activities associated with the anthocyanin pathways. In order to accurately and completely characterize gene models, genome reannotation is essential for a thorough investigation of gene function. Utilizing a vast amount of RNA-seq data from the japonica genome has been reannotated and updated to version (IC4R)-2.0. The IC4R database structure, which unifies omics data, now has richer annotations, online tools, and more user-friendly web interfaces (Sang et al. 2020).

13.5.7 Phylogenetic and Ecological Studies

The phylogenetic analysis uses conserved genomic areas and states that sequence homology and variation from homology cause divergence due to distinct or novel mutations throughout evolution and speciation (Li et al. 2022). Multiple gene sequence data from many individuals are needed for these kinds of studies, and NGS techniques offer a high throughput and affordable alternative to traditional Sanger sequencing. Research on polyploid paternity, phylogeny, population divergence, gene flow, and diversity analysis can be facilitated by accessing genomic resources for non-model species through the combination of NGS techniques and targeted sequence capture (Nunes et al. 2022). It is now feasible to investigate the evolutionary connections between late divergent species/taxa or closely related ones with the aid of NGS techniques (Zaborowska et al. 2023).

To study the ancestral connections between regionally different *O. rufipogon* accessions and cultivate variants to predict the specific selection patterns associated with paddy domestication. The *O. rufipogon* was classified according to their population structure: Or-I, Or-II, and Or-III. In conclusion, this research was able to provide answers to three concerns regarding the domestication of rice: (1) domestication progression, which included one origin type and a number of introgressions; (2) the origin of cultivated rice, which was found in Southern China; (3) Or-IIIa, the initial natural ancestor of growing rice (Huang et al. 2012).

The domesticated African rice species, *O. glaberrima*, originated from *O. barthii* approximately 3000 years ago in a specific region along the Niger River, as indicated by population genomics studies conducted on both species. These studies offered support for the independent domestication of African rice (Wang et al. 2014). A set of 533 genotypes and 950 breeding materials of rice were additionally sequenced. Integrating the aggregate yielded 6,551,358 more dependable SNPs. Six distinct groups were found when the population structure was examined: aus, tropical *japonica*, temperate *japonica*, *indica* II, and an intermediate group (Xie et al. 2015). Significantly, a comparable yet separate elimination of 113-kb within the RAPD loci in *O. glaberrima* demonstrated that the domestication of vegetative structure occurred synchronously in both Asian and African cultivable species (Wu et al. 2018).

13.5.8 Allele Mining

Allele mining helps track allele change, find novel haplotypes, and build allele-specific markers (for MAS). Compared to searching for novel alleles based on Eco-TIILING, allele mining uses sequencing data which has become more effective due to advancements in NGS technologies (Tare et al. 2023). Primarily, a genetic factor governing a desirable phenotype was identified, then primer templates were developed, and PCR was conducted using various genotypes to reflect the highest level of diversity possible for a given characteristic. The amplicons were then sequenced and checked for variations. Following their association with a genotype that exhibits higher performance, these novel alleles can be employed in crop breeding programmes due to the availability of allele-specific primers (Kaur et al. 2023). Different software was employed in allele mining to anticipate changes in amino acids and find nucleotide variations, which encode protein structure and activities. A few bioinformatics software programmes are Plant CARE, DCPD, ClustalW, and Fast PCR (Satpathy 2022).

Zhang et al. (2016) used representative SNPs to allele search for leaf rolling, a critical aspect of plant architecture, and to identify top accessions to improve leaf rolling. The 3K Rice panel provided the best SNPs at the identified loci. They differentiated the loci containing gRl4–2 fragmented in 3 sub-regions by rescanning the candidate locus using more saturated SNP markers (Zhang et al. 2016). RYMV is a prominent rice disease in Africa. A resistance survey was carried out in a reference germplasm accession including 268 *O. glaberrima*. The severe level of RYMV resistance was confirmed by the identification of 40 resistant accessions in total. This collection's resistance gene or candidate gene variability was determined using Sanger sequencing or high-depth Illumina data. In the end, one of the three RYMV resistance genes was validated or potential candidate resistance alleles were linked to all 40 resistant accessions (Pidon et al. 2020).

13.5.9 Epigenetics Studies

Epigenomics is the study of genome-wide patterns of chromatin alteration, primarily histone and DNA methylation. These regulate the accessibility of chromatin to transcriptional machinery, DNA replication, and repair, which in turn affects the activities of the genome (Kornak et al. 2021). NGS was adopted early by the epigenetics field, which led to the detailed study of genomes of numerous species and cell types and their epigenetic changes (Searle et al. 2023). Somaclonal variation caused by tissue culture has been linked to both genetic and epigenetic modifications. The extensive modifications in DNA methylation observed in regenerated under stress in rice plants support the relevant epigenetic modifications in somaclonal variation. Higher expression of a few genes linked to the abiotic stress response was correlated with decreased methylation in Nagina 22 and increased methylation in Pokkali in associated with desiccation and salinity correspondingly, according to bisulfite sequencing analysis of DNA methylation (Rajkumar et al. 2020). Utilizing WGB-Sequencing of a pair of paddy genotypes under P-starvation stress, the function of *Pup1* is regulated by methylation. They found that around an average of 14% of cytosines in the genome were methylated and that this number increased considerably under stress (Kumar et al. 2022).

In rice Hitomebore, Wang et al. (2022) found that regenerated lines (RL) had different DNA methylation patterns than the wild-type. When paired with bisulfite sequencing, Methylation Sensitive Amplification Polymorphism (MSAP) examination of 5′ CCGG locations revealed a usually lower level of methylation pattern in the RLs (Wang et al. 2022). According to Rao et al. (2023), six breeding materials, which are autotetraploid hybrids and their parents and diploid donors and their hybrids, have epigenetic alterations linked to heterosis and polyploidy. By employing WGB-sequencing and RNAseq for comparative DNA-methylome and transcriptome research, the cellular pathways governing impacts on polyploidization and hybridization were investigated. The autotetraploid lines displayed high transposable element methylation compared to the diploid donors, indicating that methylation pattern changes may influence expression of trait. The hybrid lines showed hypermethylation in comparison to their corresponding parents (Rao et al. 2023).

13.5.10 Metagenomics Analysis

Microorganisms highly accompany plants as they are found in the rhizosphere as endophytes and epiphytes, and they have positive, negative, or neutral impacts on the health of plants. Studying the diversity and dynamics of the microbial population is essential for improving our understanding of plant–microbe interactions (Khan et al. 2023). The direct extraction of DNA from environmental materials for the purpose of studying whole microbial populations is known as metagenomics. This means getting a shotgun or metagenomic library set up, sequencing it, and then doing an excessive amount of data analysis. Metagenomics provides a fresh perspective on microbial communities, as existing methods were previously unable to analyze a significant proportion of non-culturable (Offiong et al. 2023).

Different high-throughput NGS methods may be used to extensively sequence the metagenome, making it feasible to identify less prevalent microbes. The reads are assembled to form contigs after sequencing, and these contigs are compared to publicly available microbial genomes. It may be possible to assemble individual genomes for a novel microorganism based on the unaligned sequence readings (Saini et al. 2022). As demonstrated by yeast and bacterial metagenomic sequencing studies, Chinese rice wine quality and microbiota composition were correlated (Hong et al. 2016). The metagenomic DNA of root microbes revealed a significantly wide range in the *nifH, pqqC*, and *accd-DRgenes* (plant growth-promoting) when clone-based sequencing and NGS were performed (Shin et al. 2018). By using the 16S rRNA gene to analyze the genomes of two gnotobiotically produced aromatic rice cultivars, researchers have discovered the distribution of endophytic bacterial communities in rice plants, which are tissue-specific and genotype-dependent and are linked to a variety of attributes that promote plant development (Krishnamoorthy et al. 2021).

13.5.11 Single-Cell Genomics

Millions of tissues have been analyzed in the majority to get a comprehension of the genome's expression and control. Although these studies are clearly informative, they frequently fail to identify any variability that may exist within the cell population (Kambale et al. 2019). The accepted view regarding genomes is that each individual's genome is fixed in their tissues for the duration of their lives. However, over time, few mutations are added to cells with each division, leading to the formation of genomic heterogeneity (somatic variations) within the organism. These differences have a role in a number of events associated with disease and development (Evrony et al. 2021).

Conventional sequencing studies lose the genomic and transcriptome differences present in single cells since they begin with a group of cells (Chen et al. 2023). Consequently, information about a single cell's genome can be obtained using single-cell genome sequencing (SCGS). The capacity to sequence single-cell genomes is made possible by advancements in cell isolation, NGS techniques, and WG amplification. Even while whole genome amplification issues remain, the next generation of TGSTs with single-molecule sequencing capabilities will improve the speed and accuracy of SCGS (Canaguier et al. 2022). Because plant cells with cell walls are difficult to separate single-cell genomes from, a small number of SCGS experiments are conducted in plant systems (Ouyang et al. 2022).

According to single-cell transcriptome investigations, abiotic stress influences cell size and population composition and affects gene expression in a cell-type-specific way (Wang et al. 2020). A large number of unique cell-specific marker were discovered. The transcriptome architecture of important cell types of the root tip at single-cell level resolution has provided new avenues for studying cell-type functioning and the evolutionary process in plants (Liu et al. 2021). In recent times, novel insights into the molecular process of plant organ development because of the transcriptome landscape at single-cell resolution (Zhang et al. 2023b). Recently, Aoki et al. (2022) have devised a pipeline for examining the single-cell genetic variability of the root microbiome. Utilizing WG amplification, microbial cells were isolated from the soil, and their microbial sequences were obtained. These genomic sequences offered insights into the roles played by microbial organisms in the paddy ecosystem (Aoki et al. 2022).

13.5.12 Exome Sequencing

Plant genomes are frequently large, repetitive, polyploid, and complicated, which makes WGS difficult. Even though WGS has grown more economical due to advancements in NGS techniques, WGS remains more expensive than specialized sequencing methods like exome sequencing when searching for changes that affect genotypes. Approximately 2% of the genome is transcribed into proteins and read using whole-exome sequencing (WES), a streamlined and less expensive sequencing method, i.e., sequencing of a genome's exomes using next-generation technology. The exomes are enriched by utilizing nucleotides that are complementary to exome probes on a microarray or magnetic beads in solution. This procedure involves deleting the remaining part of the genome, while the exomes are retained and coupled to the probes for subsequent sequencing (Kockum et al. 2023).

To gain an extensive understanding about the effect of ion beam induced mutations on rice genome, Oono and coworkers utilized exome sequencing in rice. DNA libraries were constructed from a subset of carbon-ion beam-based artificial mutants through capturing custom probes covering 66.3 M bases of almost all exons and miRNAs sequences. In five mutant rice lines a total of 56 mutations, with 24 single nucleotide variants, 23 deletions, and five insertions were detected. Identification of a limited number of high-impact mutations makes it easier to identify potential genes causing mutant traits (Oono et al. 2020). An ion radiation beam isolated the temperature-susceptible mutant *flo11-2* from 1116 rice lines. The *flo11-2* mutant grew at 28 °C, exhibiting much greater chalkiness than the WT. Three putative genes, including *Os12g0244100*, which codes for cpHSP70-2, were identified by whole-exome sequencing of *flo11-2* mutant (Tabassum et al. 2020).

13.5.13 Whole Genome Resequencing (WGRS)

The high-depth resolution of rice reference genome and resequencing made possible by the quick advancement of sequencing technology offered unmatched ease of use for investigating the genome-wide variation in landraces (Chen et al. 2022). Dense SNP and InDel markers might be obtained at a reasonable cost through rice genomics (Hasan et al. 2021). The uniform distribution and quantity of SNP/InDel markers throughout the genome make them a promising platform for high-throughput genotyping that might yield comprehensive genetic information. These markers may be useful for gene-mapped cloning, genetic map creation, and population structure study (Long et al. 2022).

Genomic analysis of polycarpellary rice investigated by Das et al. (2021) using WGRS data created around 3.7 million bp of genomic data for further research. The readings were mapped to Nipponbare, with a length of around 101 bases. 1,096,419 variations were found by the genome-wide variant analysis, comprising 153,386 InDels and 943,033 SNPs. For a total of 11,529 identified genes, 24,920 non-synonymous SNPs were discovered (Das et al. 2021). A recent work generated markers for known agronomically significant genes using WGS 24 *japonica* lines (Ji et al. 2021). Differences in *javanica* genome sequences and the allelic variations for the grain weight *TGW2* gene were found by the resequencing of 20 landraces (Long et al. 2022).

13.6 Methods for Identification of Functional Genes

13.6.1 MutMap

It is easy to develop many mutants with chemical and physical mutagens; however, identifying the specific mutation and causing the phenotypic change can be challenging. The quick advancement of NGS has clarified and enhanced the capacity to detect nucleotide sequence changes in the genome (Zhou et al. 2018). MutMap was designed to decipher SNPs that cause differences among the genotypes of a mutant and the wild-type population. Both mutant and WT are hybridized, and self-pollination occurs in the F_1 generation. After pooling the DNA of 20 mutants, full genome sequencing was performed. Generally, the quality of the sequencing was more than 10-X coverage of genome. The sequences of the mutant were aligned with WT plant's reference sequence. After calculating the SNP index, the candidate SNPs were found to have an SNP index of 1 (Tribhuvan et al. 2018).

The causative SNPs and candidate genes were identified using bioinformatics and further Sanger sequencing analysis (Fig. 13.3). Lastly, overexpression, RNA interference, or deletion methods were used to validate the putative gene (Yuan et al. 2017). The two genes that regulate salt tolerance and pale green leaf, *OsCOA1* and *OsRR22*, were found using the MutMap technique and confirmed using transgenic and mutant plants (Takagi et al. 2015).

The MutMap-Gap strategy was employed in rice as a substitute method for conditions in which there is a significant divergence between the reference genome and the mutant at a specific site (Takagi et al. 2013). In order to create a desired reference genome, the WT-X is sequenced and matched to the reference genome, and differences in SNPs between the two are identified and then modified. Using a modified MutMap method, a New SNP was detected in gene *AGPL2* which was associated with floury endosperm in rice (Zhang et al. 2023c). The MutMap+ was developed to stop genotype crossings between wild-type and mutant organisms. The homozygous M_3 individuals undergo sequencing for SNP mapping, whereas

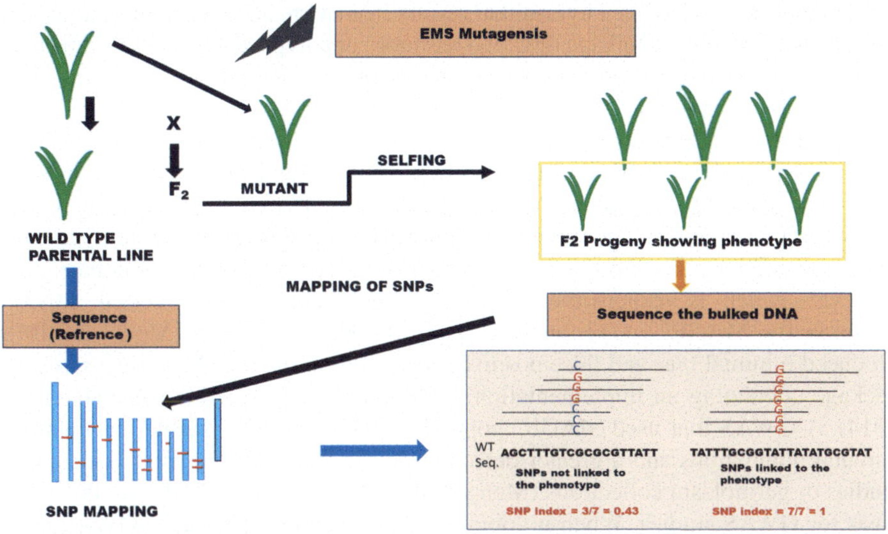

Fig. 13.3 Workflow of MutMap in Rice

the heterozygous individuals in the M_1 and M_2 self-pollinated (⊗). After aligning with a reference genome, phenotype-associated polymorphic SNPs were discovered (Fekih et al. 2013). The MutMap+ analysis was employed to identify the underlying mutation that led to the decreased Mg condition of *LMGC1*, leading to the discovery of the gene that encodes the RanBP2-type ZFn proteins (Kobayashi et al. 2023).

13.6.2 Genome-Wide Quantitative Trait Loci Mapping (GWAS)

GWAS is an established technique to find the relationship between traits and markers in humans as well as plants. Millions of SNPs have been obtained for rice genome investigations owing to the recent widespread use of numerous NGS projects. Recently, SNPs are the most desired markers for QTL scanning because of their uniform distribution throughout the genome and abundance (Istanbuli et al. 2024). Several research groups have used GWAS to explore the agronomic characteristics. In order to better understand population structure and uneven relatedness among individuals, a mixed linear model (MLM) has been adopted, which reduces false positive relationships (Uprety 2023). The local linkage disequilibrium (LD) of the SNP determines the GWAS resolution. In cultivated rice, the average LD decay has increased from 100 to 200 kb (Kim et al. 2022). Notably, the power of QTL mapping was significantly enhanced by the haplotype-level association (Liu et al. 2022). Nevertheless, genome-wide haplotype-level association mapping remains a challenging process.

Multi-parent Advanced Generation Inter-Cross population (MAGIC) was recently established to explore possible uses for varietal development in rice and

QTL mapping. MAGIC is a hybrid that results from a cross between more than six parents, with decreased linkage drag and has more genotypic diversity (Dinesh et al. 2023). A distinct lineage exists between the generations and among the lines, which can circumvent the population structure. A number of MAGIC lines which were generated by several superior parental genotypes were employed to identify QTLs for different qualities in rice (Raghavan et al. 2017). Based on NGS and/or SNP array data, the SNP markers were used to create the MAGIC genetic map (Satrio et al. 2021). MAGIC lines consist of 1328 Indica lines and their GWAS study identified several QTLs, like Sub1, which is responsible for submergence tolerance, and Xa4/Xa5 which is responsible for disease resistance (Bandillo et al. 2013). Additionally, MAGIC's QTL analysis using the Hidden Markov Model (HMM) produced minimal bias and false positive rates in the estimated QTL effect sizes. R package containing an implementation of the method is available (Verbyla et al. 2014). A GWAS that used MAGIC populations showed better resolution than bi-parental populations and a greater detection power than one that used association studies or germplasm collections (Meng et al. 2016). Currently, SNPs are the main basis for GWAS studies. When it comes to QTL mapping, bin-based GWAS may provide more accurate QTL position results and fewer false positive correlations than SNP-based GWAS (Guan et al. 2022).

13.6.2.1 QTL-Seq

For QTL mapping, F2 and/or BC1F2 are often utilized. In order to develop the advanced population, it takes five to six cropping seasons, which is a long-time period. QTL-sequencing (QTL-seq) is a contemporary implementation of traditional BSA & WGR (Majeed et al. 2022).

Recently, with advancement of NG-Seq. platforms, the amalgamation of BSA with NGS has turn out to be an prime procedure for finding genes and QTLs (Abe et al. 2012; Kadambari et al. 2018; Wambugu et al. 2018). After crossing two extremely diverse parents in a mapping population development, two extreme phenotypic bulks are created from the populations to be mapped (Fig. 13.4). Through a thorough comparison analysis of the NGS-based sequence, the QTL region is rapidly mapped. More precisely, a locus is most likely to have contributed to the phenotypic variation if the allele frequency in that area differs significantly between these two bulks. QTL-seq enhances productivity and provides a high level of genetic diversity. These methods improve QTL position resolution and help in identifying functional contributing variations in genes. Interval mapping and candidate gene discovery by sequencing have been made possible by QTL-seq. There are more widely relevant approaches which are accessible for determining main QTLs governing length and weight of seed (Yaobin et al. 2018), cold tolerance (Yang et al. 2021), salt tolerance (Lei et al. 2020), and spikelet fertility (Nubankoh et al. 2020).

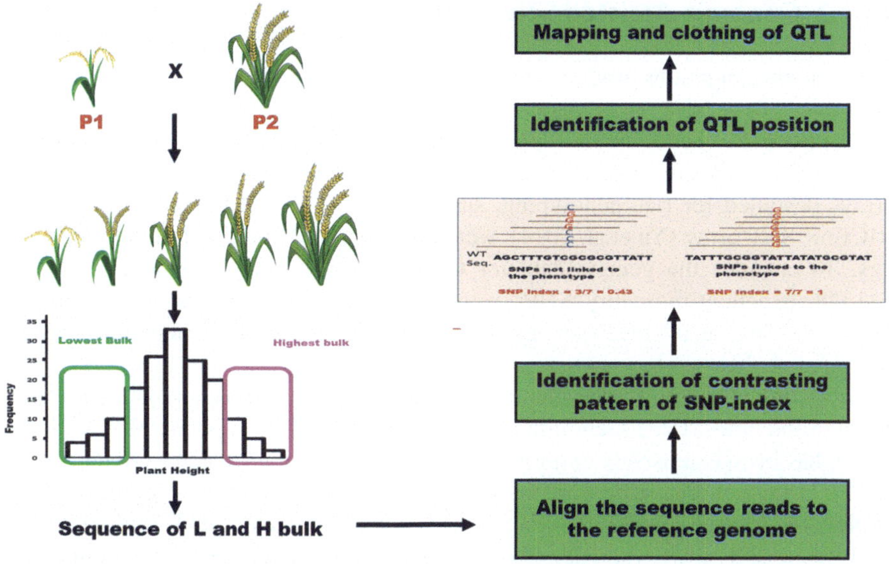

Fig. 13.4 Workflow of QTL-seq in rice

13.7 Genomic Selection for Breeding

13.7.1 Genetic Diagnosis

A crucial step in rice breeding and genetic research is choosing the parental lines. Varieties with unique genetic origins are frequently selected by breeders for their breeding programmes. For instance, a population derived from varieties with significantly distinct origins will increase the mapping capability and map several QTLs in order to screen QTLs for various agronomic attributes (Mst et al. 2023). It can be beneficial to do phylogenetic analysis with this breeding technique. Higher diversity in the parent founders may lead to additional recombination possibilities, allowing for a more accurate and efficient investigation of several desired features (Rainford and Porter 2022). Conversely, in hybrid breeding, we may use SNP analysis to conduct genetic diagnostics and choose parental lines with superior alleles to regulate yield-related characteristics like flowering date, panicle length, height, and grain structure. In production of rice, heterosis has been successfully utilized. It has been determined that several yield and yield attributing genes, including *sd1, Ghd8, ipa1, Ghd7, gs3,* and *dep1* others, are essential for hybrids and have partial heterotic effects (Huang et al. 2015). These discoveries will contribute to the development of sensible genetic designs for rice hybrids.

13.7.2 Genomic-Assisted Selection

Genomic investigations produced many SNPs as a result of several NGS projects. A high-throughput SNP array detects thousands of SNPs over the entire genome, made possible to genotype multiple samples very quickly. SNP array technologies relying on diverse concepts are available for rice (Zhou et al. 2018). Recently, SNP arrays designed for rice genotyping, such as RiceSNP50 (Chen et al. 2014) and RICE6K SNP array (Yu et al. 2014), were created using Illumina Infinium technology. To enhance the genetic data for GWAS and genomic selection, a total of 581,006 excellent-in-quality SNPs were generated. These SNPs were carefully selected to have an average spacing of 200 base pairs among each neighboring SNP (Meng et al. 2016).

Gene-linked markers are valuable tools for improving characteristics through MAS (Song et al. 2023). Genomic-assisted selection (GAS), a potential MAS technique, has been extensively employed for rice genetic enhancements (Abdeta et al. 2022). Using genome-wide SNP markers, GAS is an effective and high-throughput method to enhance agronomic parameters. It does both background and foreground analysis in SNP array test. Rice SNP arrays that may accomplish background and foreground selection are RiceSNP50 and RICE6K.

13.8 Omics Integration

The functional depiction and investigation of genes crucial to agronomic traits are the main methods for increasing rice production. The advanced technology has been used in research for a number of years in an effort to better understand molecular pathways and their mechanistic components. Genomics sheds light on the dominant or recessive genes in different rice germplasm, whereas transcriptomics provides insight into the intricate rice RNA expression networks that may be crucial to yield or stress changes. Similarly, proteomics helps identify the key proteins that increase rice quality, and metabolomics offers the essential metabolite signatures linked to rice yield enrichment and quality (Zaghum et al. 2022). Bioinformatics databases integrate information from the omics sciences to produce a comprehensive picture of the variables influencing the capacity of rice to adapt and perform better under stress or quantitatively or qualitatively. By using omics-generated datasets, rice crop improvement can be accelerated by discovering and characterizing genes (Iqbal et al. 2021).

13.9 Conclusions and Future Directions

The rice genome sequence is the simplest to decipher since sequencing techniques and methods are now well-established, reliable, and efficient. However, the sequencing is not the end; the main challenges will be annotating the sequence, identifying the precise region of every gene based on its intron/exon structure, and assigning a function to every gene. The ability to alter rice more effectively and consistently than any other grain gives it an edge over other Gramineae in this regard. This enables creation of efficient methods for functional genomics, like insertion mutagenesis and reverse genetics.

In general, the utilization of NGS technologies in rice improvement has greatly progressed our knowledge and capacities for augmenting the yield and flexibility of this indispensable crop. NGS technology has made it possible to examine the many biological processes of plants in plenty of inventive ways by enabling the use of sequencing technologies. Implementing NGS in breeding has enabled identifying and analyzing genes that regulate crucial agricultural characteristics, including yield potential and the ability to tolerate both abiotic and biotic challenges.

In conclusion, using NGS technology for rice development has significant positive social effects on society by raising agricultural productivity, sustainability, and food security. This discovery gives breeders the tools they need to create new rice varieties with increased attributes by marker-assisted and genomic selection, accelerating breeding procedures and enabling the identification of critical genes. The current literature would be supplemented by additional research on the functional validation of candidate genes, merging of multi-omics data, and exploring diversity in various agroecological sites, which would further contribute to the ongoing progress of rice improvement.

References

Abdeta TM, Biratu W, Yadesa L (2022) Genomic mapping, molecular marker and marker assisted selection in Rice: a review. Agro Bali Agric J 5:422–433. https://doi.org/10.37637/ab.v5i3.979

Abe A, Kosugi S, Yoshida K, Natsume S, Takagi H, Kanzaki H, Matsumura H, Yoshida K, Mitsuoka C, Tamiru M, Innan H, Cano L, Kamoun S, Terauchi R (2012) Genome sequencing reveals agronomically important loci in rice using MutMap. Nat Biotechnol 30:174–178. https://doi.org/10.1038/nbt.2095

Adeboye K, Semon M, Oyetunde OA, Oduwaye O, Adebambo A, Daniel I (2021) Diversity array technology (DArT)-based mapping of phenotypic variations among recombinant inbred lines of WAB638-1/PRIMAVERA under drought stress. Euphytica 217:130. https://doi.org/10.1007/s10681-021-02860-6

Ahmed N, Lévy J, Ren S, Mushtaq H, Bertels K, Al-Ars Z (2019) GASAL2: a GPU accelerated sequence alignment library for high-throughput NGS data. BMC Bioinformatics 20:520. https://doi.org/10.1186/s12859-019-3086-9

Aoki W, Kogawa M, Matsuda S, Matsubara K, Hirata S, Nishikawa Y, Hosokawa M, Takeyama H, Matoh T, Ueda M (2022) Massively parallel single-cell genomics of microbiomes in rice paddies. Front Microbiol 13:1024640

Ashraf MF, Hou D, Hussain Q, Imran M, Pei J, Ali M, Shehzad A, Anwar M, Noman A, Waseem M, Lin X (2022) Entailing the next-generation sequencing and metabolome for sustainable agriculture by improving plant tolerance. Int J Mol Sci 23:651. https://doi.org/10.3390/ijms23020651

Bandillo N, Raghavan C, Muyco PA, Sevilla MAL, Lobina IT, Dilla-Ermita CJ, Tung C-W, McCouch S, Thomson M, Mauleon R, Singh RK, Gregorio G, Redoña E, Leung H (2013) Multi-parent advanced generation inter-cross (MAGIC) populations in rice: progress and potential for genetics research and breeding. Rice 6:11. https://doi.org/10.1186/1939-8433-6-11

Canaguier A, Guilbaud R, Denis E, Magdelenat G, Belser C, Istace B, Cruaud C, Wincker P, Le Paslier M-C, Faivre-Rampant P, Barbe V (2022) Oxford nanopore and bionano genomics technologies evaluation for plant structural variation detection. BMC Genomics 23:317. https://doi.org/10.1186/s12864-022-08499-4

Chen H, Xie W, He H, Yu H, Chen W, Li J, Yu R, Yao Y, Zhang W, He Y (2014) A high-density SNP genotyping array for rice biology and molecular breeding. Mol Plant 7:541–553

Chen Z, Li X, Lu H, Gao Q, Du H, Peng H, Qin P, Liang C (2020) Genomic atlases of introgression and differentiation reveal breeding footprints in Chinese cultivated rice. J Genet Genomics 47:637–649. https://doi.org/10.1016/j.jgg.2020.10.006

Chen R, Deng Y, Ding Y, Guo J, Qiu J, Wang B, Wang C, Xie Y, Zhang Z, Chen J, Chen L, Chu C, He G, He Z, Huang X, Xing Y, Yang S, Xie D, Liu Y, Li J (2022) Rice functional genomics: decades' efforts and roads ahead. Sci China Life Sci 65:33–92. https://doi.org/10.1007/s11427-021-2024-0

Chen S, Zhou Z, Li Y, Du Y, Chen G (2023) Application of single-cell sequencing to the research of tumor microenvironment. Front Immunol 14:1285540. https://doi.org/10.3389/fimmu.2023.1285540

D'Agostino N, Li W, Wang D (2022) High-throughput transcriptomics. Sci Rep 12:20313. https://doi.org/10.1038/s41598-022-23985-1

Das SP, Jasrotia RS, Deb D, Iquebal MA, Jaiswal S, Dey N (2021) Genomic analysis of polycarpellary rice (Oryza sativa L.) through whole genome resequencing. J Plant Biochem Biotechnol 30:364–372. https://doi.org/10.1007/s13562-020-00602-8

Di Bella S, La Ferlita A, Carapezza G, Alaimo S, Isacchi A, Ferro A, Pulvirenti A, Bosotti R (2020) A benchmarking of pipelines for detecting ncRNAs from RNA-Seq data. Brief Bioinform 21:1987–1998. https://doi.org/10.1093/bib/bbz110

Dida G (2022) Molecular markers in breeding of crops: recent progress and advancements. Open Access J Microbiol Biotechnol 7:1–11. https://doi.org/10.23880/oajmb-16000244

Dinesh MS, Delvadiya IR, Duluri S (2023) Enchanting insights and exploring the relevance of the Magic population in advancing cereal crop improvement. Int J Plant Soil Sci 35:162–178. https://doi.org/10.9734/ijpss/2023/v35i173195

Dong Q, Hu B, Zhang C (2022) microRNAs and their roles in plant development. Front Plant Sci 13:824240. https://doi.org/10.3389/fpls.2022.824240

Eren K, Taktakoğlu N, Pirim I (2022) DNA sequencing methods: from past to present. Eurasian J Med 54:47–56. https://doi.org/10.5152/eurasianjmed.2022.22280

Evrony GD, Hinch AG, Luo C (2021) Applications of single-cell DNA sequencing. Annu Rev Genomics Hum Genet 22:171–197. https://doi.org/10.1146/annurev-genom-111320-090436

Fekih R, Takagi H, Tamiru M, Abe A, Natsume S, Yaegashi H, Sharma S, Sharma S, Kanzaki H, Matsumura H, Saitoh H, Mitsuoka C, Utsushi H, Uemura A, Kanzaki E, Kosugi S, Yoshida K, Cano L, Kamoun S, Terauchi R (2013) MutMap+: genetic mapping and mutant identification without crossing in Rice. PLoS One 8:e68529. https://doi.org/10.1371/journal.pone.0068529

Gao J, Dai G, Zhou W, Liang H, Huang J, Qing D, Chen W, Wu H, Yang X, Li D, Gao L, Deng G (2019) Mapping and identifying a candidate gene Plr4, a recessive gene regulating purple leaf in Rice, by using bulked segregant and transcriptome analysis with next-generation sequencing. Int J Mol Sci 20:4335. https://doi.org/10.3390/ijms20184335

Glinos DA, Garborcauskas G, Hoffman P, Ehsan N, Jiang L, Gokden A, Dai X, Aguet F, Brown KL, Garimella K, Bowers T, Costello M, Ardlie K, Jian R, Tucker NR, Ellinor PT, Harrington ED, Tang H, Snyder M, Juul S, Mohammadi P, MacArthur DG, Lappalainen T, Cummings BB (2022) Transcriptome variation in human tissues revealed by long-read sequencing. Nature 608:353–359. https://doi.org/10.1038/s41586-022-05035-y

Gu H, Liang S, Zhao J (2022) Novel sequencing and genomic technologies revolutionized rice genomic study and breeding. Agronomy 12:218

Guan W, Ke C, Tang W, Jiang J, Xia J, Xie X, Yang M, Duan C, Wu W, Zheng Y (2022) Construction of a high-density recombination bin-based genetic map facilitates high-resolution mapping of a major QTL underlying anthocyanin pigmentation in eggplant. Int J Mol Sci 23:10258. https://doi.org/10.3390/ijms231810258

Gupta N, Verma VK (2019) Next-generation sequencing and its application: empowering in public health beyond reality. Microb Technol Welf Soc 17:313–341. https://doi.org/10.1007/978-981-13-8844-6_15

Halder K, Chaudhuri A, Abdin MZ, Datta A (2023) Tweaking the small non-coding RNAs to improve desirable traits in plant. Int J Mol Sci 24:3143. https://doi.org/10.3390/ijms24043143

Hasan N, Choudhary S, Naaz N, Sharma N, Laskar RA (2021) Recent advancements in molecular marker-assisted selection and applications in plant breeding programmes. J Genet Eng Biotechnol 19:128. https://doi.org/10.1186/s43141-021-00231-1

Hasan S, Huang L, Liu Q, Perlo V, O'Keeffe A, Margarido GRA, Furtado A, Henry RJ (2022) The long read transcriptome of Rice (Oryza sativa ssp. japonica var. Nipponbare) reveals novel transcripts. Rice 15:29. https://doi.org/10.1186/s12284-022-00577-1

Homchan S, Bhadola P, Gupta Y (2021) Statistical analysis of simple sequence repeats in genome sequence: a case of Acheta domesticus (Orthoptera: Gryllidae). SPAST Abstr 1:14799

Hong X, Chen J, Liu L, Wu H, Tan H, Xie G, Xu Q, Zou H, Yu W, Wang L, Qin N (2016) Metagenomic sequencing reveals the relationship between microbiota composition and quality of Chinese Rice wine. Sci Rep 6:26621. https://doi.org/10.1038/srep26621

Hu T, Chitnis N, Monos D, Dinh A (2021) Next-generation sequencing technologies: an overview. Hum Immunol 82:801–811. https://doi.org/10.1016/j.humimm.2021.02.012

Huang X, Kurata N, Wei X, Wang Z-X, Wang A, Zhao Q, Zhao Y, Liu K, Lu H, Li W, Guo Y, Lu Y, Zhou C, Fan D, Weng Q, Zhu C, Huang T, Zhang L, Wang Y, Feng L, Furuumi H, Kubo T, Miyabayashi T, Yuan X, Xu Q, Dong G, Zhan Q, Li C, Fujiyama A, Toyoda A, Lu T, Feng Q, Qian Q, Li J, Han B (2012) A map of rice genome variation reveals the origin of cultivated rice. Nature 490:497–501. https://doi.org/10.1038/nature11532

Huang X, Yang S, Gong J, Zhao Y, Feng Q, Gong H, Li W, Zhan Q, Cheng B, Xia J, Chen N, Hao Z, Liu K, Zhu C, Huang T, Zhao Q, Zhang L, Fan D, Zhou C, Lu Y, Weng Q, Wang Z-X, Li J, Han B (2015) Genomic analysis of hybrid rice varieties reveals numerous superior alleles that contribute to heterosis. Nat Commun 6:6258. https://doi.org/10.1038/ncomms7258

Iqbal Z, Iqbal MS, Khan MIR, Ansari MI (2021) Toward integrated multi-omics intervention: Rice trait improvement and stress management. Front Plant Sci 12:741419. https://doi.org/10.3389/fpls.2021.741419

Istanbuli T, Nassar AE, Abd El-Maksoud MM, Tawkaz S, Alsamman AM, Hamwieh A (2024) Genome-wide association study reveals SNP markers controlling drought tolerance and related agronomic traits in chickpea across multiple environments. Front Plant Sci 15:1260690. https://doi.org/10.3389/fpls.2024.1260690

Jain P, Hussian S, Nishad J, Dubey H, Bisht DS, Sharma TR, Mondal TK (2021) Identification and functional prediction of long non-coding RNAs of rice (Oryza sativa L.) at reproductive stage under salinity stress. Mol Biol Rep 48:2261–2271. https://doi.org/10.1007/s11033-021-06246-8

Ji H, Shin Y, Lee C, Oh H, Yoon IS, Baek J, Cha Y-S, Lee G-S, Kim SL, Kim K-H (2021) Genomic variation in Korean japonica Rice varieties. Genes (Basel) 12:1749. https://doi.org/10.3390/genes12111749

Jin S-K, Xu L-N, Yang Q-Q, Zhang M-Q, Wang S-L, Wang R-A, Tao T, Hong L-M, Guo Q-Q, Jia S-W, Song T, Leng Y-J, Cai X-L, Gao J-P (2023) High-resolution quantitative trait locus mapping for rice grain quality traits using genotyping by sequencing. Front Plant Sci 13:1050882. https://doi.org/10.3389/fpls.2022.1050882

Kadambari G, Vemireddy LR, Srividhya A, Nagireddy R, Jena SS, Gandikota M, Patil S, Veeraghattapu R, Deborah DAK, Reddy GE, Shake M, Dasari A, Ramanarao PV, Durgarani CV, Neeraja CN, Siddiq EA, Sheshumadhav M (2018) QTL-Seq-based genetic analysis identifies a major genomic region governing dwarfness in rice (Oryza sativa L.). Plant Cell Rep 37:677–687. https://doi.org/10.1007/s00299-018-2260-2

Kambale R, Siddiqui M, Muthurajan R, Rahman H (2019) Single-cell omics approaches in plants, pp 255–275

Kaur K, Neelam K, Singh J, Malik P, Singh K (2023) Uncovering natural allelic and structural variants of OsCENH3 gene by targeted resequencing and in silico mining in genus Oryza. Sci Rep 13:830. https://doi.org/10.1038/s41598-023-28053-w

Kawahara Y, Oono Y, Wakimoto H, Ogata J, Kanamori H, Sasaki H, Mori S, Matsumoto T, Itoh T (2016) TENOR: database for comprehensive mRNA-Seq experiments in Rice. Plant Cell Physiol 57:e7. https://doi.org/10.1093/pcp/pcv179

Khan D, Kabiraj A, Biswas R, Roy RK, Bandopadhyay R (2023) Plant-microbe community dynamics in rhizosphere: reviewing the grassroots ecology towards sustainable agriculture. Indian J Agric Sci 93:247–257. https://doi.org/10.56093/ijas.v93i3.129812

Kim K-W, Nawade B, Nam J, Chu S-H, Ha J, Park Y-J (2022) Development of an inclusive 580K SNP array and its application for genomic selection and genome-wide association studies in rice. Front Plant Sci 13:1036177. https://doi.org/10.3389/fpls.2022.1036177

Kobayashi NI, Takagi H, Yang X, Nishizawa-Yokoi A, Segawa T, Hoshina T, Oonishi T, Suzuki H, Iwata R, Toki S, Nakanishi TM, Tanoi K (2023) Mutations in RZF1, a zinc-finger protein, reduce magnesium uptake in roots and translocation to shoots in rice. Plant Physiol 192:342–355. https://doi.org/10.1093/plphys/kiad051

Kockum I, Huang J, Stridh P (2023) Overview of genotyping technologies and methods. Curr Protoc 3:e727. https://doi.org/10.1002/cpz1.727

Kornak U, Bischof O, Ebert R, Taipaleenmäki H, Hesse E, Jakob F (2021) Epigenetics and noncoding RNA—Principles and clinical impact. Osteologie 30:201–210. https://doi.org/10.1055/a-1527-4585

Korsa F (2022) Effects of functional single nucleotide polymorphisms on plant phenotypes. Arch Crop Sci 5:185–192. https://doi.org/10.36959/718/619

Krishnamoorthy A, Gupta A, Sar P, Maiti MK (2021) Metagenomics of two gnotobiotically grown aromatic rice cultivars reveals genotype-dependent and tissue-specific colonization of endophytic bacterial communities attributing multiple plant growth promoting traits. World J Microbiol Biotechnol 37:59. https://doi.org/10.1007/s11274-021-03022-5

Kumar S, Seem K, Kumar S, Vinod KK, Chinnusamy V, Mohapatra T (2022) Pup1 QTL regulates gene expression through epigenetic modification of DNA under phosphate starvation stress in Rice. Front Plant Sci 13:871890. https://doi.org/10.3389/fpls.2022.871890

Lakhotia SC, Mallick B, Roy J (2020) Chapter 2—Non-coding RNAs: ever-expanding diversity of types and functions. In: Pandey R (ed) RNA-based regulation in human health and disease. Academic, pp 5–57

Lei L, Zheng H, Bi Y, Yang L, Liu H, Wang J, Sun J, Zhao H, Li X, Li J, Lai Y, Zou D (2020) Identification of a major QTL and candidate gene analysis of salt tolerance at the bud burst stage in Rice (Oryza sativa L.) using QTL-Seq and RNA-Seq. Rice 13:1–14. https://doi.org/10.1186/s12284-020-00416-1

Li W, Yang L, Meng Z, Qiu Y, Wang PS-P, Li X (2022) Phylogenetic analysis: a novel method of protein sequence similarity analysis. Int J Pattern Recognit Artif Intell 36:2258007. https://doi.org/10.1142/S0218001422580071

Liu Q, Liang Z, Feng D, Jiang S, Wang Y, Du Z, Li R, Hu G, Zhang P, Ma Y, Lohmann JU, Gu X (2021) Transcriptional landscape of rice roots at the single-cell resolution. Mol Plant 14:384–394. https://doi.org/10.1016/j.molp.2020.12.014

Liu D, Zhang X, Wu Q, Liu S, Zeng Q, Wang Q, Wang C, Li C, Singh RP, Bhavani S (2022) Combined linkage and association mapping reveals two major QTL for stripe rust adult plant resistance in Shaanmai 155 and their haplotype variation in common wheat germplasm. Crop J 10:783–792

Long W, Luo L, Luo L, Xu W, Li Y, Cai Y, Xie H (2022) Whole genome resequencing of 20 accessions of Rice landraces reveals Javanica genomic structure variation and allelic genotypes of a grain weight gene TGW2. Front Plant Sci 13:857435. https://doi.org/10.3389/fpls.2022.857435

Majeed A, Johar P, Raina A, Salgotra RK, Feng X, Bhat JA (2022) Harnessing the potential of bulk segregant analysis sequencing and its related approaches in crop breeding. Front Genet 13:944501. https://doi.org/10.3389/fgene.2022.944501

Majhi PK, Singh SK, Anandan A, Khaire AR, Korada M, Habde SV, Singh DK, Bhagvan AP, Jayasudha S, Singh A (2022) Parental polymorphism survey for evaluation and selection of contrasting parents for drought tolerance in Rice (Oryza sativa L.) by using SSR markers. Int J Environ Clim Change 12:2507–2519. https://doi.org/10.9734/ijecc/2022/v12i1131244

Mangrauthia SK, Sailaja B, Pusuluri M, Jena B, Prasanth VV, Agarwal S, Senguttuvel P, Sarla N, Ravindra Babu V, Subrahmanyam D, Voleti SR (2018) Deep sequencing of small RNAs reveals ribosomal origin of microRNAs in Oryza sativa and their regulatory role in high temperature. Gene Rep 11:270–278. https://doi.org/10.1016/j.genrep.2018.05.002

Matsumoto T, Wu J, Antonio BA, Sasaki T (2008) Development in Rice genome research based on accurate genome sequence. Int J Plant Genomics 2008:348621. https://doi.org/10.1155/2008/348621

Meng L, Guo L, Ponce K, Zhao X, Ye G (2016) Characterization of three Indica Rice multiparent advanced generation intercross (MAGIC) populations for quantitative trait loci identification. Plant Genome 9:plantgenome2015.10.0109. https://doi.org/10.3835/plantgenome2015.10.0109

Miya M, Yoshikawa T, Sato Y, Itoh J-I (2021) Genome-wide analysis of spatiotemporal expression patterns during rice leaf development. BMC Genomics 22:1–15. https://doi.org/10.1186/s12864-021-07494-5

Mohapatra S, Panda AK, Bastia AK, Mukherjee AK, Sanghamitra P, Meher J, Mohanty SP, Pradhan SK (2021) Development of submergence-tolerant, bacterial blight-resistant, and high-yielding near isogenic lines of popular variety, 'Swarna' through marker-assisted breeding approach. Front Plant Sci 12:672618

Mst F, Hossain M, Kang SG, Matin M (2023) Genetic variation, population structure, and marker-trait association of rice (Oryza sativa L.) cultivars using morphological characteristics and molecular markers

Ngernmuen A, Suktrakul W, Damchuay K, Longya A, Kate-Ngam S, Jantasuriyarat C (2019) Substantial enhancement of high polymorphic SSR marker development using in silico method from 18 available rice blast fungus genome sequences and its application in genetic diversity assessment. Biologia (Bratisl) 74:1181–1189. https://doi.org/10.2478/s11756-019-00264-5

Nubankoh P, Wanchana S, Saensuk C, Ruanjaichon V, Cheabu S, Vanavichit A, Toojinda T, Malumpong C, Arikit S (2020) QTL-seq reveals genomic regions associated with spikelet fertility in response to a high temperature in rice (Oryza sativa L.). Plant Cell Rep 39:149–162. https://doi.org/10.1007/s00299-019-02477-z

Nunes R, Storer C, Doleck T, Kawahara AY, Pierce NE, Lohman DJ (2022) Predictors of sequence capture in a large-scale anchored phylogenomics project. Front Ecol Evol 10:943361. https://doi.org/10.3389/fevo.2022.943361

Offiong N-AO, Edet JB, Shaibu SE, Akan NE, Atakpa EO, Sanganyado E, Okop IJ, Benson NU, Okoh A (2023) Metagenomics: an emerging tool for the chemistry of environmental remediation. Front Environ Chem 4:1052697. https://doi.org/10.3389/fenvc.2023.1052697

Oono Y, Ichida H, Morita R, Nozawa S, Satoh K, Shimizu A, Abe T, Kato H, Hase Y (2020) Genome sequencing of ion-beam-induced mutants facilitates detection of candidate genes responsible for phenotypes of mutants in rice. Mutat Res Mol Mech Mutagen 821:111691. https://doi.org/10.1016/j.mrfmmm.2020.111691

Ouyang S, Zhu W, Hamilton J, Lin H, Campbell M, Childs K, Thibaud-Nissen F, Malek RL, Lee Y, Zheng L, Orvis J, Haas B, Wortman J, Buell CR (2007) The TIGR Rice genome annotation resource: improvements and new features. Nucleic Acids Res 35:D883–D887. https://doi.org/10.1093/nar/gkl976

Ouyang W, Luan S, Xiang X, Guo M, Zhang Y, Li G, Li X (2022) Profiling plant histone modification at single-cell resolution using snCUT&Tag. Plant Biotechnol J 20:420–422. https://doi.org/10.1111/pbi.13768

Pidon H, Chéron S, Ghesquière A, Albar L (2020) Allele mining unlocks the identification of RYMV resistance genes and alleles in African cultivated rice. BMC Plant Biol 20:222. https://doi.org/10.1186/s12870-020-02433-0

Qian G, Ping J, Lu J, Zhang Z, Wang L, Xu D (2014) Construction of full-length cDNA library and development of EST-derived simple sequence repeat (EST-SSR) markers in Senecio scandens. Biochem Genet 52:494–508. https://doi.org/10.1007/s10528-014-9662-9

Quesnelle DC, Bendena WG, Chin-Sang ID (2023) A compilation of the diverse miRNA functions in Caenorhabditis elegans and Drosophila melanogaster development. Int J Mol Sci 24:6963. https://doi.org/10.3390/ijms24086963

Raghavan C, Mauleon R, Lacorte V, Jubay M, Zaw H, Bonifacio J, Singh RK, Huang BE, Leung H (2017) Approaches in characterizing genetic structure and mapping in a rice multiparental population. G3 (Bethesda) 7:1721–1730. https://doi.org/10.1534/g3.117.042101

Rainford P, Porter B (2022) Using phylogenetic analysis to enhance genetic improvement

Rajkumar MS, Shankar R, Garg R, Jain M (2020) Bisulphite sequencing reveals dynamic DNA methylation under desiccation and salinity stresses in rice cultivars. Genomics 112:3537–3548

Rao X, Ren J, Wang W, Chen R, Xie Q, Xu Y, Li D, Song Z, He Y, Cai D, Yang P, Lyu S, Li L, Liu W, Zhang X (2023) Comparative DNA-methylome and transcriptome analysis reveals heterosis- and polyploidy-associated epigenetic changes in rice. Crop J 11:427–437. https://doi.org/10.1016/j.cj.2022.06.011

Ratti M, Lampis A, Ghidini M, Salati M, Mirchev MB, Valeri N, Hahne JC (2020) MicroRNAs (miRNAs) and long non-coding RNAs (lncRNAs) as new tools for cancer therapy: first steps from bench to bedside. Target Oncol 15:261–278. https://doi.org/10.1007/s11523-020-00717-x

Rice Annotation Project (2008) The Rice annotation project database (RAP-DB): 2008 update. Nucleic Acids Res 36:D1028–D1033. https://doi.org/10.1093/nar/gkm978

Saini N, Kumar S, Deepak B, Sharma D (2022) High-throughput sequencing technologies in metagenomics: advanced approaches for algal research, pp 545–569

Sang J, Zou D, Wang Z, Wang F, Zhang Y, Xia L, Li Z, Ma L, Li M, Xu B, Liu X, Wu S, Liu L, Niu G, Li M, Luo Y, Hu S, Hao L, Zhang Z (2020) IC4R-2.0: rice genome reannotation using massive RNA-seq data. Genomics Proteomics Bioinformatics 18:161–172. https://doi.org/10.1016/j.gpb.2018.12.011

Satam H, Joshi K, Mangrolia U, Waghoo S, Zaidi G, Rawool S, Thakare RP, Banday S, Mishra AK, Das G, Malonia SK (2023) Next-generation sequencing technology: current trends and advancements. Biology 12:997. https://doi.org/10.3390/biology12070997

Sato Y, Antonio BA, Namiki N, Takehisa H, Minami H, Kamatsuki K, Sugimoto K, Shimizu Y, Hirochika H, Nagamura Y (2011) RiceXPro: a platform for monitoring gene expression in japonica rice grown under natural field conditions. Nucleic Acids Res 39:D1141–D1148. https://doi.org/10.1093/nar/gkq1085

Satpathy R (2022) Application of bioinformatics resources for mining of simple sequence repeats (SSRs) marker in plant genomes: an overview. Res J Biotechnol 17. https://doi.org/10.25303/1708rjbt1360143

Satrio RD, Fendiyanto MH, Supena EDJ, Suharsono S, Miftahudin M (2021) Genome-wide SNP discovery, linkage mapping, and analysis of QTL for morpho-physiological traits in rice during vegetative stage under drought stress. Physiol Mol Biol Plants 27:2635–2650. https://doi.org/10.1007/s12298-021-01095-y

Searle B, Müller M, Carell T, Kellett A (2023) Third-generation sequencing of epigenetic DNA. Angew Chem Int Ed Engl 62:e202215704. https://doi.org/10.1002/anie.202215704

Shin S-Y, Jeong JS, Lim JY, Kim T, Park JH, Kim J-K, Shin C (2018) Transcriptomic analyses of rice (Oryza sativa) genes and non-coding RNAs under nitrogen starvation using multiple omics technologies. BMC Genomics 19:532. https://doi.org/10.1186/s12864-018-4897-1

Singh SK, Majhi PK, Anandan A, Korada M, Habde SV, Khaire AR, Singh DK, Jayasudha S, Singh A, Bhagvan AP (2022) Microsatellites based parental polymorphism survey for moisture stress in Rice (Oryza sativa L.) between the parental genotypes HUR-1309 and CR Dhan 801. Emerg Life Sci Res 8:248–258

Singh A, Rami E, Upadhyay P, Gangawane AK (2023) The impact of climate change on crop production and combat strategies. In: Climate change and sustainable development. CRC Press, Boca Raton

Song L, Wang R, Yang X, Zhang A, Liu D (2023) Molecular markers and their applications in marker-assisted selection (MAS) in bread wheat (Triticum aestivum L.). Agriculture 13:642. https://doi.org/10.3390/agriculture13030642

Sun J, Wang J, Guo W, Yin T, Zhang S, Wang L, Xie D, Zou D (2021) Identification of alkali-tolerant candidate genes using the NGS-assisted BSA strategy in rice. Mol Breed 41:44. https://doi.org/10.1007/s11032-021-01228-x

Tabassum R, Dosaka T, Ichida H, Morita R, Ding Y, Abe T, Katsube-Tanaka T (2020) FLOURY ENDOSPERM11-2 encodes plastid HSP70-2 involved with the temperature-dependent chalkiness of rice (Oryza sativa L.) grains. Plant J 103:604–616. https://doi.org/10.1111/tpj.14752

Taheri S, Lee Abdullah T, Yusop MR, Hanafi MM, Sahebi M, Azizi P, Shamshiri RR (2018) Mining and development of novel SSR markers using next generation sequencing (NGS) data in plants. Molecules 23:399. https://doi.org/10.3390/molecules23020399

Takagi H, Abe A, Yoshida K, Kosugi S, Natsume S, Mitsuoka C, Uemura A, Utsushi H, Tamiru M, Takuno S, Innan H, Cano LM, Kamoun S, Terauchi R (2013) QTL-seq: rapid mapping of quantitative trait loci in rice by whole genome resequencing of DNA from two bulked populations. Plant J 74:174–183. https://doi.org/10.1111/tpj.12105

Takagi H, Tamiru M, Abe A, Yoshida K, Uemura A, Yaegashi H, Obara T, Oikawa K, Utsushi H, Kanzaki E, Mitsuoka C, Natsume S, Kosugi S, Kanzaki H, Matsumura H, Urasaki N, Kamoun S, Terauchi R (2015) MutMap accelerates breeding of a salt-tolerant rice cultivar. Nat Biotechnol 33:445–449. https://doi.org/10.1038/nbt.3188

Tare S, Tiwari S, Rathore M, Yasin M, Tripathi N, Prakash N, Tiwari PN, Tripathi M (2023) An overview of allele mining for crop improvement: TILLING and eco-TILLING. Plant Cell Biotechnol Mol Biol 26:91–103. https://doi.org/10.56557/PCBMB/2023/v24i1-28214

Thant AA, Zaw H, Kalousova M, Singh RK, Lojka B (2021) Genetic diversity and population structure of Myanmar Rice (Oryza sativa L.) varieties using DArTseq-based SNP and SilicoDArT markers. Plants 10:2564. https://doi.org/10.3390/plants10122564

Tribhuvan KU, Sandhya, Kumar K, Sevanthi AM, Gaikwad K (2018) MutMap: a versatile tool for identification of mutant loci and mapping of genes. Indian J Plant Physiol 23:612–621. https://doi.org/10.1007/s40502-018-0417-1

Tu M, Zeng J, Zhang J, Fan G, Song G (2022) Unleashing the power within short-read RNA-seq for plant research: beyond differential expression analysis and toward regulomics. Front Plant Sci 13:1038109. https://doi.org/10.3389/fpls.2022.1038109

Uprety S (2023) A prospect for genome wide association studies in crops

Verbyla AP, George AW, Cavanagh CR, Verbyla KL (2014) Whole-genome QTL analysis for MAGIC. Theor Appl Genet 127:1753–1770. https://doi.org/10.1007/s00122-014-2337-4

Wambugu P, Ndjiondjop M-N, Furtado A, Henry R (2018) Sequencing of bulks of segregants allows dissection of genetic control of amylose content in rice. Plant Biotechnol J 16:100–110. https://doi.org/10.1111/pbi.12752

Wang C, Han B (2022) Twenty years of rice genomics research: from sequencing and functional genomics to quantitative genomics. Mol Plant 15:593–619. https://doi.org/10.1016/j.molp.2022.03.009

Wang M, Yu Y, Haberer G, Marri PR, Fan C, Goicoechea JL, Zuccolo A, Song X, Kudrna D, Ammiraju JSS, Cossu RM, Maldonado C, Chen J, Lee S, Sisneros N, de Baynast K, Golser W, Wissotski M, Kim W, Sanchez P, Ndjiondjop M-N, Sanni K, Long M, Carney J, Panaud O, Wicker T, Machado CA, Chen M, Mayer KFX, Rounsley S, Wing RA (2014) The genome sequence of African rice (Oryza glaberrima) and evidence for independent domestication. Nat Genet 46:982–988. https://doi.org/10.1038/ng.3044

Wang Y, Huan Q, Chu X, Li K, Qian W (2020) Single-cell transcriptome analyses recapitulate the cellular and developmental responses to abiotic stresses in rice. 2020.01.30.926329

Wang N, Yu Y, Zhang D, Zhang Z, Wang Z, Xun H, Li G, Liu B, Zhang J (2022) Modification of gene expression, DNA methylation and small RNAs expression in Rice plants under in vitro culture. Agronomy 12:1675. https://doi.org/10.3390/agronomy12071675

Wing RA, Purugganan MD, Zhang Q (2018) The rice genome revolution: from an ancient grain to Green Super Rice. Nat Rev Genet 19:505–517. https://doi.org/10.1038/s41576-018-0024-z

Woronik A, Kiontke K, Jallad RS, Herrera RA, Fitch DHA (2022) Laser microdissection for species-agnostic single-tissue applications. J Vis Exp. https://doi.org/10.3791/63666

Wu Y, Zhao S, Li X, Zhang B, Jiang L, Tang Y, Zhao J, Ma X, Cai H, Sun C, Tan L (2018) Deletions linked to PROG1 gene participate in plant architecture domestication in Asian and African rice. Nat Commun 9:4157. https://doi.org/10.1038/s41467-018-06509-2

Xia L, Zou D, Sang J, Xu X, Yin H, Li M, Wu S, Hu S, Hao L, Zhang Z (2017) Rice expression database (RED): an integrated RNA-Seq-derived gene expression database for rice. J Genet Genomics 44:235–241. https://doi.org/10.1016/j.jgg.2017.05.003

Xie W, Wang G, Yuan M, Yao W, Lyu K, Zhao H, Yang M, Li P, Zhang X, Yuan J, Wang Q, Liu F, Dong H, Zhang L, Li X, Meng X, Zhang W, Xiong L, He Y, Wang S, Yu S, Xu C, Luo J, Li X, Xiao J, Lian X, Zhang Q (2015) Breeding signatures of rice improvement revealed by a genomic variation map from a large germplasm collection. Proc Natl Acad Sci U S A 112:E5411–E5419. https://doi.org/10.1073/pnas.1515919112

Xie Y, Chan L-Y, Cheung M-Y, Li M-W, Lam H-M (2023) Current technical advancements in plant epitranscriptomic studies. Plant Genome 16:e20316. https://doi.org/10.1002/tpg2.20316

Yan T, Sun M, Su R, Wang X, Lu X, Xiao Y, Deng H, Liu X, Tang W, Zhang G (2023) Transcriptomic profiling of cold stress-induced differentially expressed genes in seedling stage of Indica Rice. Plants 12:2675. https://doi.org/10.3390/plants12142675

Yang L, Wang J, Han Z, Lei L, Liu HL, Zheng H, Xin W, Zou D (2021) Combining QTL-seq and linkage mapping to fine map a candidate gene in qCTS6 for cold tolerance at the seedling stage in rice. BMC Plant Biol 21:1–14. https://doi.org/10.1186/s12870-021-03076-5

Yaobin Q, Peng C, Yichen C, Yue F, Derun H, Tingxu H, Xianjun S, Jiezheng Y (2018) QTL-Seq identified a major QTL for grain length and weight in Rice using near isogenic F2 population. Rice Sci 25:121–131. https://doi.org/10.1016/j.rsci.2018.04.001

Yu H, Xie W, Li J, Zhou F, Zhang Q (2014) A whole-genome SNP array (RICE6K) for genomic breeding in rice. Plant Biotechnol J 12:28–37. https://doi.org/10.1111/pbi.12113

Yuan JH, Li JH, Yuan JJ, Jia KL, Li SF, Deng CL, Gao WJ (2017) The application of MutMap in forward genetic studies based on whole-genome sequencing. Yi Chuan Hered 39:1168–1177. https://doi.org/10.16288/j.yczz.17-095

Zaborowska J, Perry A, Cavers S, Wachowiak WM (2023) Evolutionary targets of gene expression divergence in a complex of closely related pine species. J Syst Evol 61:198–212. https://doi.org/10.1111/jse.12896

Zaghum MJ, Ali K, Teng S (2022) Integrated genetic and omics approaches for the regulation of nutritional activities in Rice (Oryza sativa L.). Agriculture 12:1757. https://doi.org/10.3390/agriculture12111757

Zhang Q, Zheng T, Hoang L, Wang C, Nafisah JC, Zhang W, Xu J, Li Z (2016) Joint mapping and allele mining of the rolled leaf trait in Rice (Oryza sativa L.). PLoS One 11:e0158246. https://doi.org/10.1371/journal.pone.0158246

Zhang B, Fei Y, Feng J, Zhu X, Wang R, Xiao H, Zhang H, Huang J (2022) RiceNCexp: a rice non-coding RNA co-expression atlas based on massive RNA-seq and small-RNA seq data. J Exp Bot 73:6068–6077. https://doi.org/10.1093/jxb/erac285

Zhang H, Chen G, Xu H, Jing S, Jiang Y, Liu Z, Zhang H, Wang F, Hu X, Zhu Y (2023a) Transcriptome analysis of Rice embryo and endosperm during seed germination. Int J Mol Sci 24:8710. https://doi.org/10.3390/ijms24108710

Zhang S, Zhu C, Zhang X, Liu M, Xue X, Lai C, Xuhan X, Chen Y, Zhang Z, Lai Z, Lin Y (2023b) Single-cell RNA sequencing analysis of the embryogenic callus clarifies the spatiotemporal developmental trajectories of the early somatic embryo in Dimocarpus longan. Plant J Cell Mol Biol 115:1277–1297. https://doi.org/10.1111/tpj.16319

Zhang L, You R, Chen H, Zhu J, Lin L, Wei C (2023c) A new SNP in AGPL2, associated with floury endosperm in Rice, is identified using a modified MutMap method. Agronomy 13:1381. https://doi.org/10.3390/agronomy13051381

Zhou X, Bai X, Xing Y (2018) A Rice genetic improvement boom by next-generation sequencing. Curr Issues Mol Biol 27:109–126. https://doi.org/10.21775/cimb.027.109

Zhou J, Yang Y, Lv Y, Pu Q, Li J, Zhang Y, Deng X, Wang M, Wang J, Tao D (2022a) Interspecific hybridization is an important driving force for origin and diversification of Asian cultivated Rice Oryza sativa L. Front Plant Sci 13:932737. https://doi.org/10.3389/fpls.2022.932737

Zhou D, Zhou X, Sun C, Tang G, Liu L, Chen L, He H, Xiong Q (2022b) Transcriptome and metabolome analysis provides insights into the heterosis of yield and quality traits in two hybrid Rice varieties (Oryza sativa L.). Int J Mol Sci 23:12934. https://doi.org/10.3390/ijms232112934

Ziv N, Shuster BM, Siegal ML, Gresham D (2017) Resolving the complex genetic basis of phenotypic variation and variability of cellular growth. Genetics 206:1645–1657. https://doi.org/10.1534/genetics.116.195180

Regulatory Framework of Plant Variety Protection for Modernized Plant Breeding Approaches

14

S. N. Vasudevan, Thota Joseph Raju, S. K. Pooja, B. T. Krishnaprasad, and Geetha Govind

Abstract

Seed regulation is being governed by seed legislation, whereas the legal framework for protecting intellectual property rights (IPR) on plant varieties is provided by the Plant Variety Protection (PVP) law. Many countries consider plant varieties as intellectual property and provide legal protection for them. Protection is offered to plant varieties which are developed through conventional as well as modern breeding methodologies, viz. marker-assisted selection, genomic selection, and genome editing. This book chapter summarizes genesis of IPR; PVP; Plant Breeder's Rights (PBR); International Union for the Protection of New Plant Varieties (UPOV)/World Trade Organization (WTO) guidelines; Genetic Use Restriction Treaty (GURT); distinctness, uniformity, and stability (DUS) criteria; testing procedure; national and international laws in IPR; modern plant breeding technologies (MPBT) alternative to genetically modified organisms (GMO); gene patenting in rice; and salient features of the Protection of Plant Varieties and Farmers' Rights Act (PPVFRA).

Keywords

IPR · UPOV · WTO · PVP regulation · PPVFRA · DUS · MPBT · Gene patent

---

S. N. Vasudevan · T. J. Raju (✉)
Department of Seed Science and Technology, College of Agriculture, Hassan, University of Agricultural Sciences, Bangalore, Karnataka, India

S. K. Pooja · B. T. Krishnaprasad · G. Govind
Department of Agricultural Biotechnology, College of Agriculture, Hassan, University of Agricultural Sciences, Bangalore, Karnataka, India

A. Singh et al. (eds.), *Climate-Smart Rice Breeding*,
https://doi.org/10.1007/978-981-97-7098-4_14

14.1 Introduction

Agriculture serves as the fundamental basis of society, and plant breeding is a key component in the domain of agriculture. Higher agriculture production is reinforced by global population expansion and the increasing requirements for food, fodder, animal protein, fiber, and fuel (Hemathilake and Gunathilake 2022). Farmers also need cultivars that are suited to the growing challenges posed by changing climates, including biotic and abiotic stresses besides nutritional security. Expediting genetic enhancement by creating novel germplasms and technologies is crucial for attaining sustainable growth in production. The need and complexity of developing new and enhanced crop varieties, while also safeguarding the environment, is currently at its peak (Wang et al. 2018; Hasegawa et al. 2018). Rice, an essential dietary component for over 50% of the world's population, holds significant importance in ensuring food security and supporting agricultural economies worldwide (Ashraf et al. 2024). The practice of plant breeding involves the development of novel crop varieties with improved characteristics and has played a crucial role in improving rice output and its ability to withstand challenges (He et al. 2019; Benavente and Gimenez 2021; Huang and Liu 2022). Modern breeding approaches, including novel plant breeding techniques (NPBTs), have become influential instruments for expediting rice enhancement in recent years (Fiaz et al. 2021). Although NPBTs hold significant promise, their incorporation into current intellectual property right (IPR) frameworks and Plant Variety Protection (PVP) regimes poses unique hurdles.

Intellectual property (IP) safeguards producers of novel plant varieties against persons/organizations if they use protected variety without the consent of the proprietor of the new variety breeding materials for getting due recognition and compensation for their meticulous work and monetary expenditures (ITPGRFA 2001). A fruitful IP ecosystem fosters scientific investigation and innovation, leading to improved agricultural output and preservation of genetic diversity. Investments in research and development need substantial financial resources. It is imperative for corporate sector plant breeders and seed producers to have profitable outcome from their investment. Continuing investments and research are advantageous for farmers, growers, and society as it results in the development of novel varieties (Buttel and Belsky 2024). Effective and robust IP protection promotes breeding programs which are essential for maintaining the environment while meeting the food, feed, fiber, and fuel demands. Patents and Plant Breeder's Rights (PBR), which are alternatively known as PVP or Plant Variety Rights (PVR), are essential for fostering extensive innovation in the domain of agricultural sciences. Protection as an incentive for innovation is balanced with access to allow others to further enhance plant varieties in the most effective IP system (Cooke and Reeves 2003). Therefore, PBR is the most effective method of plant variety protection. The breeder's exemption is a provision within the PBR that permits access and utilization of commercial varieties protected under the PBR in order to breed further and constitutes a fundamental component of the PBR system (Balaravi 2010). This equilibrium can be maintained when patents are applied to the domain of plant breeding by exercising caution regarding the delineation of patentable subject matter, extent and caliber of patent assertions, validity of duration of patent protection, and exemptions to the patent right that pertains to breeding and research. As the domain of plant breeding continues to evolve and incorporate new technologies, the necessity for a robust and adaptable regulatory framework for PVP has become increasingly evident. Precision breeding and genome editing possess significant potential for developing new crop varieties with improved traits, but they also raise PVP system issues (Nerkar et al. 2022). For nearly six decades, novel plant varieties are safeguarded by PVR on meeting distinctness, uniformity, and stability (DUS) criteria. With the increasing demand for a wider range of plant varieties, highlighting the growing obstacles to registering new varieties, an immediate reform of the system is necessary. A modernized PVP framework is needed to accommodate these advances and protect plant breeder's ideas. In this regard, the highlights of regulatory framework for novel varieties and the concept of genomic DUS are discussed here.

14.2 Genesis

14.2.1 Plant Variety Protection (PVP)

Agricultural research in India, particularly in developing novel plant varieties, has fallen to the government and public sector institutes. However, India had no plant variety protection laws as it was thought to be not. In 1930, the US Plant Patent Act (PPA) granted property rights to privately developed asexually reproducing plants. The drive for the Plant Variety Protection Act (PVPA), 1970, was created by advances in breeding techniques, safeguarding sexually reproducing plants. The US Patent and Trademark Office ruled in 1985 that plants under breeder's rights are patentable under industrial patent law, allowing the utility of patents (Chawla 2010). Thus, US utility patents could include genetically modified plants and other plant varieties. In 1993, there were broad patent claims that covered all genetically modified cotton plants, regardless of the method used to create them. This unusual attempt to claim patent protection for an entire species was met with opposition. However, India's accession to the Trade Related Intellectual Property Rights Agreement (TRIPS) in 1994 made it necessary to enact such legislation. According to Article 27.3 (b) (1996) of this agreement, member countries are obligated to ensure the safeguarding of plant varieties by either a patent or an effective *sui generis system* (combination of both). Since member countries might tailor laws to their own systems, India chose to do so. The 1970 Indian Patent Act prohibits patenting agriculture and horticultural invention; hence, the Indian government created the *sui generis system* for plant variety protection in 2001 by enacting "The Protection of Plant Varieties and Farmers' Rights Act" (PPVFRA) in Parliament, integrating breeders' (Venkatesh and Pal 2013), farmers', and village communities' rights and addressing equitable benefit sharing. This regulation provides flexibility in protecting genera/species, levels, and duration in comparison to similar laws in other nations. In 2002, the Government of India (GOI) passed the Biological Diversity Act to preserve biological materials from exploitation. According to this

Act, if foreign countries want to utilize our material either for research or trade, they should obtain prior consent from the National Biological Diversity Authority of India. This Act protects farmers' and communities' rights and natural materials, but not breeders' sovereign rights on their variations. Transgenic plant varieties are protected under the PPVFR Act, provided permission from the Genetic Engineering Approval Committee (GEAC) to ensure environmental safety. PVP encourages funding for scientific research and development aimed at producing novel and improved plant varieties. Breeders are given monetary incentives and the exclusive right to market and sell their protected plant varieties; this structure, which covers international borders, serves to facilitate international trade. It assists in making a greater number of plant varieties more readily accessible, giving farmers and consumers a greater choice (UPOV 1991). However, it limits access to genetic resources, especially for researchers and small-scale breeders and potentially raises farmer costs by limiting farmer's ability to conserve and replant seeds from protected varieties (ITPGRFA 2001). The time-limited nature of PVP allows variety to enter the public domain, potentially affecting the availability of certain varieties. PVP could also result in the concentration of power in the hands of large corporations, potentially leading to market monopolies and increased seed prices (ETC Group 2015). PVP can be difficult and expensive to get, especially for researchers or small-scale breeders, which makes it harder for them to safeguard their varieties (UPOV 1991). Indigenous group's traditional knowledge and genetic resources are utilized without proper recognition or remuneration (Shiva 1997), which raises questions about bio piracy.

14.2.2 Plant Breeding Technologies: Beginning and Evolution

The progression of plant breeding techniques has extended across thousands of years, constituting an ongoing endeavor to enhance and revolutionize approaches for breeding. During traditional selection and domestication, agricultural societies began selecting and saving seeds from wild plants with desirable traits. Through unintentional selection and later purposeful cultivation, humans gradually domesticated plants like wheat, barley, and maize. Subsequently, controlled crossbreeding emerged as a prevalent technique, enabling breeders to deliberately merge the genetic characteristics of two parent plants with desirable qualities. The landrace selection method led to the creation of crop varieties that are precisely adapted to local conditions. Later, Mendelian principles helped plant breeders comprehend inheritance patterns and produce more predictable results. The green revolution, which occurred during the mid-twentieth century, resulted in a significant rise in agricultural output and played a crucial role in reducing worldwide food shortages. In the twentieth century, scientists used radiation or chemicals to cause plant mutations to create genetic variation. The emergence of recombinant DNA technology in the late twentieth century allowed scientists to transfer useful/novel genes from various species into crop plants, creating genetically modified plants. Prioritizing gene targeting for site-specific mutagenesis, intragenesis, and cisgenesis has sparked a discussion among scientists, breeders, and consumers about creating consumer-friendly genetically modified organisms (GMOs) (Kumar 2023). The utilization of marker-assisted breeding techniques accelerated the efficiency and precision of plant breeding. In the twenty-first century, CRISPR-Cas9-enabled target modification of specific genes without introducing foreign genes is revolutionizing the creation of novel crop varieties. Continued advancements in biotechnology, genomics, and bioinformatics are enhancing the efficiency and precision of plant breeding. Emerging techniques like RNA interference (RNAi) and synthetic biology offer new possibilities for tailoring crop plants to meet evolving agricultural and environmental challenges.

14.3 Intellectual Property Right Protection Organizations

To know the impact of IPRs on plant varieties and agriculture products, one must understand PBRs and how they differ from patents. In the USA, PVP is named for seed propagated and "plant patent" for all other vegetative propagated types including fruits and flower crops. The *sui generis* system was India's unique plant variety protection method. India's PPV&FR Act, 2001, is compliant with the International Union for the Protection of New Plant Varieties (UPOV) system and the TRIPS agreement. India holds an observer status in the UPOV (Chawla 2010).

14.3.1 International Plant Variety Protection (PVP)

IP becomes increasingly important as IPR gains respect in the global forums. Companies, institutions, and industries seek global IPR protection and have organization and signed accords and treaties to protect their invention. In this regard, the General Agreement on Tariffs and Trade (GATT) was established in 1947 with the goal of sustainable reduction of tariffs and trade barriers and elimination of preferences on a reciprocal and mutually advantageous basis. It was updated in eight rounds from 1947 to 1995 of which Uruguay's round (eight round, 1986–1994) introduced trade liberalization in agriculture. As a result, trade in agriculture was doubled during 1995. As the GATT emphasized only on trade in goods, it was effectively replaced by the World Trade Organization (WTO) during 1995. This created a new structure for monitoring trade service and resolving conflicts faster. TRIPS is the largest multinational IP agreement under the WTO, enforced on January 1, 1995, as there was a lack of multilateral framework principle on IPR and to frame measures for international trade of counterfeit goods in the international market by providing common protection and enforcement of IPR to protect the international economy. This agreement sets and describes intellectual property law standards and protection for each and individual IPs (Kumar 2023). The UPOV coordinated the application of PBR across countries, and an intergovernmental organization based in Geneva (Switzerland) was established in 1961 and was updated in 1972, 1978, and 1991 (International Union for the Protection of New Varieties of Plants (UPOV) n.d.). Article 27.3 (b) (1996) allows parties to exclude non-microorganisms, plants, and animals from patentability. India and many other countries, such as Thailand, lack a patenting mechanism to safeguard plants. TRIPS require member nations to safeguard their plant varieties through either patent protection or *sui generis system* or a combination of both (Cullet 1999). By fostering the development of novel plant varieties for society, the UPOV promotes an effective PVP system (Jordens 2005). Union member states grant breeders of new plant varieties an exclusive property right based on consistent rules under the UPOV Convention. Variety protection requires the uniqueness from well-known varieties, uniformity, stability, innovativeness, and not commercialized before the application date. The 1991 UPOV Convention Act increased pressure on developing nations to join and grants breeder's exclusive monopoly rights but offered farmers no rewards. The controversial 1991 UPOV Act grants the breeder exclusive harvest rights, extending the breeder's monopoly to farmers' harvests. These traits make the law unsuitable for emerging nations where breeders can directly control the trading of processed food, ornaments, and other commodities. Farmers cannot use harvested crops as planting material under the 1991 Act. The implementation will limit the reuse of seed to countries that meet standards. After multiple UPOV amendments, plant protection measures are now like patents or exclusive monopoly rights (UPOV 2005, 2013). The UPOV Convention encourages plant breeding by granting breeders of new plant varieties an IPR (breeder's right). The Convention needs a breeder's permission to propagate and harvest protected variety material under specified situations (Helfer 2003).

The Food and Agriculture Organization (FAO) promotes the International Treaty on Plant Genetic Resource for Food and Agriculture (ITPGRFA) conservation and usage for sustainable agriculture and provides debate and information sharing for international standard formulation and technical assistance for building national policies and legislation to embrace international regulatory frameworks (Helfer 2003). The ITPGRFA helps member states exchange and conserve crop genetic resources (Oguamanam 2014) to promote climate-smart agriculture research and food security. This treaty allows IPRs on new plant varieties obtained from material derived from the multilateral system (MLS) to re-compensate plant breeders for their efforts and triggers the sharing of monetary and nonmonetary benefits for farmers and protection of farmers' rights especially in developing countries. PVP in the International Seed Federation (ISF) promotes seed, plant variety, and technological intellectual property rights. The ISF standardizes seed licensing and buyer–seller contracts, promotes international seed health standards for pest and disease control, recognizes inventors, and provides farmers with high-quality seed varieties for food security and sustainability. It works with the International Seed Testing Association (ISTA), Organization for Economic Co-operation and Development (OECD), and UPOV to enhance seed sector development and strengthening. The FAO's Global Partnership Initiative for Plant Breeding Capacity Building (GIPB) improves plant breeding through knowledge institutions and extensively examined worldwide plant breeding and distribution through seed system.

India proposes replacing the 1966 Seeds Act with the 2004 Seeds Bill. The amendments of the Plant Variety Act, Biodiversity Act, and Patents Act over the decades led to IP protection. The participation of multinational corporations (MNC) and progressive adoption of transgenic seeds required a new Act. The measure seeks to "regulate the seed market and ensure quality seeds" (Balaravi 2010). The proposed changes will unify the Seed Law with other Seed Laws worldwide and expose the Indian seed industry to big business, which may necessitate biosafety requirements and new labelling and liability processes for GM seeds and nanotechnology. According to a 1998 Seed Policy Review Group in India, a long-awaited shakeup and reform of the Indian Seed Laws, including new seed legislation (New Seeds Bill, 2004), will become increasingly necessary and require regulatory adjustments in the twenty-first century. The Convention on Biological Diversity (CBD) was founded in 1992 and implemented on December 29, 1993, to conserve biological resources worldwide. Two protocol agreements augmented the convention, namely, the Colombian Cartagena protocol (implemented in 2003), which addressed biotechnology development, transfer, benefit sharing, and biosafety, and the Nagoya protocol (2010, Japan) implemented in 2014, which promoted knowledge and biological resource access. The Central Government of India created the National Biodiversity Authority (NBA) in 2003 to enforce the Biological Diversity Act (2002) which facilitates the regulation and advises the government on biological resource conservation, sustainable use, and traditional knowledge and also helps the state government to choose biodiversity-sensitive heritage sites under Section 37 (1).

14.3.2 Plant Protection Rights

Hybridization: Protection by law for seed producers against unauthorized use of novel varieties was difficult to enforce before the 1970s due to difficulties of recognizing the parent of a new commercial variety. Previously, only hybrid seeds, which do not breed true when replanted, were protected. Developing nations like India lack sufficient protection for parent seed lines from misuse beyond a few years, making hybridization less effective. Several countries during the 1930s offer clone protection through PBRs, which, unlike plant patents, do not prevent sexual reproduction or the use of germplasm to create new cultivars. Improvements in breeding technologies stimulated the PVPA, 1970 (Helfer 2003).

Plant breeders' rights (PBR): The IPR protection for breeders, originators, or owners of a variety is known as PBR. Understanding PBRs and how they differ from patents is crucial to comprehend the effect of IPRs on plant varieties and agricultural production. To encourage commercial plant breeding for agriculture and society, efforts must be made to allow breeders to earn from their crops (Mishra 1999). A breeder's rights certificate granted under the PPV&FR Act, 2001, provides exclusive privileges to the breeder, successor, agent, or licensee to manufacture, market, distribute, import, or export their own variety. On the other hand, it is advantageous in promoting private research, leading to more beneficial types for farmers; competitions will rise among the seed-producing companies to make available the best planting materials at an affordable price to the farming community. Farmers in India and the seed industry will get access to international plant varieties from nations with established IPR regulations. Increased genetic variability will enable Indian plant breeders to create superior varieties.

Farmer's rights: Farmers' rights under the Act define the farmer privilege under Section 39 (I) (i)–(iv). The Food and Agriculture Organization (FAO) acknowledges the significant role of farming communities in creating, selecting, and preserving genetic material and knowledge. Although knowledge is not patentable, information of "folk varieties" and genetic diversity are essential for current plant breeding and biotechnology research. Therefore, efforts were made to help these farmers under the PPV&FR Act, wherein it gives farmers benefits and importance such as the right to register traditional varieties free of cost and to store, use, re-sow, trade, share, and sell farmer's variety/farm produce without a brand name or branded with the breeder's registered name. Farmers are protected from innocent violation if they are unaware of breeder rights (Section 42 (1)). Farmers who conserve genetic resources of landraces and wild relatives of economic plants and improve them through selection and preservation are recognized and rewarded by the National Gene Fund, if their material is used as a donor in registrable varieties (Shanker et al. 2000). According to the Act, farmers or groups of farmers can seek compensation if a variety or propagating material fails to perform as claimed by the breeder under specified conditions.

Researcher's right: Researchers have access to protected varieties for legitimate study (Section 30). This Section allows individuals to use registered varieties for experiments or research and to use them as initial sources for creating new varieties, with breeder authorization required for repeated use as a parental line.

Community rights: It provides compensation for their contributions to the development of new varieties as given by the PPVFR Authority (Section 41 (1)).

14.3.2.1 Genetic Use Restriction Technologies (GURT)

Using the GURT principle by plant breeders to strengthen the protection of newly developed plant varieties allows the control of gene expression of an organism, further allowing constraints on the use of the organism or trait. At particular stages or generations of the crop, regulating the expression of the gene can be achieved by two types of GURTs: variety-GURT causes the seeds of the affected plant variety to be sterile, whereas trait-GURT results in the expression of a selected trait by a mechanism for trait expression into the variety, which can only be turned on or off with specific chemical inducer treatment (Goeschl and Swanson 2003).

14.3.3 Regulatory Framework Approaches in the Convention Countries

The European Union PVP follows UPOV, 1991. The Community Plant Variety Right (CPVR) offers the same IPR to applicants in all EU countries based on a single application. CPVRs can only be transferred or terminated uniformly inside the EU. As an alternative, it supplements European plant protection regulations. The same plant variety cannot be protected under both community and national regimes. The CPVR protects new botanical genera, species and hybrids provided that the varieties meet UPOV Convention requirements (Helfer 2003). The Australian PBR accepted UPOV, 1991, and protected plant varieties under the 1994 PBR Act. The PBR office oversees it (unlike IP Australia) which grants patents. Based on desired protection and capacity to achieve requirements, applicants can select between two protective systems. Faster and cheaper PBRs are available and good for short-term protection and noncommercial use without rights. The USA patented plant varieties through TRIPS and UPOV by PVP certificates. The UPOV is implemented in the USA under the PVP Act, 1970, and updated in 1994 to protect F_1 hybrids and tuber-propagated plants. The USA widely adopts the practice of patenting plant varieties, demonstrating its prevalence (Shanker et al. 2000; Brahmi et al. 2001; Phillips and Boone 2016; Plant Variety Protection and Intellectual Property Rights n.d.; Plant Variety Protection Office (PVPO) of the United States Department of Agriculture (USDA) n.d.). The Value for Cultivation and Usage (VCU) test is now mandatory along with DUS testing before a variety may be registered in the EU. Establishing comprehensive and uniform global methods is essential for undertaking plant variety testing. These procedures are fundamental for improving the accessibility of improved plant varieties and the PVP system. Therefore, it is crucial for the UPOV, as an intergovernmental entity, to actively participate and provide significant contributions in these areas. Once a PVP certificate reaches its official expiration date (ex PVP) and the active variety patent becomes invalid, the original variety is no longer protected by intellectual property rights and becomes available for public use.

14.4 Protection of Plant Varieties and Farmers' Rights Act (PPVFRA), 2001

14.4.1 Implementation of the PVPFR Act

This Act lays the foundation to establish an effective system for protecting plant varieties along with farmer and plant breeder rights. The Ministry of Agriculture and Farmers Welfare, GOI, has established the PVP&FR Authority in New Delhi. Breeders and farmers can register novel plant varieties with the PVP&FR Authority. This Act allows plant breeders (or licensees/agents/successors) exclusive rights to manufacture, market, sell, import, export, and distribute new plant varieties with a "Certificate of Registration" from the PPV&FR Authority. The plant varieties that can be registered in India include new varieties, extant varieties (which are commonly known and notified under Section 5 of the Seeds Act, 1966), farmer's varieties, and essentially derived varieties (EDVs), and this Act has protected Indian farmers and plant breeders. To implement the provisions of the Act, the Department of Agriculture and Co-operation, Ministry of Agriculture, established the PPV&FR Authority on November 11, 2005. Salient features of the Act are highlighted in Fig. 14.1, important sections in Table 14.1, registration forms in Table 14.2, and registration fee in Table 14.3 under the PPVFR Act, 2001.

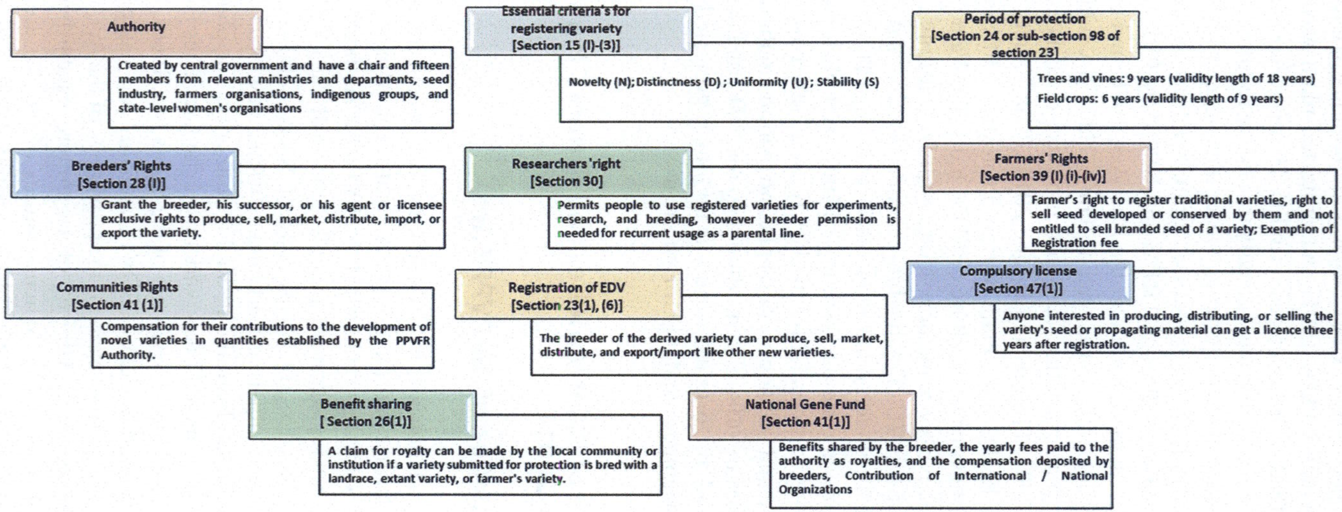

Fig. 14.1 Salient features of the PPVFR Act, 2001

Table 14.1 List of important sections under the PPVFR Act, 2001 (https://plantauthority.gov.in/plant-variety-forms 2024)

Sl. no.	Title	Section
1	Breeder	Section 2 (c)
2	EDV[a]	Section 2 (i)
3	Extant variety	Section 2 (j)
4	Farmer	Section 2 (k)
5	Farmer variety	Section 2 (l)
6	Variety	Section 2 (za)
7	Establishment of PPVFR authority	Section 3
8	Function of authority	Section 8
9	National registration plant variety	Section 13
10	Application for registration	Section 14
11	Duration of new plant variety (18 years) Duration of new extant variety (15 years) Other crops	Section 24–2 (i) Section 24–2 (ii) Section 24–2 (iii)
12	Registration to confer rights	Section 28
13	Surrender of registration	Section 33
14	Farmer's right	Section 39
15	Exemption form fee	Section 44
16	Infringement	Section 64

[a] Essentially derived varieties

Table 14.2 List of application forms to register under the PPVFR Act, 2001 (https://plantauthority.gov.in/plant-variety-forms 2024)

Sl. no.	Form	Title
1	Form I	For registration of new variety and extant variety
2	Form II	For registration of EDVs
3	Application form	For registration of farmers
4	PV-3	Notice of opposition
5	PV-6	Renewal of registration of plant variety
6	PV-9	Registration as an agent or license
7	PV-28	Grant of compulsory license
8	PV-33	Request for certified copies of entries in the plant varieties' register or for inspection of such entry

14.4.2 Important Sections and Salient Features of the PPVFR Act, 2001

The Act has 11 chapters and is divided into 97 clauses; the first chapter has the title and the definitions used in the context of the Act, whereas the last chapter has the miscellaneous clauses. The other nine chapters deal with PPVFR authority, registration of plant varieties, duration, effect of registration, benefit sharing, surrender and revocation of certificates, farmer's rights, compulsory license, plant varieties'

Table 14.3 Registration fees for different varieties under the PPVFR Act, 2001 (https://plantauthority.gov.in/plant-variety-forms 2024)

Sl. no.	Type of variety	Individual (Rs.)	Educational institute (Rs.)	Commercial entity (Rs.)	Annual maintenance fee (Rs.)
1	New variety	7000	10,000	50,000	2000
2	Extant variety (about which there is common knowledge)	7000	10,000	50,000	2000
3	Extant variety (notified under section of the Seeds Act, 1966)	2000	2000	2000	2000
4	EDVs	7000	10,000	50,000	Not specified
5	Farmer's variety	No fee	No fee	No fee	10

protection appellate tribunal, finance, accounts, audit, infringement, offences, and penalties (Table 14.1). Information on protection and examination of varieties (new varieties, extant varieties, farmer varieties, varieties of common knowledge, essentially derived varieties); authority; eligibility; period of protection; rights of farmers, breeders, researchers, and community; benefit sharing, compulsory licensing; and registration through national gene fund constitution through the Act are listed in salient features (Fig. 14.1). To register the variety for protection under the Act, check with the application form and other plant variety (PV) forms (Table 14.2) and registration fees for individual, educational institute, and commercial (Table 14.3).

14.5 Plant Variety Protection (PVP) for Conventional to New Breeding Approaches

Breeders often introduce traits into diverse backgrounds that have evolved to various environmental situations. This results in screening hundreds of thousand plants growing in various environments. Currently, marker-assisted selection can be used to select desirable plants at an early stage and during off-season. This involves screening and selecting progeny with desirable traits (trait markers) and desirable genetic backgrounds (background/genome-wide markers). The UPOV's PVP framework for granting IPR for new varieties is based on novelty (N), distinctness (D), uniformity (U), and stability (S) under the specified environment. DUS standards must be met to safeguard new varieties. The present PVP system uses morphological description of plant varieties. The DUS test is conducted in a field or greenhouse over two successive seasons in at least two locations. Several morphological traits are recorded for the new and comparable varieties during this period, as part of general knowledge (https://www.resource/en/dus_guidance.html n.d.). Field and greenhouse assessments for the phenotype-based DUS test are time-consuming and expensive. Discerning them only based on morphological characteristics becomes increasingly challenging as these attributes can be influenced by environmental conditions. Phenotypic DUS traits followed in rice are given in Fig. 14.2. EDVs share essential qualities with the initial variety, except from their unique traits. To detect EDV from fraud, phenotype-based DUS tests typically fall short due to crucial morphological features distinguishing EDV from the initial variety. These drawbacks can be overcome by marker-based and sequencing-based tests (Yu and Chung 2021). The International Union for the Protection of New Varieties of Plants (UPOV) suggested three ways for integrating molecular marker technology into DUS testing (UPOV INF/18/1 2011). Molecular markers can only be utilized in distinctness, uniformity, and stability (DUS) testing if they are directly related to the DUS features, as per the UPOV criteria. Molecular markers are widely used in the identification of different rice varieties. Several successive reports on the usage of SSR markers in rice DUS testing (Pourabed et al. 2015) followed by the development of 40 SNP markers which can effectively differentiate between *indica* and *japonica* rice have been reported (Zheng et al. 2022). In addition, collection of KASP markers is available to quickly genotype and distinguish basmati rice cultivars (Steele et al. 2021).

14.5.1 Modernized Methods for Proving DUS

14.5.1.1 Molecular Markers-Based DUS Test

The working group on biochemical and molecular techniques and DNA profiling (BMT) at UPOV includes expertise in DUS testing and plant breeding (Heckenberger et al. 2005). The UPOV-BMT aims to expand practical features for plant variety registration and enhance PBR protection such as using molecular markers directly linked to DUS traits (Cockram et al. 2012). Marker–trait relationships across various germplasm accurately predict trait expression (Cooke and Reeves 2003).

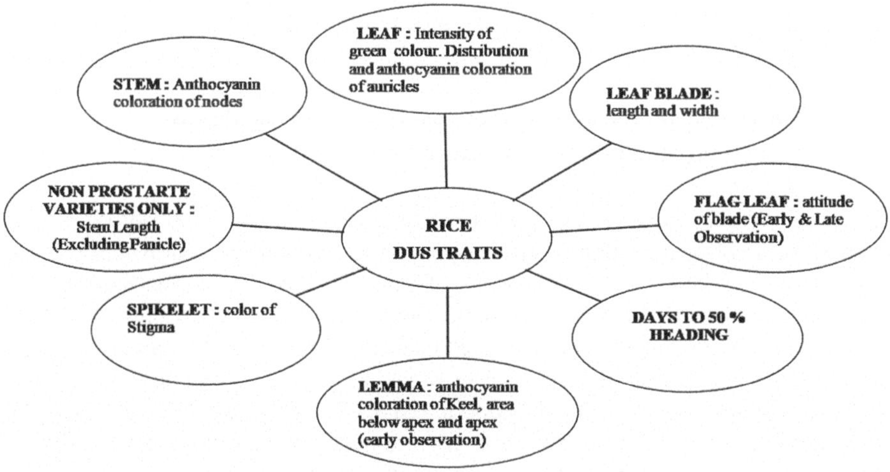

Fig. 14.2 List of DUS traits as per UPOV for rice

14.5.1.2 Sequencing-Based DUS Test

To find causal variants faster and more cost-effectively than whole-genome sequencing (WGS), targeted next-generation sequencing (NGS) sequences a specified genomic area. DUS testing may benefit from targeted NGS instead of standard marker technologies (Jones et al. 2013; Zhang et al. 2020). A noteworthy example of targeted NGS is CRISPR-Cas9-enriched long-read sequencing (LRS). CRISPR-Cas9 can be used to isolate a region of interest, followed by enrichment and LRS to remove long, intact DNA molecules from the trait of interest. This enables the target region to be isolated without amplification and studied utilizing LRS technology (Lopez-Girona et al. 2020). LRS data provides a thorough perspective of the region of interest, identifying intricate structural variation and repetitive elements that other DNA-sequencing technologies struggle to uncover.

Plant breeders can utilize novel breeding technologies (NBTs) to manipulate gene sequences by introducing desirable alleles or removing undesirable ones. Genome editing technology has the potential to profoundly transform the field of plant breeding, leading to the emergence of novel plant varieties. Genome editing can modify the genotypic distances across types by a small number of base pair nucleotides. If new genome-edited variety is protected by PBR, it will lead to disagreement because genome-edited varieties retain all the essential traits of the original variety except for a few specific base pair nucleotides. Consequently, it would be eligible for an EDV claim. The herbicide resistance of the initial variety is a distinctive characteristic in DUS testing, which adds significant agronomic benefit to this minor modification. Precise legal definitions of EDV are essential to hinder or deter the application of NBT in breeding operations (Lopez-Girona et al. 2020). It is imperative to redefine the concept of EDV and establish standardized regulations that are universally accepted by the public and international breeders for future endeavors. However, none of these have been adopted by the UPOV, and molecular markers are only accepted if they totally correlate with DUS traits and encourage potential, new opportunities, and rapid adoption for next-generation DUS testing. An essential priority in the development, evaluation, and release of new crop varieties should be the revision of the definition of plant variety.

14.5.2 Challenges in Plant Variety Protection Regulatory Framework for New Breeding Techniques

The rapid advancement of modern plant breeding technologies (MPBTs) has revolutionized the field of agriculture, offering promising solutions to improve crop traits. However, the regulatory framework for PVP has fallen behind these advancements, posing several challenges that hinder the development and commercialization of MPBT-derived crops. According to a ruling by the Delhi High Court in India, transgenic plants would likewise be safeguarded under plant variety rights. No transgenic food crop has been approved for commercial cultivation in India. This is because of delays in safety approvals and intense opposition from nongovernmental organizations (NGOs), biodiversity conservation groups, farmers, and public. On the other side, genome-edited plants (categorized as SDN-1 and SDN-2) are exempted from the regulation and risk assessment under Rule 20 of the Rules 1989 (Department of Biotechnology Ministry of

Science and Technology Government of India 2022) demonstrated to contain no exogenous DNA, but India lacks a policy regarding synthetic biology.

- **Regulatory inconsistency and uncertainty**: One of the most pressing challenges is the lack of harmonized regulations for MPBTs across different countries and regions. The regulatory landscape is fragmented, with varying interpretations and applications of existing rules, leading to confusion, uncertainty, and increased costs for developers and investors (ETC Group 2015). This inconsistency creates barriers to trade and hinders the global adoption of MPBT innovations.
- **Overly broad definition of GMOs**: Many existing regulatory frameworks define genetically modified organisms (GMOs) broadly, encompassing MPB-derived crops though they may not introduce foreign genes. This classification subjects MPBT products to the same rigorous and costly regulatory procedures as traditional GMOs, despite their potential benefits and reduced risks. This regulatory burden discourages investment in MPBT research and development.
- **Overlapping intellectual property rights (IPRs)**: A plant variety can receive protection through both a patent and Plant Variety Protection (PVP). The patent may encompass the novel genetic engineering methodologies employed in the creation of the variety, whereas PVP ensures the preservation of the unique traits of the plant itself. A trademark may be linked to a safeguarded plant variety, especially if the variety possesses a distinct appellation or brand awareness. This overlapping safeguard helps to deter unlawful utilization of the variety's appellation or trademark. Copyright can be applicable to several resources associated with a safeguarded plant variety, including photos, descriptions, and marketing materials. This safeguard guarantees that the original artistic works connected with the variety are not violated. Having several overlapping IPRs offers enhanced protection against the illicit utilization and commercial exploitation of a safeguarded plant variety (Overlapping intellectual property rights in plant variety protection: challenges and opportunities n.d.). This extensive safeguard encourages investment in the field of plant breeding and promotes innovation. Concurrent IPRs can result in disputes between individuals or entities that possess different IP rights, especially in relation to licensing and commercialization contracts. Breeders and holders of plant varieties may face increased administrative burdens when dealing with different intellectual property rights. Efficient methods and standardized intellectual property regimes help alleviate these challenges. Concurrent IPRs can foster equitable competition by preventing any one business from having exclusive control over a specific plant type. This promotes the development of new ideas and varieties.
- **Lack of scientific risk assessment**: The risk assessment processes for MPBTs often fail to consider the unique characteristics of these techniques and their potential environmental and human health impacts. Traditional GMO risk assessment protocols, designed for organisms containing foreign genes, may not be applicable to MPBT-derived crops. This lack of tailored risk assessment methodologies hinders the objective and science-based evaluation of MPBT innovations. Compared to transgenic crops, intragenic/cisgenic crops require less regulation (Vasudevan et al. 2023).
- **Public perception and consumer concerns**: Public perception and consumer concerns regarding MPBTs pose additional challenges. Misconceptions about

the safety and potential risks of MPBTs can lead to resistance to their adoption and hinder public acceptance. Addressing these concerns through effective communication and education is crucial to build public trust in MPBT innovations.

- **Intellectual property protection**: The current PVP framework may not adequately protect the intellectual property rights of MPBT developers. The absence of clear guidelines for patenting MPBT-derived crops and traits can create uncertainty and discourage investment in research and development.
- **Overall cost estimation of PVP (including DUS testing)**: Breeding operation acceptance and molecular technology vary with public vs. industry institutes find key field crops vs. vegetables. A breeder from a well-funded institute can obtain all DUS characteristics and molecular markers during line development and may only need money for PVP registration and maintenance, while a poor breeder may struggle to acquire DUS characteristics and PVP cost estimation (Zaid and Knox 2020).

14.5.3 Way Forward and Future Needs for Modern Breeding Approaches

A number of actions can be taken to address these issues and promote the appropriate development and use, harmonize regulations, and categorize MPBTs separately. Encourage public knowledge, improve PVP frameworks to offer exact guidelines for patenting crops and traits derived via MPBT. Advanced genotyping and sequencing technology can be applied in variety registration through chip and array technology by thousands of single nucleotide polymorphism (SNP) markers used in breeding programs (Sun et al. 2020). Double haploids (anther culture) is a valuable technique in precision breeding for developing new plant varieties (International Union for the Protection of New Varieties of Plants (UPOV) n.d.). It allows for the production of homozygous haploid plants quickly by inducing androgenesis and diploidization. To enhance the callusing capacity of anthers, different media compositions are being developed (Dwivedi et al. 2015). Due to statistical advances and increased computer capability, breeding programs are gradually employing genome-wide, low-pass sequencing technologies (Wang et al. 2020). By utilizing advanced genotyping and sequencing technologies with sophisticated statistical methods and powerful computational capabilities, ex PVP lines have the potential to serve as a novel source of germplasm for any breeding effort. Furthermore, it has created fresh prospects for identifying and utilizing the genes that regulate agronomic performance in both traditional breeding initiatives and genome editing initiatives. Examining the extension of intellectual property rights outside Plant Variety Protection (PVP) inside the UPOV system, such as through variety patents, utility patents, trade secrets, and exploring how both public and private sectors employ ex PVP lines in their breeding and research initiatives (Yu and Chung 2021).

14.5.4 Gene Patents in Rice: Its Status and Update

To safeguard rice crops under UPOV, 1991 (PPVFRA, 2001 and 2005), and grant patent through the Indian Patent Act, 1970, which bans plants from patentability (gene patenting), however, the synthetic gene, vector, and method of crop species transformation are patentable if distinct from the native gene (Correa et al. 2020). The total number of applications received during 2022–2023 by PV authorities in India is 250. Out of 100 applications received for cereals, the maximum application (50) is for rice (Protection of Plant Varieties and Farmers' Rights Authority 2022–2023). Rice gene patents are covered under abiotic stress tolerance (cold, drought, flooding, salt, cadmium, arsenic, diamide, iron deficiency), biotic stress (striped rice borer, bacterial blight, blast, leaf streak, gray mold, rice stripe virus, glyphosate or glufosinate ammonium tolerance), and agronomic features (yield, grain weight, root length, plant height, chlorophyll content) (Ricroch et al. 2022). There are a couple of registered rice germplasm possessing tolerance to various traits (Table 14.4; https://icar-nrri.in/patent-ipr 2024) which can be used as a prominent source in rice breeding programs and some of the approved/nonregulated biotechnological rice events (gene patents) presented in Table 14.4.

Table 14.4 List of gene patent in rice

Individual trait and gene	Country	Patent no.	References
Magnaporthe oryzae polynucleotide associated with blast resistance and uses	India	397653398	https://nipb.icar.gov.in/patents (2024)
Polynucleotide fragments for the expression of genes in plants in response to pathogens and wounding	India	460126	
Nucleotide sequence of rice responsible for resistance to *Magnaporthe oryzae* and uses thereof	India	346774	
"A single copy gene based 50K SNP rice DNA chip and uses thereof"	India	339438	
Novel allele (Rc) alleles in rice encoding bHLH-type transcription factors for enhancing red pigmentation. Provided novel PCR primers for identifying rice germplasm indicating the potential for red and non-red pigmentation in rice pericarp and seed coat	USA	9556446	https://patents.justia.com/patents-by-us-classification/800/320.2 (2024)
Cultivar, method for differentiating plant cultivars, and method for causing earlier maturing of rice individual Rice cultivar 'Koshihikari kazusa no. 6' or rice cultivar 'Habataki' in the third chromosome of the rice individual	USA	9029669	
Method for increasing photosynthetic carbon fixation in rice	USA	20150118385	
Rice gene (GS3), exerting primary control over grain length and grain weight	USA	20150082496	
Pathogen-inducible promoter from rice and uses thereof	India	331964	https://nipb.icar.gov.in/patents (2024)

14.6 Conclusion

The field of biology is experiencing fast advancements in scientific innovation, notably in current plant breeding techniques. The transition from conventional to random mutation, accomplished by incorporating chemical or radiation mutation breeding, to the more precise point-specific mutation facilitated by novel breeding techniques poses a challenge for regulatory systems to promptly adapt. It provides valuable insights into the difficulties of addressing this regulatory deficiency for PVP. Contemporary methods of plant breeding comprise a wide variety of procedures, each characterized by its own specific strategies and assessments. The selection of a breeding methodology is contingent upon the particular breeding goals, accessible resources, and temporal limitations. Conventional breeding involves the process of crossing and selection, whereas mutation breeding provides genetic variety by inducing mutations. On the other hand, marker-assisted selection (MAS) uses molecular markers to monitor the inheritance of desired characteristics. Molecular (genotypic) testing of DUS alone cannot replace the traditional (phenotypic) DUS testing, but both of these analyses combined can increase the efficiency of DUS testing. The lack of regulatory clarity surrounding products of NBTs is not a result of scientific apprehensions, but rather the intrusion of political influence in the regulatory clearance procedure. The primary factors contributing to ambiguity around the regulatory licensing of novel plant breeding types are unrelated to scientific considerations, as previously mentioned. According to the credible sources, if the regulation of gene-edited technologies were to be based only on scientific risk assessment criteria, these items would be approved safely. However, due to political intervention in the process of granting regulatory approval, policymakers can enable the responsibility in the development and commercialization of MPBTs by tackling these difficulties and implementing a supportive regulatory strategy. This will unlock the potential of MPBTs to revolutionize agriculture and contribute to sustainable food and nutritional security.

References

Article 27(3)(b) (1996) In: Agreement on Trade Related Aspects of Intellectual Property Rights (TRIPS), World Intellectual Property Office (WIPO) Publication No. 223(E). ISBN:92-805-0640-4

Ashraf H, Ghouri F, Baloch FS, Nadeem MA, Fu X, Shahid MQ (2024) Hybrid rice production: a worldwide review of floral traits and breeding technology, with special emphasis on China. Plants 13(5):578. https://doi.org/10.3390/plants13050578

Balaravi S (2010) Fault lines in the 2010 seeds bill. Econ Polit Wkly 45(32):12–16. https://www.jstor.org/stable/20764379

Benavente E, Gimenez E (2021) Modern approaches for the genetic improvement of rice, wheat and maize for abiotic constraints-related traits: a comparative overview. Agronomy 11(2):376. https://doi.org/10.3390/agronomy11020376

Brahmi P, Saxena S, Dhillon BS (2001) The protection of plant varieties and farmers rights bill. Curr Sci 86(3):392–398. https://www.jstor.org/stable/24108733

Buttel FH, Belsky J (2024) Biotechnology, plant breeding, and intellectual property: social and ethical dimensions science, technology and human values. JSTOR 12(1):31–49. http://www.jstor.org/stable/688815

Chawla HS (2010) Introduction to plant biotechnology. Science Publishers, USA
Cockram J, Jones H, Norris C, O'Sullivan DM (2012) Evaluation of diagnostic molecular markers for DUS phenotypic assessment in the cereal crop, barley (*Hordeum vulgare* ssp. vulgare L.). Theor Appl Genet 125(8):1735–1749. https://doi.org/10.1007/s00122-012-1950-3
Cooke RJ, Reeves JC (2003) Plant genetic resources and molecular markers: variety registration in a new era. Plant Genet Resource 1(2–3):81–87. https://doi.org/10.1079/PGR200312
Correa CM, Correa JI, De Jonge B (2020) The status of patenting plants in the global south. J World Intellect Prop 1:121–146. https://doi.org/10.1111/jwip.12143
Cullet P (1999) Revision of the TRIPS agreement concerning the protection of plant varieties lessons from India concerning the development of a *sui generis* system. J World Intellect Prop 2:617
Department of Biotechnology Ministry of Science and Technology Government of India (2022) Guidelines for the safety assessment of genome edited plants. New Delhi, pp 5–7
Dwivedi SL, Britt AB, Tripathi L, Sharma S, Upadhyaya HD, Ortiz R (2015) Haploids: constraints and opportunities in plant breeding. Biotechnol Adv 33(6):812–829. https://doi.org/10.1016/j.biotechadv.2015.07.001
ETC Group (2015) Putting the cartel before the horse…and farm, seeds, soil, peasants, etc
Fiaz S, Wang X, Khan SA, Ahmar S, Noor MA, Riaz A, Ali K, Abbas F, Mora-Poblete F, Figueroa CR, Alharthi B (2021) Novel plant breeding techniques to advance nitrogen use efficiency in rice: a review. GM Crops Food 12(2):627646. https://doi.org/10.1080/21645698.2021.1921545
Goeschl T, Swanson T (2003) The development impact of genetic use restriction technologies: a forecast based on the hybrid crop experience. Environ Dev Econ 8(1):149–165. https://doi.org/10.1017/S1355770X03000081
Hasegawa T, Fujimori S, Havlík P, Valin H, Bodirsky BL, Doelman JC, Fellmann T, Kyle P, Koopman JF, Lotze-Campen H, Mason-D'Croz D (2018) Risk of increased food insecurity under stringent global climate change mitigation policy. Nat Clim Chang 8(8):699–703. https://doi.org/10.1038/s41558-018-0230-x
He Y, Yan L, Ge C, Yao XF, Han X, Wang R, Xiong L, Jiang L, Liu CM, Zhao Y (2019) PINOID is required for formation of the stigma and style in rice. Plant Physiol 180(2):926–936. https://doi.org/10.1104/pp.18.01389
Heckenberger M, Bohn M, Frisch M, Maurer H, Melchinger A (2005) Identification of essentially derived varieties with molecular markers: an approach based on statistical test theory and computer simulations. Theor Appl Genet 111(3):598–608. https://doi.org/10.1007/s00122-005-2052-2
Helfer LR (2003) Intellectual property rights and the international treaty on plant genetic resources for food and agriculture. Proceedings of the annual meeting (American Society of International Law) 97(2–5):33–35. https://doi.org/10.1017/S0272503700059723
Hemathilake DMKS, Gunathilake DMCC (2022) Agricultural productivity and food supply to meet increased demands. In: Bhat R (ed) Future foods. Academic Press, pp 539–553. isbn:9780323910019. https://doi.org/10.1016/B978-0-323-91001-9.00016-5
https://icar-nrri.in/patent-ipr. Accessed 20 May 2024
https://nipb.icar.gov.in/patents. Accessed 20 May 2024
https://patents.justia.com/patents-by-us-classification/800/320.2. Accessed 20 May 2024
https://plantauthority.gov.in/plant-variety-forms. Accessed 20 May 2024
https://www.resource/en/dus_guidance.html
Huang Z, Liu W (2022) Hybrid rice production. In: Modern techniques of rice crop production. Springer, Berlin, pp 629–645
International Union for the Protection of New Varieties of Plants (UPOV). https://www.upov.int/
ITPGRFA (2001) International treaty on plant genetic resources for food and agriculture
Jones H, Norris C, Smith D, Cockram J, Lee D, O'Sullivan DM, Mackay I (2013) Evaluation of the use of high-density SNP genotyping to implement UPOV model 2 for DUS testing in barley. Theor Appl Genet 126(4):901–911. https://doi.org/10.1007/s00122-012-2024-2

Jordens R (2005) Progress of plant variety protection based on the international convention for the protection of new varieties of plants (UPOV convention). World Pat Inf 27:232–243. https://doi.org/10.1016/j.wpi.2005.03.004

Kumar A (2023) Silent features of TRIPS (trade related intellectual property rights) agreements. Int J Innov Anal 3(1):31–38. issn:2583-0295

Lopez-Girona E, Davy MW, Albert NW, Hilario E, Smart MEM, Kirk C, Thomson SJ, Chagne D (2020) CRISPR-Cas9enrichment and long read sequencing for fine mapping in plants. Plant Methods 16:121. https://doi.org/10.1186/s13007-020-00661-x

Mishra JP (1999) Biotechnology and intellectual property rights. Yojana, pp 15–20

Nerkar G, Devarumath S, Purankar M, Kumar A, Valarmathi R, Devarumath R, Appunu C (2022) Advances in crop breeding through precision genome editing. Front Genet 13:880195. https://doi.org/10.3389/fgene.2022.880195

Oguamanam C (2014) Farmers rights and the intellectual property dynamic. In: David M, Halbert D (eds) Agriculture in sage handbook of intellectual property. Sage, London, pp 238–257

Overlapping intellectual property rights in plant variety protection: challenges and opportunities. https://uclawsf.edu/jd-concentrations/intellectual-property-law/

Phillips R, Boone B (2016) Plant variety protection: a global study. Edward Elgar Publishing, Cheltenham

Plant Variety Protection and Intellectual Property Rights. https://www.wipo.int/edocs/mdocs/sme/en/wipo_ip_bis_ge_03/wipo_ip_bis_ge_03_11-main1.pdf

Plant Variety Protection Office (PVPO) of the United States Department of Agriculture (USDA). https://www.ams.usda.gov/services/plant-variety-protection

Pourabed E, Noushabadi MRJ, Jamali SH, Alipour NM, Zareyan A, Sadeghi L (2015) Identification and DUS testing of rice varieties through microsatellite markers. Int J Plant Genomics 2015:965073. https://doi.org/10.1155/2015/965073

Protection of Plant Varieties and Farmers' Rights Authority, New Delhi, Annual report 2022–2023, pp 8–10. https://plantauthority.gov.in/

Ricroch AE, Martin-Laffon J, Rault B, Pallares VC, Kuntz M (2022) Next biotechnological plants for addressing global challenges: the contribution of transgenesis and new breeding techniques. New Biotechnol 66:25–35. https://doi.org/10.1016/j.nbt.2021.09.001. Epub 2021 Sep 16. PMID: 34537403.

Shanker R, Kochhar AS, Archak S, Gautam PL (2000) Plant variety protection: lessons from cross-country perspective, policy brief no. 11. National Centre for Agricultural Economics and Policy Research (ICAR), New Delhi

Shiva V (1997) Biopiracy: the plunder of nature and knowledge

Steele K, Tulloch MQ, Burns M, Nader W (2021) Developing KASP markers for identification of basmati rice varieties. Food Anal Methods 14:663–673. https://doi.org/10.1007/s12161-020-01892-3

Sun C, Dong Z, Zhao L, Ren Y, Zhang N, Chen F (2020) The Wheat 660K SNP array demonstrates great potential for marker assisted selection in polyploid wheat. Plant Biotechnol J 18(6):1354–1360. https://doi.org/10.1111/pbi.13361

UPOV (1991) International convention for the protection of new varieties of plants (1991 Act)

UPOV (2005) UPOV report on the impact of plant variety protection. UPOV Publication 353. http://www.upov.int/en/about/pdf/353_Executive_Summary

UPOV (2013). http://www.upov.int. Accessed 20 May 2024

UPOV INF/18/1 (2011) Guidance on the use of biochemical and molecular makers in the examination of distinctness, uniformity and stability (DUS). International Union for the Protection of New Varieties of Plants, Geneva. https://www.upov.int/edocs/infdocs/en/upov_inf_18

Vasudevan SN, Pooja SK, Raju TJ, Damini CS (2023) Cisgenics and intragenics: boon or bane for crop improvement. Front Plant Sci 14(1275145):1–11. https://doi.org/10.3389/fpls.2023.1275145

Venkatesh P, Pal S (2013) Determinants and valuation of plant variety protection in India. J Intellect Property Rights 18:448–456

Wang J, Vanga SK, Saxena R, Orsat V, Raghavan V (2018) Effect of climate change on the yield of cereal crops: a review. Climate 6(2):41. https://doi.org/10.3390/cli6020041

Wang N, Yuan Y, Wang H, Yuan T, Wang H, Yu D, Liu Y, Zhang A, Gowda M, Nair SK (2020) Applications of genotyping-by-sequencing (GBS) in maize genetics and breeding. Sci Rep 10(16308):1–12. https://doi.org/10.1038/s41598-020-73321-8

Yu JK, Chung YS (2021) Plant variety protection: current practices and insights. Genes (Basel) 12(8):127. https://doi.org/10.3390/genes12081127

Zaid AM, Knox AM (2020) Plant variety protection and novel breeding techniques. In: The economics of plant variety protection. CABI, Wallingford, pp 211–232

Zhang S, Li B, Chen Y, Shaibu AS, Zheng H, Sun J (2020) Molecular-assisted distinctness and uniformity testing using SLAF-sequencing approach in soybean. Genes (Basel) 11(2):175. https://doi.org/10.3390/genes11020175

Zheng XH, Ye JH, Cheng CP, Wei XH, Ye XF, Yang YL (2022) *Xiangeng* identification by SNP markers in *Oryza sativa* L. Acta Agron Sin 48(2):342–352. https://doi.org/10.3724/SP.J.1006.2022.02085

The manufacturer's authorised representative in the EU is Springer Nature Customer Service Centre GmbH, Europaplatz 3, 69115 Heidelberg, Germany. If you have any concerns regarding our products, please contact ProductSafety@springernature.com

Printed and bound by CPI Group (UK) Ltd, Croydon, CR0 4YY
01/10/2026
02225941-0010